Environmental Science - A Ground Zero Observation on the Indian Subcontinent

Abhijit Mitra • Sufia Zaman

Environmental Science - A Ground Zero Observation on the Indian Subcontinent

Springer

Abhijit Mitra
Department of Marine Science
University of Calcutta
Kolkata, West Bengal, India

Sufia Zaman
Department of Oceanography
Techno India University
Kolkata, West Bengal, India

ISBN 978-3-030-49133-8 ISBN 978-3-030-49131-4 (eBook)
https://doi.org/10.1007/978-3-030-49131-4

This Springer imprint is published by the registered company Springer Nature Switzerland AG
The registered company address is: Gewerbestrasse 11, 6330 Cham, Switzerland

Acknowledgement

The purpose of this book is to shed light on various issues of the environment on which our survival depends. The wheels of civilization can perfectly move if the frictions posed by pollution, population, over exploitation of resources can be minimized with appropriate technologies and policies. This book is by no means a final work, as both technologies and policies are dynamic in nature. The contents of the book should be viewed as part of an ongoing process, of a continuing effort to understand a set of complex interrelated global issues. The contents of the book have been enriched by the contributions of Dr. Prosenjit Pramanick, Dr. Shankhadeep Chakraborty, Dr. Subhasmita Sinha, Dr. Roopali Roy Chowdhury, Dr. Pritam Mukherjee, Ms. Nabonita Pal, Ms. Arpita Saha, Ms. Sudeshna Biswas, Mr. Joystu Dutta and Ms. Ankita Mitra through their rigorous field work. Several innovative programmes that constitute the annexure sections of the present book may serve as road map in solving several environmental issues.

We are thankful to Dr. Pardis Fazli of Dept. of Biological and Agricultural Engineering, University Putra, Selangor, Malaysia, for her effort in representing our data in graphical forms.

Finally Dr. Abhijit Mitra expresses his gratefulness to his wife Shampa, daughter Ankita and mother Manjulika whose inspirations and encouragements acted as booster to complete the manuscript. The advice of late Dhanesh Chandra Mitra, father of Dr. Abhijit Mitra, to create a strong footprint in life still inspires Dr. Mitra.

Dr. Sufia Zaman expresses her deepest gratitude to her mother, Mrs. Ayesha Zaman, for her unconditional love and practical day-to-day support, and father, Mr. Salim-uz-Zaman, who gave her immense moral support. Dr. Zaman also acknowledges the support of her beloved husband Dr. Sahid Imam Mallick and her baby girl Shanza Mallick. Dr. Zaman wishes to accord her deep sense of gratitude to her family members including her uncle (Mr. Pradip Kumar Mitra) and aunt (Late Mrs. Kanika Mitra), younger sister (Ms. Sharmilee Zaman), brother-in-law Kazi Dr. Sazzad Manir, her in-laws and her beloved grandmother (Mrs. Shibani Dhar) for their encouragement and inspiration throughout the strenuous period of manuscript preparation.

Contents

About the Authors

Abhijit Mitra Associate Professor and former Head, Dept. of Marine Science, University of Calcutta (India), has been active in the sphere of Oceanography since 1985. He obtained his PhD as NET qualified scholar in 1994. Since then he joined Calcutta Port Trust and WWF (World Wide Fund for Nature-India) in various capacities to carry out research programmes on environmental science, biodiversity conservation, climate change and carbon sequestration. Dr. Mitra is also serving as the Director of Research at Techno India University, West Bengal. He has to his credit about 540 scientific publications in various National and International journals and 40 books of postgraduate standards. Dr. Mitra is presently the member of several committees like PACON International, IUCN, SIOS, Mangrove Society of India, The West Bengal National University of Juridical Sciences, etc., and has successfully completed about 17 projects on biodiversity loss in fishery sector, coastal pollution, aquaculture, alternative livelihood, climate change and carbon sequestration. Dr. Mitra also visited as faculty member and invited speaker several universities in Singapore, Kenya, Oman and the USA. In 2008, Dr. Mitra was invited as visiting fellow at the University of Massachusetts at Dartmouth, USA, to deliver a series of lecture on Climate Change. Dr. Mitra also successfully guided 39 PhD students. Presently his research areas include environmental science, mangrove ecology, sustainable aquaculture, alternative livelihood, climate change and carbon sequestration.

Sufia Zaman presently serving as Head, Department of Oceanography, Techno India University, West Bengal, started her career in the field of Marine Science since 2001. She worked in the region of Indian Sundarbans and has wide range of experience in exploring the floral and faunal diversity of the Sundarbans. She has published 5 books on carbon sequestration, 235 scientific papers and contributed chapters in several books on biodiversity, environmental science, aquaculture and livelihood development. Dr. Zaman is presently a member of Fisheries Society of India. She is also running projects on carbon sequestration by mangroves of Indian Sundarbans. She is the recipient of DST Women Scientist and Jawaharlal Memorial Doctoral fellowship awards. Her areas of research include aquaculture, fish nutrition, phytoplankton diversity, climate change mangrove ecology and alternative livelihood. Dr. Zaman is also the first researcher in the maritime state of West Bengal (India) who initiated trial experiments on iron fertilization and subsequent enhancement of primary (phytoplankton) and secondary (fish) productions in the brackish water ponds of Indian Sundarbans with financial assistance from Department of Science and Technology, Government of India. Dr. Zaman is also providing consultancy on green technology to several industries, NGOs and corporate sectors.

Chapter 1
Introduction to Environment

Abstract A general overview of environment along with its components and resources are highlighted in this chapter. The methods to conserve resources of the planet Earth is an important section of this chapter, where *in situ* and *ex situ* conservation strategies have been explained clearly with examples. The annexure at the end of the chapter focuses on innovative alternative livelihood through which over-exploitation can be avoided and employment can be generated.

Keywords Natural resources · Renewable and non-renewable resources · *In situ* and *ex situ* conservation · Alternative livelihood

1.1 Definition and Overview of Environment

The term environment is derived from the French word 'Environ' which means 'surrounding'. Our surrounding includes biotic components like human beings, birds, reptiles, amphibians, fishes, plants, microbes, etc. and abiotic factors such as light, atmospheric pressure, air, water, soil, etc. Environment is an aggregation of several components and variables, which surrounds all the living organisms of the planet Earth. Environment includes air, water, land and all living creatures and the intricate inter-relationships that exist between them. The natural environment consists of four interlinking systems, namely, the hydrosphere, the lithosphere, the atmosphere and the biosphere. Each of these four compartments/systems is discussed here in brief.

Hydrosphere
Hydrosphere includes all water bodies such as lakes, ponds, rivers, streams, estuaries, seas, oceans, etc. Hydrosphere functions in a cyclic nature, which is termed as hydrological cycle or water cycle.

Lithosphere
Lithosphere means the mantle of rocks constituting the Earth's crust. The Earth is a cold spherical solid planet of the solar system, which spins in its axis and revolves

A. Mitra, S. Zaman, *Environmental Science - A Ground Zero Observation on the Indian Subcontinent*, https://doi.org/10.1007/978-3-030-49131-4_1

around the sun at a certain constant distance. Lithosphere mainly, contains soil, earth rocks, mountain, etc. Lithosphere is divided into three layers, namely, crust, mantle and core (outer and inner).

Atmosphere
The layer of the air that envelopes the Earth is known as the atmosphere. This layer contains gases like oxygen, nitrogen, inert gases, carbon dioxide, etc. This layer protects the solid earth and human beings from the harmful radiations of the sun. There are five concentric layers within the atmosphere, which can be differentiated on the basis of temperature and each layer has its own characteristics. These include the troposphere, stratosphere, mesosphere, thermosphere and exosphere.

Biosphere
It is the layer which consists of the life and several living forms. Thus it consists of plants, animals and microorganisms, ranging from the tiniest microscopic organism to the largest blue whales in the sea. Several ecosystems and habitats compose the biosphere of the planet Earth. Thus biosphere can be considered as a layer packed with a wide spectrum of biodiversity. The richness of the habitats in biosphere is a function of environmental conditions like availability of light, nutrients, rainfall, optimum soil conditions, etc.

1.2 Renewable and Non-renewable Resources

Resources are very important for the survival and prosperity of human civilization. Human beings are exploiting resources both from the land and water since long evolutionary period of time. This has posed an adverse impact on the resource bank of the planet Earth. On this background, the concept of renewable resource has gained much priority.

The resources which are being continuously consumed by man but are renewed by nature constantly are called as renewable resources. These resources are inexhaustible because they cannot be exhausted permanently. Renewable resources are also called as 'Non-Conventional' sources of energy. Examples of renewable resources include solar energy, wind energy, tidal energy, hydro power, geothermal energy and biofuels. In the southern states of Indian sub-continent, wind-based energy is widely used to run the wheels of civilization (Fig. 1.1).

Non-renewable Resources
The non-renewable resources do not replenish and cannot be renewed. It takes thousands of years of time to form the non-renewable resources which exist inside the Earth in the form of coal, fossil fuels, etc. Examples of non-renewable resources are coal, mineral ores, metal ores, crude oil and nuclear energy. A comparative account of renewable and non-renewable resources is presented in Table 1.1.

Fig. 1.1 Wide use of wind energy in the southern parts of India

Table 1.1 A comparative account of renewable and non-renewable resources

Point of comparison	Renewable resource	Non-renewable resource
Presence	The renewable resources are mostly present in the atmosphere of the Earth	The non-renewable resources are typically found in the underground layers of the Earth
Replaceable	The renewable resources are replaced by nature itself in a very short period	The non-renewable resources cannot be replaced by nature during the time of human life span
Availability	The renewable energy resources are plentiful and abundant in nature	The non-renewable resources are scarce resources and not available in an abundant manner in nature
Cost	The renewable resources are obtained free of cost or at very less cost in nature	The non-renewable resources are very costly and not easily available
Impact on environment	The renewable resources do not affect the environment of the Earth and do not cause any climate changes in the atmosphere	The non-renewable resources seriously affect the environment and cause climate changes in the environment like burning of fossil fuels cause greenhouse gas emission

(continued)

Table 1.1 (continued)

Point of comparison	Renewable resource	Non-renewable resource
Pollution	The renewable resources do not cause pollution in the environment and do not release any pollutants into the environment	The non-renewable resources pollute the Earth by releasing various types of pollutants into the air, water, soil, etc. when fossil fuels are burned
Impact on atmosphere	The renewable resources are called 'clean and green' energy sources because they do not produce any adverse impact to the environment	The non-renewable resources release greenhouse gases into the atmosphere which leads to global warming
Impact on health	The renewable resources do not cause any health problems to the living beings of the Earth	The non-renewable resources adversely affect the health of the living beings by releasing smoke, radiations, carcinogenic or cancer-causing elements into the environment
Impact on nature	The use of renewable resources promotes the balance in the nature and natural habitat of the Earth	The use of non-renewable resources disrupts the balance in nature which is due to digging the earth to take out coal, minerals, fuels, etc.

1.3 Conservation of Natural Resources

The term natural resource implies any substance of the environment which is used to achieve our objective and fulfil our desire. For example, we require bricks, cement, iron, wood, etc. to construct a building. Again we need water, fishes, aquatic plants, etc. to initiate an aquaculture pond. **A resource can be defined as 'any natural or artificial substance, energy or organism, which is used by human being for its welfare'. The resources are usually used to upgrade the human civilization.** The resources can be two types: (a) natural resources and (b) artificial resources.

All that nature has provided, such as soil, air, water, minerals, coal, sunshine (sunlight), animals, plants, etc., are known as natural resources. The resources, which have been developed by human beings during the growth of civilization, are called artificial resources. For example, electricity generated from power plants, products developed from industries, fishes grown in the aquaculture ponds and oyster farmed in the coastal waters are artificial resources.

The classification of resources present in the planet Earth may be divided and sub-divided into several groups and sub-groups (Table 1.2).

Exhaustible Resources

There are some resources, which are available in limited quantities and are going to be depleted as a result of continuous use. These are called **exhaustible resources**. For example, the reservoir of coal in the Earth is limited, and one day there will be no more coal available for our use and same may happen for petroleum also. Fishes also come under the category of exhaustible resources. Today population explosion is a burning issue in the planet Earth. This has forced the growing population to

Table 1.2 Classification of resources

Type of resource	Division
Natural	Exhaustible – these are of two types, namely, renewable (water, forest, etc.) and non-renewable (coal, petroleum, iron, biological species)
	Inexhaustible – includes solar energy, wind, rainfall and tidal energy)
Artificial	Electricity

exploit natural resources without considering their sustainability. However, for the sake of our future generation and to sustain the mankind, it is of utmost importance to conserve the natural resources.

Conservation is the proper management of a natural resource to prevent its exploitation, destruction or degradation. Conservation is the sum totals of activities, which can derive benefits from natural resources but, at the same time, prevent excessive use that may lead to destruction or degradation.

Inexhaustible Resources

The resources which cannot be exhausted by human consumption are called **inexhaustible resources.** These include energy sources like solar radiation, wind power, water power (flowing streams) and tidal power and substances like sand, clay, air, water in oceans, etc.

The main focus of conservation is:

(i) To maintain the ecological balance for supporting life
(ii) To preserve different kinds of species (biodiversity)
(iii) To make the resources available for present and future generations (sustainability)
(iv) To ensure survival of human race
(v) To regulate the biogeochemical cycles

On this background, the need for conservation of natural resources was felt long back in the Indian culture.

In Indian sub-continent natural resources are conserved in the form of sacred groves/forests, sacred pools and lakes, sacred species, etc., e.g. the River Ganges. In our country the conservation of natural forests is known from the time of Lord Ashoka. Sacred forests are forest patches of different dimensions dedicated by the tribals to their deities and ancestral spirits. Cutting down trees, hunting and other human interferences were strictly prohibited in these forests. This practice is widespread particularly in peninsular, central and eastern India and has resulted in the protection of a large number of plants and animals. Similarly, several water bodies, e.g. Khecheopalri Lake in Sikkim, were declared sacred by people, thus, protecting aquatic flora and fauna. **Worshipping** certain plants like banyan, peepal, tulsi, etc. has not only preserved them but also encouraged their plantation. History recalls numerous instances where people have laid down their lives for protecting trees.

The Chipko movement in India is one of the best examples. This movement was started by women in Gopeshwar village in Garhwal in the Himalayas. They stopped the felling of trees by hugging them when the lumbermen arrived to cut them. This saved about 12,000 square kilometres of sensitive water catchment area. Similar movements also occurred in some other parts of the country.

Today in the twenty-first century, the importance of biodiversity has been felt in every sector starting from healthcare to recreation. Hence, conservation of biodiversity has been given priority all over the world.

There are two basic strategies for conservation of biodiversity, namely, (i) **in situ** conservation and (ii) **ex situ** conservation:

(i) **In situ** conservation includes the protection of plants and animals within their natural habitats or in protected areas. Protected areas are areas of land or sea dedicated to protection and maintenance of biodiversity. For example, National Parks, Wildlife Sanctuaries, Biosphere Reserves, etc.
(ii) **Ex situ (off site)** conservation is the conservation of plants and animals outside their natural habitats. These include Botanical Gardens, Zoo, Gene Banks, DNA Banks, Seed Banks, Pollen Banks, Seedling, Tissue Culture etc.

Conservation of plants is of utmost importance for the sake of mankind. Many plants have high commercial values, e.g. they are the sources of timber, fuel, crops, fruits, medicinal ingredients etc. Such plants are often exploited to meet the needs of the people and in most of the cases the essence of conservation is lost. In recent times great focus has been given on the medicinal properties of the plants where leaf extracts, barks, fruits and root extracts are used to cure several diseases. It has been found that such medicinal plants are often over exploited and destroyed in mass by the people. Many industrial houses also appoint local people to collect these resources and thus leading to mass destruction of the species. In such cases strict policy should be developed to cultivate the species as a backup of the resources. A very important example in this context is the cultivation of *Suaeda maritima* and *Salicornia brachiata* in certain pockets of Indian Sundarbans to get the leaves and stems of these herbs throughout the year as these vegetative parts of the species are now used to prepare snacks like samosa for the local people (Vide Annexure 1A for full details). Although, the entire concept is yet in the embryonic stage, but as it will proceed through a business chain, mass exploitation of these resources will start. Hence, awareness programmes have been initiated through several seminars and workshops to aware the people of all ranks of the society to conserve the species (Figs. 1.2, 1.3, 1.4 and 1.5).

Cultivation of the species *S. maritima* has also been initiated involving the local people of Sundarbans to protect the species (Fig. 1.6). Hence, awareness generation involving industrialists, academicians, consumers, local inhabitants, researchers and policymakers is an effective approach to conserve the natural resources.

The International Conference held on 17–18th January 2020 on 'Climate Change: Impact, Management, Law and Policy Formulation' is an eye opener in this context where researchers, academicians, policymakers, lawyers and judges exchange their ideas to conserve natural resources in the backdrop of climate change (Fig. 1.7).

Fig. 1.2 A view of snacks box containing samosa prepared from *Suaeda maritima*

Fig. 1.3 The inner materials of the snacks box showing the food products along with the specimen species (*S. maritima*) – an attempt to aware the participants about the salient features of the plant species

Fig. 1.4 Display and distribution of *S. maritima*-based food products in seminars and workshops

1.4 Take Home Messages

(A) Environment is an aggregation of several variables, which surrounds all the living organisms of the planet Earth. Environment includes air, water, land and all living creatures and the intricate inter-relationships that exist between them. The natural environment consists of four interlinking systems, namely, the hydrosphere, the lithosphere, the atmosphere and the biosphere.

(B) The resources which are being continuously consumed by man but are renewed by nature constantly are called as renewable resources. These resources are inexhaustible because they cannot be exhausted permanently. Renewable resources are also called as 'Non-Conventional' sources of energy. Examples of renewable resources include solar energy, wind energy, tidal energy, hydro power, geothermal energy and biofuels.

(C) The non-renewable resources do not replenish and cannot be renewed. It takes thousands of years of time to form the non-renewable resources which exist inside the earth in the form of coal, fossil fuels, etc. Examples of non-renewable resources are coal, mineral ores, metal ores, crude oil and nuclear energy.

Fig. 1.5 View of the participants receiving the knowledge of species conservation

(D) Conservation is the proper management of a natural resource to prevent its exploitation, destruction or degradation. Conservation is the sum totals of activities, which can derive benefits from natural resources but at the same time, prevent excessive use that can lead to destruction or degradation.

(E) Mass awareness involving people of all ranks of the society is needed to carry forward the essence of conservation. This can be achieved through seminars and workshops and binding the approach of conservation with legal framework.

Fig. 1.6 Cultivation of *S. maritima* in the supra-littoral zone of Indian Sundarbans

1.5 Brain Churners

This chapter provides the readers a basic idea on environment. The chapter has immense importance on the backdrop of natural resource reservoir of the planet Earth, which is gradually depleting with the rise of human population. It is expected that the readers will find interest in answering the questions asked on various types of natural resources and their conservation. The section will also help the readers to face various competitive examinations as environment has been given thrust in recent times. The readers can consider this section as a sort of 'self-appraisal approach' as correct answers to all the questions are filled in black in the Answer Box at the end of this section.

1. Every year which day is celebrated as World Environment Day?

    (a) 7th June
    (b) 5th July
    (c) 5th June
    (d) None of these

Fig. 1.7 Flyer displaying The International Conference held on 17–18th January 2020 on 'Climate Change: Impact, Management, Law and Policy Formulation' at Kolkata (India) in which the researchers, legal people and policymakers joined hands to conserve the natural resources as an adaptation to climate change

2. Who is not associated with the field of environment?
   - (a) Mother Teresa
   - (b) Medha Patkar
   - (c) Sunder Lal Bahuguna
   - (d) Anil Agarwal
3. Ecology is the study of relationship of
   - (a) Organisms and environment
   - (b) Man and environment
   - (c) Soil and water
   - (d) Man and child
4. Biotic environment includes
   - (a) Producers
   - (b) Consumers

(c) Decomposers
(d) All the above

5. The group of organisms which convert light into food is called
   (a) Autotrophs
   (b) Heterotrophs
   (c) Decomposers
   (d) Omnivores
6. Which one of the following is not biodegradable
   (a) Vegetables
   (b) Fruits
   (c) Earthworm
   (d) Aluminium foil
7. In our country, the Van Mahotsava Day is observed on
   (a) 2nd October
   (b) 1st December
   (c) 10th August
   (d) 1st July
8. Plants are green because of presence of pigment called
   (a) Glucose
   (b) Nitrogen
   (c) Chlorophyll
   (d) Oxygen
9. Air is composed of gases, water vapours and
   (a) Dust particles
   (b) Rainfall
   (c) Snowfall
   (d) Light
10. Chief source of energy in environment is
    (a) Fire
    (b) Moon
    (c) Sun
    (d) Stars
11. Which of the following is not a primary contributor to the greenhouse effect?
    (a) Carbon dioxide
    (b) Carbon monoxide
    (c) Chlorofluorocarbons
    (d) Methane gas

12. Process through which plants reproduce is
    (a) Eating
    (b) Evaporation
    (c) Pollination
    (d) Condensation
13. 71% of Earth surface is covered with
    (a) Land
    (b) Air
    (c) Water
    (d) Coal
14. World Forest Day is celebrated every year on
    (a) 3rd March
    (b) 3rd April
    (c) 21st April
    (d) 21st March
15. The ozone layer in the stratosphere acts as an efficient filter for
    (a) Solar UV-B rays
    (b) X-rays
    (c) Gamma rays
    (d) UV-A rays
16. Ozone layer is found in
    (a) Ionosphere
    (b) Troposphere
    (c) Stratosphere
    (d) All the above
17. CFC stands for
    (a) Chlorofluorocarbons
    (b) Carbon fluorine carbon
    (c) Compact fluorocarbons
    (d) None of the above
18. The total air volume can be determined by
    (a) Anemometer
    (b) Spirometer
    (c) Vane anemometer
    (d) None of the above
19. Stratosphere is devoid of weather phenomenon because
    (a) All of the atmosphere ozone is present in this layer
    (b) It is free from moisture and dust

(c) It has a maximum temperature of 20C
(d) The harmful radiation is absorbed by ozone in this layer

20. The blanket of gases and vapours around is known as

(a) Atmosphere
(b) Stratosphere
(c) Ionosphere
(d) Troposphere

Answer Sheet

S. No.	a	b	c	d
1.			■	
2.	■			
3.	■			
4.				■
5.	■			
6.				■
7.				■
8.			■	
9.	■			
10.			■	
11.		■		
12.			■	
13.			■	
14.				■
15.	■			
16.			■	
17.	■			
18.		■		
19.		■		
20.	■			

Annexure 1A: Food Products from Mangrove Associate Species: An Avenue for Alternative Livelihood and Conservation

Introduction

Development of food products from untapped natural resources plays an important role in the upliftment of rural nutrition and rural economy. Nutrition is one of the most critical issues to address the health sector of coastal population and island dwellers. The coastal zone is enriched with a wide spectrum of natural resources of which mangroves are noted for their ecosystem services. Mangroves are widely distributed in the coastal zone particularly at the land–sea interface. Lot of literatures are available on development of food products from mangroves as well as seaweeds and their microbial assay (Pramanick et al. 2015a, b, 2016a, b, 2017a, b). However, no research works have been conducted on the development of food products from mangrove associate species although they are important sources of bioactive substances and have high nutritional value.

A large number of people have the misconception that mangrove ecosystems are unproductive swamps and act as abode of harmful insects, mosquitoes, snakes and need to be cleared for the sake of public health. However, in recent years the concept has radically changed. In addition to several ecosystem benefits of mangroves like erosion control, protection against storms and tidal surges, livelihood generation, carbon sequestration, etc., mangroves are also used as part of food and consumed by humans in many places of the world. Tender leaves of *Acacia* sp. are used as substitute for tamarind in chutneys. *Bruguiera gymnorrhiza* fruits are edible. Fruits of *Sonneratia caseolaris* are edible in raw form and are also cooked. Fruit juice of *S. caseolaris* is also consumed by people. Fruit jelly, has been prepared from *S. apetala* fruit. Ice creams, cookies and bread have been prepared from mangrove associate seaweeds *Enteromorpha intestinalis* (Mitra et al. 2016; Pramanick 2017).

Suaeda maritima is a salt marsh plant which is considered as a mangrove associate species. It is salty in taste in raw form and hence it is cooked with other types of vegetables to reduce the salty taste (Tanaka 1976). In many pockets of the world preferably in South Asian countries, this species is used in salad, curry with crabs and various other types of chilli-based preparations. In South Indian states, it is pickled in vinegar and used for cooking (Patra et al. 2011). This mangrove associate species contain antioxidants and is non-toxic in nature. Polyphenolic compounds may be the major bioactive components in the species which are responsible for antioxidation and antiproliferation. Natural antioxidants have been proved to retard the process of tumour growth because of different redox status between normal cells and cancer cells (Nair et al. 2007).

In addition to *Suaeda* sp., another species selected for the present study is *Salicornia brachiata*, which is widely distributed in the supra-littoral zone of the marine and estuarine systems. *Salicornia* has been used for decades for edible purposes. The wide application of the species in salads is well-known. The species is also the source of salt that is commonly used in cooking. *Salicornia* is mulled to be the right candidate for reclamation of barren lands, salt flats and seashores. In short, they can be deemed for seawater agriculture. It is suggested that as global warming threatens to submerge more landmass, and freshwater is depleting, a shift to saline crop might be a viable option (Katschnig et al. 2013).

On this background, the present research programme aims to develop two very popular Indian food products, namely, samosa and kachori from mangrove associate species *Suaeda maritima* and *Salicornia brachiata* collected from Indian Sundarbans.

The success of any experimental approach is monitored on the basis of a comparative study between control and experimental samples in terms of few parameters (variables) that are relevant indicators of human health. In the present programme, we have used the values of protein, carbohydrate, fat, fibres, Ca, Na and K as indicators of such comparative study. These indicators (compounds/elements) are intricately related to human health.

Materials and Methods

(A) **Sampling**

Plant samples were collected from the supra-littoral zone of Jharkhali during November 5 to 9, 2018. The samples were thoroughly washed with double-distilled water and sterilized to make them germ free. The fresh plants (except the roots) were grinded/crushed and mixed with potato (in case of samosa) and sattu (in case of kachori) in the ratio 1:1.

(B) **Analysis of Biochemical Constituents**

1. Estimation of Protein

The food samples (mixed with the selected mangrove associate species) as well as the control were separately homogenized in 10% cold trichloroacetic acid TCA (10 mg: 5 ml) and were centrifuged at 5000 rpm for 10 min. Supernatant was discarded and pellets were saved. Pellets were again suspended in 5 ml of 10% cold TCA and recentrifuged for 10 min. Supernatant was again discarded and the precipitate was dissolved in 10 ml of 0.1 N NaOH. 0.1 ml of this solution was used for protein estimation. In 1 ml of the sample, total protein content was estimated using the protocol of Lowry et al. (1951). A stock solution (1 mg/ml) of bovine serum albumin was prepared in 1 N NaOH; five concentrations (0.2, 0.4, 0.6, 0.8 and 1 ml) from the working standard solution were taken in series of test tubes. In another set of test tubes, 0.1 and 0.2 ml of the sample extracts were taken, and the volume was raised up to 1 ml in all the test tubes. To each test sample, 5 ml of freshly prepared alkaline solution was added at room temperature and left undisturbed for a period of 10 min. Subsequently, to each of these mixture tubes 0.5 ml of Folin–Ciocalteu reagent was rapidly added and incubated at room temperature for 30 min until the blue colour develops. The absorbance was read at 750 nm, and the amount of total protein present in the samples was calculated and expressed in percentage.

2. Estimation of Fat

The total fat in the samples (prepared food products) was estimated by Soxhlet method using petroleum ether (80 °C) as per the standard procedure (AOAC 2000).

3. Estimation of Carbohydrate

The carbohydrate concentration in the food samples was measured by Anthrone method (Yemm and Willis 1954). 100 mg of each of the samples was weighed and taken into a boiling tube, hydrolysed by keeping it in a boiling water bath for 3 h with 5 ml of 2.5 N HCl, and the mixture was cooled at room temperature. By using sodium carbonate, it was neutralized and the volume was made up to 100 ml. Then it was centrifuged and supernatant was collected. 0.5 ml of aliquots was used for estimation. The working standards were prepared by taking 0, 0.2, 0.4, 0.6, 0.8 and 1 ml of the working standard glucose solutions and '0' served as blank. 0.1 and 0.2 ml of the test sample solutions were taken in two separate test tubes. In all the test tubes, the volume was marked up to 1 ml with water, and blank was set with 1 ml of water. 4 ml of Anthrone reagent was added to each tube, mixed well and kept it in water bath for 10 min. Cooled the contents rapidly and the absorbance was read at 630 nm, and the amount of total carbohydrate present in the sample was calculated.

4. Estimation of Fibres

Fat-free samples were digested with 1.25 N HCl for 2 h. Then the digested mixture was filtered through a filter paper after cooling, and acid-free residue was further digested with 1.25 N NaOH for 2 h. The final residue after alkali digestion was oven-dried up to constant weight and calculated for the crude fibre in each sample.

% Fibre = (mass of residue after alkali digestion/mass of grinded dried sample) × 100

5. Estimation of Calcium

The samples (samosa and kachori) were obtained in solid form; they were digested using a mixture of concentrated nitric acid, sulphuric acid and hydrogen peroxide. An accurately weighed 5.0 g of the edible portion of each of the food stuffs was put in a heat-resistant beaker where 8 ml of concentrated sulphuric acid and 10 ml of concentrated nitric acid was added. The beaker was then placed on a hot plate and warmed cautiously until the reaction subsided. It was then heated vigorously until the solution begins to darken owing to incipient charring.

To avoid such charring, a 2 cc aliquot of concentrated nitric acid was constantly added at any time the solution began to darken. This treatment was continued until the solution stopped to darken on prolonged heating. At this point, the solution was allowed to cool and diluted with 10 ml of double-distilled water and boiled to fuming. This dilution and boiling to fuming process was repeated twice again but with 5 cc of double-distilled water.

At this point, the persistent colour(s) of the solution was cleared by the addition of about 2–4 ml of hydrogen peroxide with drops of nitric acid again. The solution was heated to fuming state each time hydrogen peroxide was added until the residue was colourless or no further reduction of pale yellow colour was obtained.

The solution was cooled with about 10 ml of distilled water and evaporated to fuming again. This continued until there was no more fuming; then the solution was made up to mark in 100 ml volumetric flask.

A 25 ml aliquot of each digest was pipette into a beaker and 1 M NaOH solution was added to adjust the pH to 12–13. Two drops of solochrome dark blue was then added and immediately titrated against a 0.01 M EDTA solution to the blue endpoint. For every determination and evaluation, a standard curve of mass of calcium (mg) versus amount of EDTA (millimoles) was plotted, which finally produced the results for calcium in the unknown samples (here samosa and kachori).

6. Estimation of Sodium

   Sodium in the selected samples was analysed by flame photometer.

7. Estimation of Potassium

   Flame photometric method was used to determine the potassium concentration in the food samples (samosa and kachori).

Result

Figure 1.8, 1.9, 1.10 and 1.11 reflect the results of proximate analysis and elemental composition of samosa and kachori available in the market (considered as control).

On mixing crushed fresh *Salicornia* and *Suaeda* with kachori and samosa in the ratio 1:1, an alteration in the values of protein, carbohydrate, fat, total fibre, Ca, Na and K was observed, which may be attributed to biochemical composition of the selected mangrove associate species.

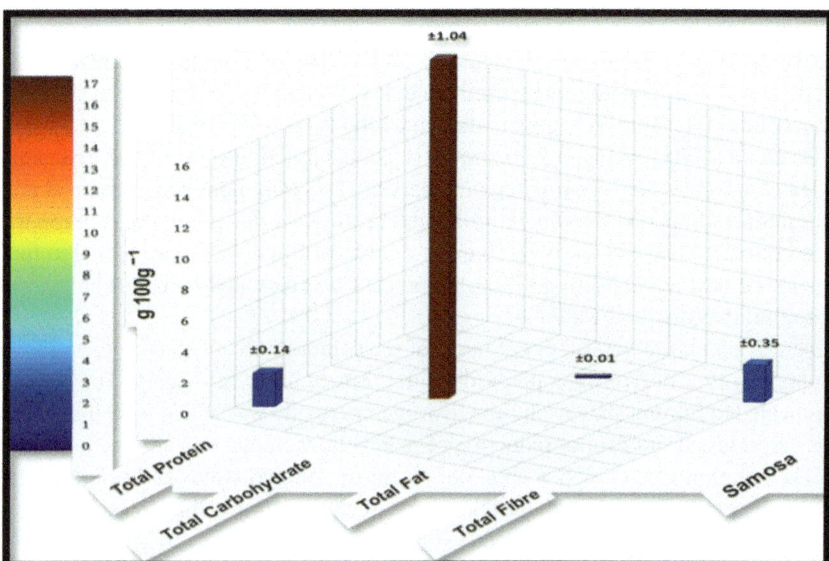

Fig. 1.8 Proximate analysis of samosa without mixing mangrove associate species (control)

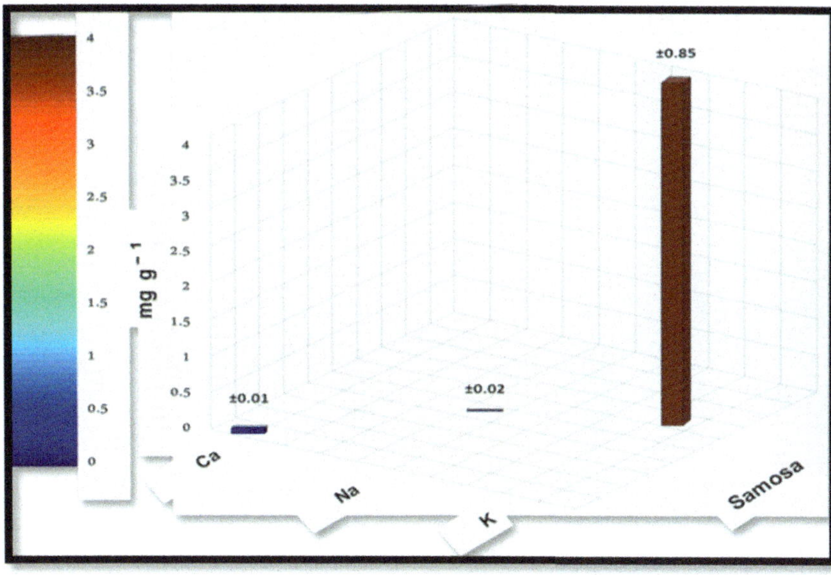

Fig. 1.9 Elemental analysis of samosa without mixing mangrove associate species (control)

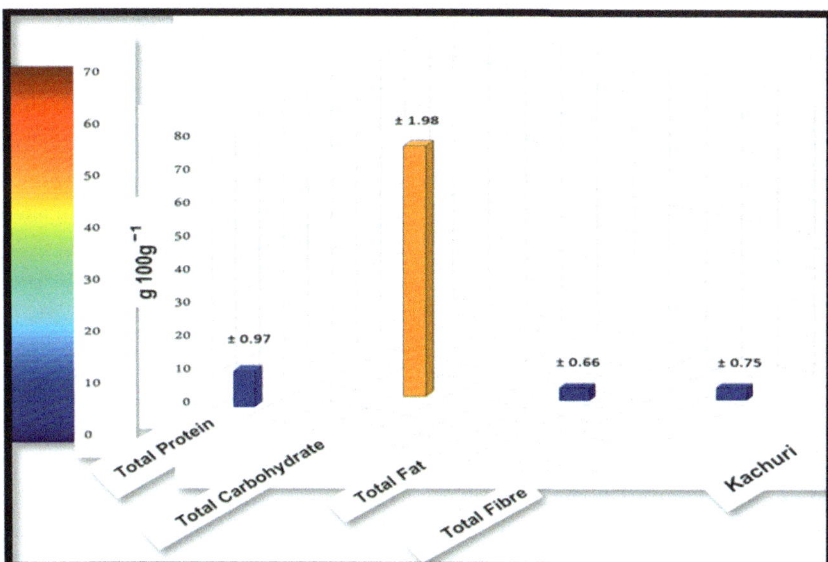

Fig. 1.10 Proximate analysis of kachori without mixing mangrove associate species (control)

Discussion

Suaeda maritima is a salt marsh mangrove associate plant growing in mangrove forest, which can grow naturally. It is a low-cost vegetable with high nutritional

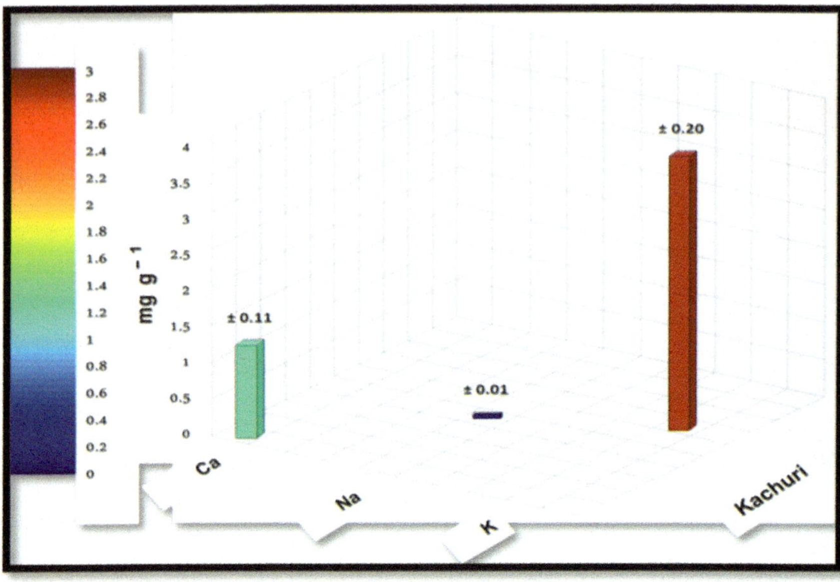

Fig. 1.11 Elemental analysis of kachori without mixing mangrove associate species (control)

value. Food products prepared from this mangrove associate species can be a new source of income of the people in the community.

Salicornia is also a mangrove associate species, which is commonly known as pickleweed, glasswort, sea beans, sea asparagus, crow's foot greens and samphire and is a halophyte, belonging to Amaranthaceae family (Singh et al. 2014). In fact, *Salicornia* name has originated from the Latin word meaning 'salt'. Studies report that some species, for example, *Salicornia europaea* show tolerance towards salinity as high as 3% NaCl (Yamamoto et al. 2009). This fleshy plant is found at the edges of wetlands, marshes, seashores and mudflats, actually on most alkaline flats (Smillie 2015). The aerial parts of this plant are consumed in salads or processed into pickles, beverages, etc.

Both *Salicornia* and *Suaeda* are rich in Na, Ca and K, which are of vital importance for human health. Considering this, the present programme was undertaken in November 2018 to prepare samosa and kachori from these two halophytes collected from the Jharkhali region of Indian Sundarbans.

It is observed that *Salicornia*-based samosa showed a decrease of protein, carbohydrate, fibre and K by 36.51%, 6.21%, 4.31% and 28.39%, respectively. However, fat, Ca and Na increased abruptly by 40%, 400% and 150%, respectively (Fig. 1.12).

In case of *Salicornia*-based kachori, the protein, carbohydrate, fat and fibre decreased by 59.80%, 2.25%, 8.20% and 2.66%, respectively. Ca, Na and K increased by 3.91%, 116.67% and 1.72%, respectively (Fig.1.13).

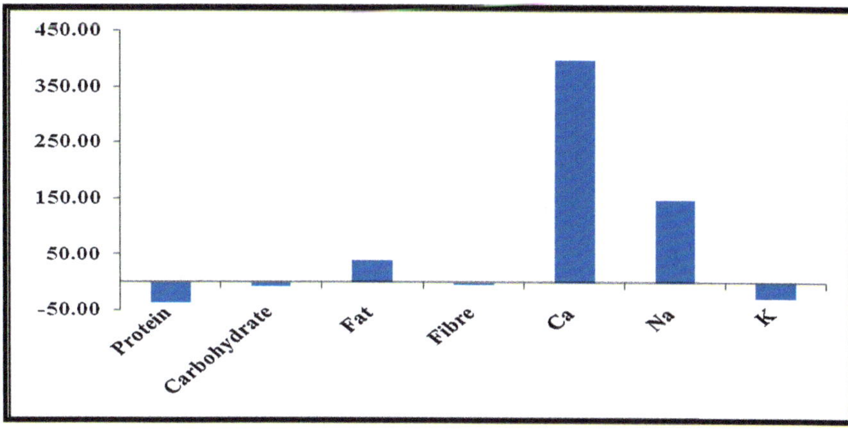

Fig. 1.12 Percent alteration of biochemical composition in final product (samosa prepared from *Salicornia brachiata*) compare to control

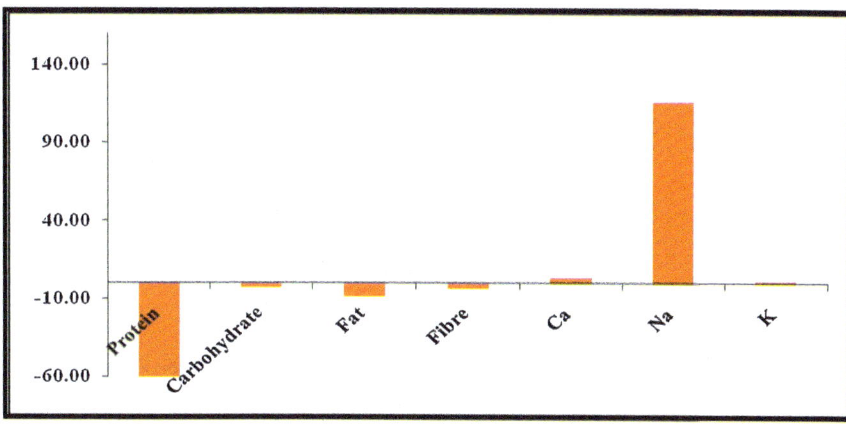

Fig. 1.13 Percent alteration of biochemical composition in final product (kachori prepared from *Salicornia brachiata*) compare to control

Suaeda-based samosa clearly showed an increase of protein, fat, fibre, Ca and Na by 4.23%, 140%, 37.80%, 525% and 200%, respectively. However, carbohydrate and K decreased by 35.50% and 22.11%, respectively (Fig. 1.14).

In case of *Salicornia*-based kachori, the carbohydrate, fibre, Ca, Na and K increased by 2.19%, 2.66%, 10.16%, 183.33% and 5.84%, respectively. 53.81% and 5.99% was the decrease in protein and fat, respectively (Fig. 1.15).

The findings in the present research programme show a promising market for mangrove associate-based food products, not only because of their rich mineral value but also for establishing future cottage industries for the island dwellers and coastal population of the country.

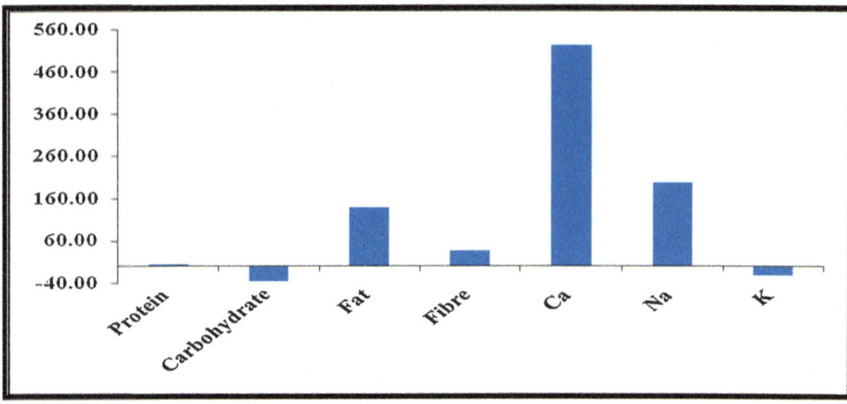

Fig. 1.14 Percent alteration of biochemical composition in final product (samosa prepared from *Suaeda maritima*) compare to control

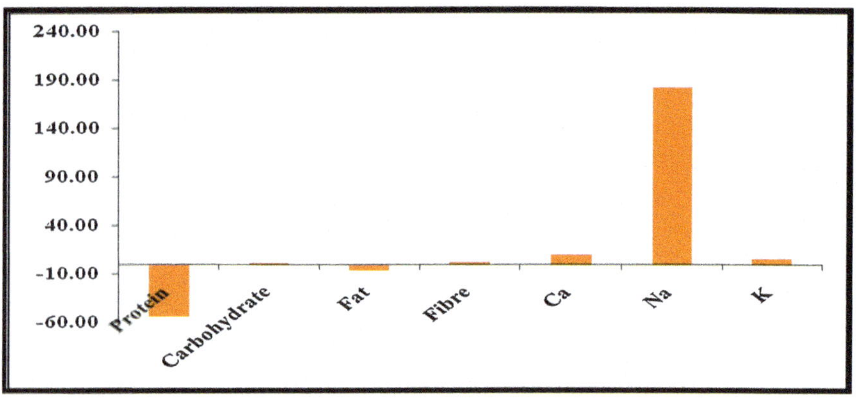

Fig. 1.15 Percent alteration of biochemical composition in final product (kachori prepared from *Salicornia brachiata*) compare to control

References

AOAC (2000) Official methods of analysis, 17th edn. The Association of Official Analytical Chemists, Gaithersburg

Katschnig D, Broekman R, Rozema J (2013) Salt tolerance in the halophyte *Salicornia dolichostachya* Moss: growth, morphology and physiology. Environ Exp Bot 92:32–42

Lowry OH, Rosebrough NJ, Farr AL, Randall RJ (1951) Protein measurement with the Folin phenol reagent. J Biol Chem:193–265

Mitra A, Sundaresan J, Bera D, Zaman S (2016) Seaweed resource utilization: an adaptation to climate change. Published by CSIR-National Institute of Science Communication And Information Resources (NISCAIR), New Delhi, p 120. ISBN 978-81-7236-349-9

Nair S, Li W, Kong AT (2007) Natural dietary anti-cancer chemopreventive compounds: redox-mediated differential signaling mechanisms in cytoprotection of normal cells versus cytotoxicity in tumor cells. Acta Pharmacol Sin 28:459–472

Patra KJ, Dhal KN, Thatoi NH (2011) *In vitro* bioactivity and phytochemical screening of *Suaeda maritima* (Dumort), a mangrove associate from Bhitarkanika India. Asian Pac J Trop Med 4:727–734

Pramanick P (2017) Ph.D thesis entitled "Study on the ecology and biochemical aspects of the major seaweeds of Indian Sundarbans with few edible products". Submitted to Techno India University, West Bengal, p 153

Pramanick P, Zaman S, Rudra T, Guha A, Mitra A (2015a) Heavy metals in a dominant seaweed species from the islands of Indian Sundarbans. Int J Life Sci Pharm Res 5(2):64–71

Pramanick P, Zaman S, Trivedi S, Bera D, Mitra A (2015b) Seasonal variation of heavy metals in *Catenella repens* in Indian Sundarbans. Am J Biol Pharm Res 2(4):149–152

Pramanick P, Zaman S, Rudra T, Guha A, Mitra A (2016a) Comparative account of heavy metal level in two species of Chlorophyceae inhabiting Indian Sundarbans. Int J Res Chem Environ 6(2):65–68

Pramanick P, Bera D, Banerjee K, Zaman S, Mitra A (2016b) Seasonal variation of proximate composition of common seaweeds in Indian Sundarbans. Int J Life Sci Sci Res 2(5):570–578

Pramanick P, Bera D, Banerjee K, Zaman S, Mitra A (2017a) Spatial variation of heavy metals in seaweeds inhabiting Indian Sundarbans, Research Article 5. In: MItra A, Sundaresan J, Banerjee K, Agarwal SK (eds) Environmental coastguards - understanding mangrove ecosystem and carbon sequestration, Climate change series 3. Published by CSIR-National Institute of Science Communication and Information Resources (NISCAIR), New Delhi, pp 207–214. ISBN: 978-81-7236-352-9

Pramanick P, Zaman S, Bera D, Biswas S, Mitra A (2017b) Nutritional characteristics of *Enteromorpha compressa* in Indian Sundarbans. J Sci Eng Health Manag 1(2):45–47. Published by Techno India University, West Bengal

Singh D, Buhmann AK, Flowers TJ, Seal CE, Papenbrock J (2014) *Salicornia* as a crop plant in temperate regions: selection of genetically characterized ecotypes and optimization of their cultivation conditions. AoB Plants. https://doi.org/10.1093/aobpla/plu071

Smillie C (2015) *Salicornia* spp as a biomonitor of Cu and Zn in salt marsh sediments. Ecol Indic 56:70–78

Tanaka T (1976) Tanaka's cyclopedia of edible plants of the world. Keigaku Publishing Co., Tokyo

Yamamoto K, Oguri S, Chiba S, Momonoki YS (2009) Molecular cloning of acetylcholinesterase gene from *Salicornia europaea* L. Plant Signal Behav 4:361–366

Yemm E, Willis AJ (1954) The estimation of carbohydrate in plant extracts by Anthrone. Biochem J 57:508–514

Chapter 2
Components of the Earth

Abstract The anatomy of the planet Earth is the basic backbone of the chapter. The different natural processes (both constructive and destructive) that shape the morphology of the Earth have been explained with ground zero observations. The various agents that play roles in changing the configuration of the Earth is highlighted with special reference to super cyclone AILA that caused considerable alteration of Sundarban estuaries during 2009 in terms of hydrological parameters. A bird's eye view of the different types of coral is a bonus to the readers of the chapter.

Keywords Constructive natural processes · Destructive natural processes · Sediment · Coral · Super cyclone AILA

2.1 Layers of the Earth

The Earth is made up of three main layers, namely, crust, mantle and core (Fig. 2.1). Each of these three layers is discussed separately.

(A) **Crust**

The outermost layer of the Earth is referred to as the crust. It is like the peel of an orange. It is very thin compared to the other layers (mantle and core). The crust is only about 3–5 miles (8 km) thick under the oceans (oceanic crust) and about 25 miles (32 km) thick under the continents (continental crust). Thus crust can be sub-divided into oceanic crust and continental crust depending on its location. It is composed mainly of silicate rocks (about 85%). Continental crust is mostly granite, while oceanic crust is basalt. Both are igneous (fire formed) rocks. Igneous rocks go through several alterations and can erode to form sedimentary rocks. Both igneous and sedimentary rocks may be reheated and exposed to extreme pressure and become metamorphic rocks or be subducted and completely melt again.

A. Mitra, S. Zaman, *Environmental Science - A Ground Zero Observation on the Indian Subcontinent*, https://doi.org/10.1007/978-3-030-49131-4_2

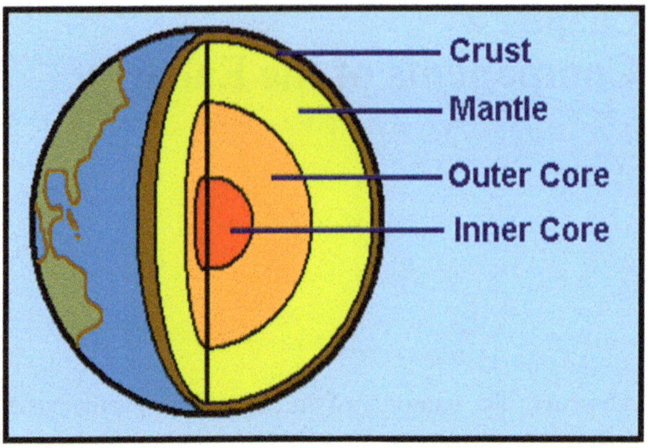

Fig. 2.1 Layers of the Earth

(B) **Mantle**

The mantle is about 1800 miles (2900 km) thick and is separated into upper and lower sections. Large convective cells in the upper mantle circulate heat and drive plate tectonic processes of the crust. The crust and the solid portion of the upper mantle compose a layer called the lithosphere. The lithosphere is divided into many plates that move in relation to each other due to tectonic forces. It essentially floats atop a semi-liquid layer known as the asthenosphere. It is about the consistency of asphalt. This layer allows the solid lithosphere to move around since the asthenosphere is much weaker than the lithosphere. Temperatures here range from 2000° F towards the crust to 4000° F towards the lower layers of mantle.

(C) **Core**

The last and innermost layer of the Earth is the core, which is separated into a liquid outer core and the solid inner core. The inner core is composed of almost entirely iron with a small amount of nickel and has temperatures and pressures so great that the metals are squeezed together and are not able to move like a liquid but instead vibrate in place as a solid. The inner core begins about 4000 miles (6437 km) beneath the crust and is about 800 miles (12,287 km) thick. The temperatures may reach 9000 degrees F and the pressures are 45,000,000 pounds per square inch. That is 3,000,000 times the air pressure felt at the sea level. The outer core is also composed of iron and nickel, which flows directionally controlling the Earth's magnetic field.

2.2 Constructive Natural Processes

The processes that build landforms through tectonic and depositional processes are termed as constructive processes. Constructive processes produce surface irregularities such in the form of mountains, plateaus, faults, folds, volcanic cones, depressions, etc. These processes are considered as endogenic processes, i.e. process which originate from within the Earth. Constructive processes include tectonic processes that include movements at plate boundaries, earthquakes, orogeny, deformation, emergence, submergence and volcanic activity. Thus the processes for building new land masses are referred to as constructive forces. Three of the major constructive forces are crustal deformation, volcanic eruptions and deposition of sediment. Crustal deformation occurs when the shape of land (or crust) is changed or deformed. One of the main causes is movement of the Earth's plates. When the plates collide or push towards each other, pressure builds. This can cause two things to happen. The rock can either fold or fault. Volcano can be defined as a vent in the Earth's crust through which lava, rock fragments, hot vapour and gases are ejected. In other words, a volcano is the Earth's way of letting off steam. The superheated particles that eject out of a volcano come from deep below the Earth's surface where temperatures can become so hot that rock actually melts.

Magma is the term used to describe this hot molten rock from deep within the Earth. A volcano begins to form when magma, which is less dense than the rock it originated from, rises towards the Earth's surface. This liquid rock collects in chambers called 'magma chambers', where pressure builds due to expanding steam and gases associated with the magma. As pressure reaches a peak within these chambers, magma finds its way through a vent or fissure in the Earth's surface, resulting in a volcanic eruption and the expulsion of the hot molten rock. Volcanoes are classified by their level of activity. If a volcano is deemed to be an active volcano, then it is a volcano that has erupted at least once in the past 10,000 years. There are about 1500 active volcanoes on planet Earth, and this includes those under the world's oceans.

Volcanoes can also erupt under the ocean water. This is because many active volcanoes, both underwater and on land, occur due to plate tectonics. Openings at the plate boundaries where these plates meet each other allow hot molten magma to escape in a volcanic eruption, even if these plate boundaries are underwater. Not all volcanoes occur along plate boundaries, but many of the world's most active volcanoes are in these locations, e.g. the Ring of Fire is an area of the Pacific Ocean Basin that is tectonically active due to the presence of plate boundaries. This area is home to many active volcanoes, including Mount St. Helens, which erupted in 1980 in Washington State. It was the most destructive volcano eruption in the United States history.

If a volcano is not active, it could be a dormant volcano, which means that it is a volcano that has not erupted in the past 10,000 years but has the potential to erupt in the future. Mt. Rainier, which is also located in Washington State, is an example of a dormant volcano.

Though destructive processes may sound as though they are always negative and dangerous for humans, but new landforms are also created by earthquake and volcanoes and even in the marine system as seen in case of Lakshadweep. Lakshadweep Islands are coral islands in the Indian waters of Arabian Sea (Fig. 2.2). Several species of corals participate in forming the coral islands (*Vide* Annexure 2A).

Darwin suggested the formation of coral islands highlighting the simultaneous subsidence of volcanic formation and growth of fringing reefs to develop ring reefs.

Deposition is the geological process in which sediments, soil and rocks are added to a landform or land mass. Wind, ice, and water, as well as sediment flowing via

Fig. 2.2 Coral island at Lakshadweep – sample of a dead coral in hand

gravity, transport previously eroded sediment, which at the loss of enough kinetic energy in the fluid, is deposited, building up layers of sediment.

Deposition occurs when the forces responsible for sediment transportation are no longer sufficient to overcome the forces of gravity and friction, creating a resistance to motion and this is known as the null-point hypothesis. Deposition can also refer to the buildup of sediment from organically derived matter or chemical processes. For example, chalk is made up partly of the microscopic calcium carbonate skeletons of marine plankton, the deposition of which has induced chemical processes (diagenesis) to deposit further calcium carbonate. Similarly, the formation of coal begins with deposition of organic material, mainly from plants in anaerobic conditions.

Deposition of sediment in the coastal zone and estuarine system creates a suitable habitat for coastal vegetation particularly mangroves. The intertidal mudflats, which are caused by gradual deposition of sediments, sustain a rich genetic resource with diverse types of flora and fauna (Fig. 2.3).

2.3 Destructive Natural Processes

Processes that break down landforms through weathering, erosion and mass wasting are termed as destructive processes. Destructive processes are involved in continuously removing the surface irregularities/relief features that are created by constructive processes through denudation (including both weathering and erosion) activities. These processes are considered as exogenic processes, i.e. process which originate on the Earth surface.

Fig. 2.3 Sediment deposition in the midst of estuary

Destructive processes include weathering which is the disintegration of rocks by mechanical, chemical and biological agents. Erosion is the removal and transportation of weathered material by water, wind, ice or gravity and mass wasting which is the rapid downslope movement of materials by gravity.

In mangrove-dominated deltaic Sundarbans situated in the lower Gangetic delta erosion is caused by regular tidal movements and wave actions (Fig. 2.4). The magnitude of erosion becomes extremely high during natural disasters, like super cyclone, which is a normal feature in this deltaic complex.

The complete uprooting of the mangroves and subsequent erosion caused by the super cyclone AILA is presented in Fig. 2.5. This super cyclone hit the lower Gangetic delta during 24th–25th May 2009 which had an average wind speed of 120 km/hr. and even changed the hydrological characteristics of the estuary (see Annexure 2B for details).

2.4 Take Home Messages

(A) The Earth is made up of three main layers, namely, crust, mantle and core. The outermost layer of the Earth is the crust. It is like the peel of an orange. It is very thin compared to the other layers (mantle and core). The mantle is about 1800 miles (2900 km) thick and is separated into upper and lower sections.

Fig. 2.4 Erosion caused by regular tides and wave actions

Fig. 2.5 Massive erosion caused by the super cyclone AILA during 24th–25th May 2009

Large convective cells in the upper mantle circulate heat and drive plate tectonic processes of the crust. The last and innermost layer of the Earth is the core, which is separated into a liquid outer core and the solid inner core.

(B) The processes that build landforms through tectonic and depositional processes are termed as constructive processes. Constructive processes produce surface irregularities in the form of mountains, plateaus, faults, folds, volcanic cones, depressions, etc. These processes are considered as endogenic processes, i.e. process which originate from within the Earth.

(C) Deposition occurs when the forces responsible for sediment transportation are no longer sufficient to overcome the forces of gravity and friction, creating a resistance to motion, and this is known as the null-point hypothesis. Deposition can also refer to the buildup of sediment from organically derived matter or chemical processes.

(D) Processes that break down landforms through weathering, erosion and mass wasting are termed as destructive processes. Destructive processes are involved in continuously removing the surface irregularities/relief features such created by constructive processes through denudation (including both weathering and erosion) activities. These processes are considered as exogenic processes, i.e. process which originate on the Earth surface.

Destructive processes include weathering which is the disintegration of rocks by mechanical, chemical and biological agents. Erosion is the removal and transportation of weathered material by water, wind, ice or gravity and mass wasting which is the rapid downslope movement of materials by gravity.

2.5 Brain Churners

This section is an evaluation of the concept gained by the readers after critically studying the components of the Earth and the various natural processes that shape the Earth. The section will help the readers to face various competitive examinations as the Earth Science-related topics have been given thrust in recent times. The readers can consider this section as a sort of 'self-appraisal approach'. The correct answers to all the questions are filled in black in the Answer Box at the end of this section.

1. The thickness of the crust of the Earth is

   (a) 20 km
   (b) 8 km
   (c) 30 km
   (d) 5 km

2. Continental crust is mostly
   (a) Basalt
   (b) Muddy
   (c) Granite
   (d) None of the above
3. Oceanic crust is
   (a) Basalt
   (b) Granite
   (c) Silicate rocks
   (d) None
4. The thickness of the mantle is
   (a) 4000 km
   (b) 8000 km
   (c) 2000 km
   (d) 2900 km
5. Magma is the term used to describe
   (a) Rocks on the surface
   (b) Molten rocks within the Earth
   (c) Sediment of the ocean
   (d) Soil on the pond
6. Ring of Fire is located in the
   (a) Pacific Ocean Basin
   (b) Indian Ocean Basin
   (c) Atlantic Ocean Basin
   (d) Bay of Bengal Ocean Basin
7. Lakshadweep island is located in the
   (a) Bay of Bengal
   (b) Atlantic Ocean
   (c) Pacific Ocean
   (d) Arabian Sea
8. Mangroves are found in the
   (a) Coastal zone
   (b) Deep ocean
   (c) Rocky shores
   (d) None of the above
9. An example of destructive natural process is
   (a) Sedimentation
   (b) Erosion

(c) Evaporation
(d) Transpiration

10. AILA refers to

(a) Tidal action
(b) Super cyclone
(c) Wave action
(d) Volcanic eruption

11. Core is the

(a) Inner most layer of the Earth
(b) Outer layer of the Earth
(c) Middlemost layer of the Earth
(d) None of the above

12. The outer core is composed of

(a) Fe and Ni
(b) Hg and Cr
(c) Co and Zn
(d) None of the above

13. Endogenic process originates

(a) Outside the Earth
(b) Within the Earth
(c) From outer space
(d) None of the above

14. The crust of the Earth is mostly

(a) Siliceous
(b) Carbonaceous
(c) Sedimentary
(d) None of the above

15. Deposition of the sediment is

(a) Constructive process
(b) Destructive process
(c) Heating process
(d) Cooling process

16. The crust and the solid portion of the upper mantle is composed of a layer called

(a) Biosphere
(b) Lithosphere
(c) Atmosphere
(d) Ionosphere

17. Indian Sundarbans is situated in the
    (a) Pacific Ocean Basin
    (b) Indian Ocean Basin
    (c) Lower Gangetic Delta
    (d) Upper Gangetic Delta

18. Super cyclone causes
    (a) Erosion of embankment
    (b) Formation of island
    (c) Ecological succession
    (d) None of the above

19. Erosion is the rapid down slope movement of materials by
    (a) Gravity
    (b) Extraterrestrial force
    (c) Desalinization
    (d) Conductivity

20. Mt. Helens is a/an
    (a) Active
    (b) Dormant
    (c) Erosional zone
    (d) None of the above

Answer Sheet

S. No.	a	B	c	d
1.		■		
2.			■	
3.	■			
4.				■
5.		■		
6.	■			
7.				■
8.	■			
9.		■		
10.		■		
11.	■			
12.	■			
13.	■			
14.	■			
15.	■			
16.		■		
17.			■	
18.	■			
19.	■			
20.	■			

Annexure 2A: Common Types of Corals

Sl. No.	Genus	Representative
1.	*Pocillopora*	 *Pocillopora damicornis*
2.	*Madracis*	 *Madracis kirbyi*
3.	*Acropora*	 *Acropora hemprichii*
4.	*Montipora*	 *Montipora aequituberculata*

(continued)

Sl. No.	Genus	Representative
5.	*Astreopora*	 *Astreopora myriophthalma*
6.	*Pavona*	 *Pavona varians*
7.	*Pachyseris*	 *Pachyseris rugosa*
8.	*Siderastrea*	 *Siderastrea savignyana*
9.	*Pseudosiderastrea*	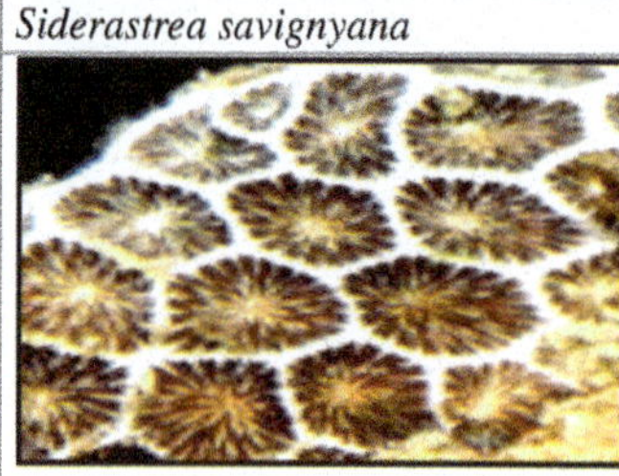 *Pseudosiderastrea tayami*

(continued)

Sl. No.	Genus	Representative
10.	*Coscinaraea*	 *Coscinaraea monile*
11.	*Psammocora*	 *Psammocora contigua*
12.	*Cycloseris*	 *Cycloseris cyclolites*
13.	*Goniopora*	 *Goniopora planulata*
14.	*Porites*	 *Porites lichen*

(continued)

Sl. No.	Genus	Representative
15.	*Favia*	 *Favia speciosa*
16.	*Favites*	 *Favites halicora*
17.	*Goniastrea*	 *Goniastrea pectinata*
18.	*Platygyra*	 *Platygyra sinensis*
19.	*Leptoria*	 *Leptoria phrygia*

(continued)

Sl. No.	Genus	Representative
20.	*Hydnophora*	 *Hydnophora microconos*
21.	*Leptastrea*	 *Leptastrea transversa*
22.	*Cyphastrea*	 *Cyphastrea microphthalma*
23.	*Echinopora*	 *Echinopora lamellosa*
24.	*Plesiastrea*	 *Plesiastrea versipora*

(continued)

Sl. No.	Genus	Representative
25.	*Galaxea*	 *Galaxea fascicularis*
26.	*Merulina*	 *Merulina ampliata*
27.	*Acanthastrea*	 *Acanthastrea echinata*
28.	*Lobophyllia*	 *Lobophyllia corymbosa*
29.	*Symphyllia*	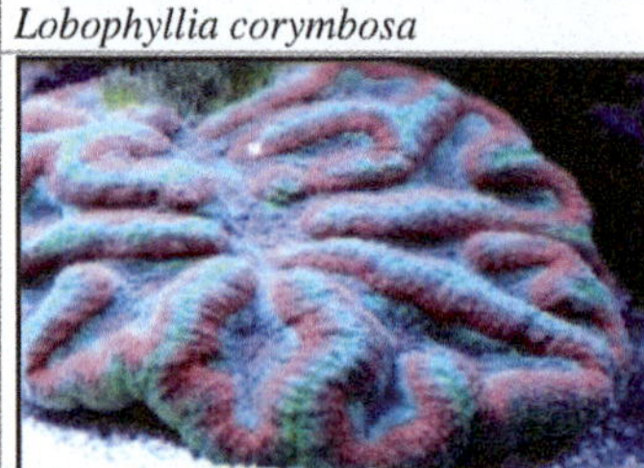 *Symphyllia radians*

(continued)

Sl. No.	Genus	Representative
30.	*Mycedium*	 *Mycedium elephantotus*
31.	*Polycyathus*	 *Polycyathus verrilli*
32.	*Heterocyathus*	 *Heterocyathus aequicostatus*
33.	*Heteropsammia*	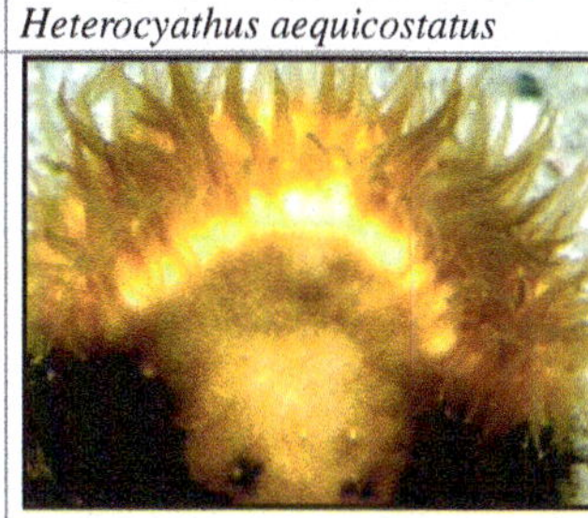 *Heteropsammia* sp.
34.	*Tubastrea*	 *Tubastrea aurea*

(continued)

Sl. No.	Genus	Representative
35.	*Dendrophyllia*	 *Dendrophyllia coarctata*
36.	*Turbinaria*	 *Turbinaria peltata*

Annexure 2B: Changing Hydrological Face of Indian Sundarban Estuary due to Super Cyclone AILA

Introduction

In tropical and some subtropical areas, organized cloud clusters form in response to perturbations in the atmosphere. If a cloud cluster forms in an area sufficiently away from the Equator, then Coriolis accelerations are not negligible and an organized, closed circulation may occur. A tropical system with a developed circulation, but with wind speeds of less than 17.4 metres/second (i.e. 63 km/h or 39 mph), is termed a tropical depression. Given that conditions are favourable for continued development (basically warm surface waters, little or no wind shear and a high pressure area aloft), this circulation can intensify to the point where sustained wind speeds exceed 17.4 metres/second, at which time it is termed a tropical storm. If development continues to the point where the maximum sustained wind speed equals or exceeds 33.5 metres/second (121 km/h or 75 mph), the storm is termed a cyclone (Indian Ocean and Bay of Bengal), typhoon (Western Pacific) or hurricane (Atlantic and Eastern Pacific).

The India Meteorological Department (IMD) classified cyclone on the basis of sustained wind speed in to six major types (Table 2A.1). Considering the speed of AILA (110 km/h) on May 25, 2009, in the Gangetic delta stretch, it can be designated as severe cyclonic storm (SCS). AILA was formed in the central Bay of

Table 2A.1 Tropical cyclone classifications (all winds are 10-min averages)

Storm type	Abbreviation	Wind Speed (km/h)
Super cyclone	SC	>221
Very severe cyclonic storm	VSCS	119–221
Severe cyclonic storm	SCS	88–118
Cyclonic storm	CS	63–87
Cyclonic depression	CDP	62 or less
Cyclonic disturbance during monsoon	CD	Not specified

Bengal as the net output of several factors. Around May 20, 2009, monsoon initiated at Andaman. Under its influence, the southerly surge over the region increased. It resulted in increase in the horizontal pressure gradient and the north–south wind gradient over the region. Hence the lower level horizontal convergence and relative vorticity increased gradually over the southeast Bay of Bengal. This condition triggered the development of the upper air cyclonic circulation extending up to mid-tropospheric level on May 21 over the southeast Bay of Bengal and associated convective cloud clusters persisted over the region. Under the influence of the cyclonic circulation, a low pressure area developed over the southeast Bay of Bengal on 22nd May morning. It laid over east central and adjoining west central Bay of Bengal on 22nd evening. It concentrated into a depression and lay centred at 1130 h IST of 23rd near Lat. 16.5° N/Long 88.0° E about 600 km south of Sagar Island, the largest island in Indian Sundarbans out of 102 island clusters.

According to INSAT imageries, a low-level circulation developed over South Bay of Bengal on May 21, 2009, at 0830 h IST. It developed into a Vortex with centre 11.5° N/ 85.5° E and intensity T1.0 at 1730 h. IST on the same day. It gained intensity of T1.5 corresponding to depression with centre 16.5° N/88.0° E at 1130 h IST of 23rd May. It was the shear pattern at the time of cyclogenesis with maximum convection lying to the southwest of the system centre.

Considering the environmental factors for cyclogenesis, the wind shear between the layers (150–300) hPa and (700–925) hPa was 05–10 knots on 21st and 22nd May according to METEOSAT observations. It became 10–20 knots on 23rd May. However, it continued to be 05–10 knots in the northeast sector of the system. The sea surface temperature (SST) was warmer (about 28 °C) over central and north Bay of Bengal, being 0.5 to 10 °C above normal. There was maximum lower level convergence to the southeast of the system centre. Similarly, the upper level divergence and the lower level relative vorticity were higher around the system centre. The system could gain upper level divergence as the upper tropospheric ridge roughly ran along 17° in association with an anti-cyclonic circulation located near lat. 17° N and long. 94° E. The quickscat-derived wind speed was about 10–15 knots on 21st and 22nd. It became 15–20 knots on 23rd. However, the windsat observations indicated 25–30 knots wind on 23rd in association with the system. The wind speed was relatively stronger in the southeast sector due to strong southerly surge of the monsoon current. All these observations indicate that the environmental factors were favourable for intensification of the system.

The cyclone retained its intensity for about 15 h after it hit the landmasses as it was close to the Bay of Bengal. It laid centred over the Gangetic delta for a considerable period of time, ascertaining the availability of moisture. Due to occurrence of AILA, there was intrusion of saline water from Bay of Bengal into the Hooghly–Matla estuarine system. The present study was conducted on May 18, 2009 (before the AILA event, and even before the initiation of monsoon at Andaman), and May 29, 2009, just 4 days after AILA in 12 different stations in and around the Matla River. The study was extended further 10 days after the incidence to observe the behaviour of the selected hydrological parameters.

Materials and Methods

The entire network of the investigation encompassed the monitoring of hydrological parameters on May 18, 2009 (pre-AILA phase, designated as Phase A), May 29 (post-AILA period, referred to as Phase B) and June 4 (extended post-AILA period, referred to as Phase C) in 12 sampling stations in and around the Matla estuarine system of Indian Sundarbans (Table 2A.2). A GPS was used to fix the coordinates of sampling sites.

The surface water salinity was recorded by means of an optical refractometer (Atago, Japan) in the field and cross-checked in laboratory by employing Mohr–Knudsen method (Strickland and Parsons 1972). The correction factor was found out by titrating the silver nitrate solution against standard seawater (IAPO standard seawater service Charlottenlund, Slot Denmark, chlorinity = 19.376‰). Our method was applied to estimate the salinity of standard seawater procured from NIO and a standard deviation of 0.02% was obtained for salinity. The average accuracy for salinity (in connection to our triplicate sampling) is ±0.28 psu. Glass bottles of

Table 2A.2 Location of the sampling stations

Station	Station no.	Geographical location	
		Longitude	Latitude
Canning	Stn. 1	88°41′16.20″	22°18′40.25″
Sandeshkhali	Stn. 2	88°48′17.62″	22°16′33.76″
Sonakhali	Stn. 3	88°42′24.76″	22°12′34.46″
Gosaba	Stn. 4	88°49′08.99″	22°10′57.92″
Amlamethi	Stn. 5	88°44′37.02″	22°03′08.92″
Sajnekhali	Stn. 6	88°48′17.60″	22°16′33.79″
Kultali	Stn. 7	88°38′39.19″	22°04′36.15″
Chotomollakhali	Stn. 8	88°54′26.71″	22°10′40.00″
Satjelia	Stn. 9	88°52′49.51″	22°05′17.86″
Kumirmari	Stn. 10	88°57′06.47″	22°11′26.46″
Pakhiralaya	Stn. 11	88°48′29.00″	22°07′07.23″
Netidhopani	Stn. 12	88°49′00.64″	21°55′36.10″

125 ml were filled to overflow from collected water samples and Winkler titration was performed for the determination of dissolved oxygen. The pH was recorded with a portable pH metre (Hanna, USA), which has an accuracy of ±0.1.

Results and Discussion

Tropical cyclones and storms are more common in the Bay of Bengal. They severely affect the eastern coast of India as compared to that of the Arabian Sea. According to Koteswaram (1984), there were about 346 cyclones that include 133 severe ones in the Bay of Bengal, whereas the Arabian Sea had only 98 cyclones including 55 severe ones between the years 1891 and 1970. These cyclones with tremendous speed hit the coastline and inundate the shores with strong tidal wave, severely destroying and disturbing coastal resources. The intrusion of seawater in to the upstream riverine zone through estuaries, creeks and inlets has high probability to alter the chemical composition of the aquatic phase, which is a subject of present discussion.

The present study recorded the increase of surface water salinity by 15.65%, 19.13%, 18.70%, 18.78%, 18.34%, 16.56%, 24.18%, 22.22%, 21.79%, 14.32%, 20.45% and 23.80% at Canning (Stn. 1), Sandeshkhali (Stn. 2), Sonakhali (Stn. 3), Gosaba (Stn. 4), Amlamethi (Stn. 5), Sajnekhali (Stn. 6), Kultali (Stn. 7), Chotomollakhali (Stn. 8), Satjelia (Stn. 9), Kumirmari (Stn. 10), Pakhiralaya (Stn. 11) and Netidhopani (Stn. 12), respectively, and this increase is significant at 1% level (Tables 2A.3 and 2A.4). Salinity of water can greatly affect organisms

Table 2A.3 Variations of hydrological parameters during different phases before and after AILA

Station	Surface water salinity (Phase A)	Surface water salinity (Phase B)	Surface water salinity Phase C)	pH (Phase A)	pH (Phase B)	pH (Phase C)	DO (Phase A)	DO (Phase B)	DO (Phase C)
Stn. 1	10.29	11.90	11.05	8.26	8.26	8.27	4.86	4.33	4.91
Stn. 2	10.82	12.89	11.93	8.26	8.26	8.26	4.84	4.04	4.97
Stn. 3	20.21	23.99	21.56	8.28	8.29	8.28	5.11	4.97	5.09
Stn. 4	21.78	25.87	22.18	8.28	8.29	8.29	5.49	4.54	5.67
Stn. 5	22.74	26.91	24.02	8.30	8.30	8.30	5.37	4.99	5.38
Stn. 6	21.92	25.55	22.01	8.28	8.29	8.30	4.91	5.03	4.99
Stn. 7	22.46	27.89	23.77	8.30	8.30	8.30	4.79	4.11	5.35
Stn. 8	22.00	26.89	23.10	8.30	8.30	8.30	5.05	4.90	5.96
Stn. 9	22.17	27.00	24.58	8.30	8.30	8.30	5.76	5.12	5.98
Stn. 10	21.99	25.14	22.10	8.29	8.30	8.30	5.92	4.75	5.12
Stn. 11	22.05	26.56	23.17	8.30	8.30	8.30	5.60	4.89	5.17
Stn. 12	25.67	31.78	27.80	8.30	8.31	8.30	5.85	4.10	5.44

Phase A, pre-AILA period (18.05.2009); Phase B, post-AILA phase (29.05.2009); Phase C, Extended post-AILA phase (05.06.2009)

Units of surface water salinity and DO are psu and ppm, respectively

Table 2A.4 ANOVA for surface water salinity

Source of variation	SS	df	MS	F	P-value	F_{crit}
Stations	620.81	11	56.44	66.70	1.94×10^{-8}	2.82
Pre- and post-AILA situation *(Phases A and B)*	97.08	1	97.08	114.74	3.70×10^{-7}	4.84
Error	9.31	11	0.85			
Total	727.20	23				

Table 2A.5 ANOVA for surface water pH

Source of variation	SS	df	MS	F	P-value	F_{crit}
Stations	0.01	11	0	35.97	5.17×10^{-7}	2.82
Pre- and post-AILA situation *(Phases A and B)*	0	1	0	7.86	0.02	4.84
Error	0	11	1.33×10^{-5}			
Total	0.01	23				

Table 2A.6 ANOVA for dissolved oxygen (DO)

Source of variation	SS	df	MS	F	P-value	F_{crit}
Stations	2.35	11	0.21	1.68	0.20	2.82
Pre- and post-AILA situation *(Phases A and B)*	2.52	1	2.52	19.90	0	4.84
Error	1.39	11	0.13			
Total	6.26	23				

surviving in the system. In the *Chattonella marina*, it has been stated that the salinity of the water can contribute to the growth rate of harmful algal blooms in that area. These blooms may pose an adverse effect on secondary production especially by killing local fish (Liu et al. 2007). While salinity has been known to have a positive impact on the growth of algal blooms, high salinity has been known to stunt algae growth, which can affect the productivity of the system.

The pH value also increased in some stations (e.g. 0.12% increase at Sonakhali, Gosaba, Sajnekhali, Kumirmari and Netidhopani) as confirmed by the ANOVA (Table 2A.5). High pH can affect the benthic community by way of precipitating the heavy metals from the aquatic phase to underlying sediment compartment (Mitra and Choudhury 1992, 1993a, b).

The DO level did not vary significantly between the stations ($F_{obs} = 1.682 < F_{crit} = 2.818$) in the study area. However, after the AILA the DO level in the surface water of the selected stations decreased within the range 2.74% to 29.91% (exception: 2.44% increase at Sajnekhali), and this decrease was significant at 1% level as revealed from ANOVA (Table 2A.6). The low DO observed during the AILA period (Phase B) may be attributed to high salinity, which may result in the mortality of aquatic life or affect their metabolic process and rate. Depletion in dissolved oxygen (and resulting decrease in water quality) can cause major shifts in the kinds of aquatic organisms found in water bodies, and evidences of such shifts, if any, need to be properly documented for such situation originating from tidal surges and intrusion of seawater.

Table 2A.7 ANOVA for salinity of post-AILA (Phase B) and extended post-AILA (Phase C)

Source of variation	SS	df	MS	F	P-value	F_{crit}
Stations	644.77	11	58.62	98.06	2.44×10^{-9}	2.82
Phase B and Phase C situation	51.33	1	51.33	85.88	1.57×10^{-6}	4.84
Error	6.56	11	0.60			
Total	702.68	23				

Table 2A.8 ANOVA for pH of post-AILA (Phase B) and extended post-AILA (Phase C)

Source of variation	SS	df	MS	F	P-value	F_{crit}
Stations	0	11	0	23.67	4.54×10^{-6}	2.82
Phase B and Phase C situation	-9.1×10^{-3}	1	-9.1×10^{-3}	-5.0×10^{-8}	0	4.84
Error	0	11	1.82×10^{-5}			
Total	0	23				

Table 2A.9 ANOVA for DO of post-AILA (Phase B) and extended post-AILA (Phase C)

Source of variation	SS	df	MS	F	P-value	F crit
Stations	2.11	11	0.19	1.78	0.18	2.82
Phase B and Phase C situation	2.84	1	2.84	26.36	0`	4.84
Error	1.19	11	0.11			
Total	6.14	23				

We conducted a similar study in the same locations 10 days after the AILA to evaluate the status of hydrological parameters and observed a recovery trend in all cases. A decrease in aquatic salinity is observed (7.14% at Canning, 7.45% at Sandeshkhali, 10.13% at Sonakhali, 14.26% at Gosaba, 10.74% at Amlamethi, 13.86% at Sajnekhali, 14.77% at Kultali, 14.09% at Chotomollakhali, 8.96% at Satjelia, 12.09% at Kumirmari, 12.76% at Pakhiralaya and 12.52% at Netidhopani) and the decrease is significant as confirmed through ANOVA (Table 2A.7). This is a clear indication of restoration of the water quality (in terms of salinity) after the AILA incidence. The relatively high salinity in the extended period (phase C) in comparison to pre-AILA condition (Phase A) is a normal trend of the geographical locale as pointed out earlier by several workers (Mitra et al. 2009). Such increase will continue till the end of June, and with the onset of monsoon the salinity value will drop. However the decreased percentage of Phase C in relation to post-AILA condition (Phase B) is a clear indication of the gradual restoration of the system, which could have been more effective if the dilution of the system, could be increased through fresh water discharge. This is, however, not feasible in this geographical locale as huge siltation of Bidyadhari basin has cut-off the fresh water supply of Bhagirathi-Ganga to this sector. The restoration of pH and DO through increased level was also confirmed through ANOVA (Tables 2A.8 and 2A.9).

The effect of AILA was severe resulting in the death of people, damages to properties and alterations in the physico-chemical characteristics of the soil and water. Embankments were destroyed, agricultural fields and freshwater ponds lost their

Table 2A.10 Damage spectrum of AILA

Affected	West Bengal	Sundarbans (Indian part)
Villages	28,359	4249
Population	670,000	254,000
Human lives lost	137	70
Livestock lost	71,196	70,811
Crop area affected (ha)	290,000	125,000

productivity due to intrusion of saline water and many people became homeless. 70 island dwellers died in the Indian Sundarbans region. Over 8000 people could not be traced and about a million became homeless in India and Bangladesh. The embankments, which act as the line of defence of the villages, were completely smashed by tidal surges. Out of 3500 km long stretch of embankment, 400 km were completely broken and another 565 became highly vulnerable. The agricultural fields were submerged by sea water, and all standing crops were destroyed. The fresh water ponds were transformed into brackish water bodies, and several fish species of saline water were recovered from these ponds. The entire spectrum of damages is summarized in Table 2A.10.

The geographical and geological setting of Indian Sundarbans has made the area extremely vulnerable to cyclones and tidal surges. Dube et al. (2004) created models to simulate the storm surges from past cyclones in the head of the Bay of Bengal (Orissa, West Bengal and Bangladesh) and discussed the formation and effect of each of these storms that were modelled. The extent of storm surges in the Bay of Bengal region and the causes behind extreme sea levels were also analysed by several workers (Dube et al. 1997). The main reasons behind the vulnerability are:

- Coastal waters (shallow bathymetry extending tens of kilometres offshore)
- Convergence of the bay
- High astronomical tides
- Thickly populated low-lying islands
- Favourable cyclone tracks impacting perpendicular to coastline
- Innumerable inlets and river systems

The situation of deltaic Sundarbans coincides with most of these points. Being located at the apex of Bay of Bengal, the islands of the deltaic complex (102 in numbers) are criss-crossed with networks of creeks and inlets. The geological formation of Sundarbans is of comparatively recent origin. Till a few thousand years back, the whole tract was under the sea. The deposition of debris and formation of Sundarbans delta occurred recently with the change of main course of River Ganga from the Bhagirathi to Padma towards the east between the twelfth and fifteenth century A.D. Basically this was the result of Bengal basin suffering from neotectonic movement and an easterly tilt. During the sixteenth century, the flow of Ganga shifted almost totally eastwards into River Padma (now in Bangladesh) and the Matla/Bidyadhari rivers which had formed innumerable network of creeks in the delta got completely cut-off from the sweet water sources. This is another major reason behind increased vulnerability in and around Matla estuary to cyclone induced waves as the pressure of fresh water supply cannot prevent the saline intrusion to upstream zone. The Gangetic delta of Indian sub-continent is a cyclone prone zone, and the event of AILA is an eye opener to think of a strong action plan, so that the damage can be minimized in future. In this context, it is extremely essential to dredge the Bidyadhari channel so that the fresh water from western sector (Bhagirathi-Ganga-Hooghly channel) sourced from Farakka barrage may be diverted in the central Indian Sundarbans in and around the Matla estuary. This will not only reduce the salinity intrusion during extreme tidal surges and tropical storms but may also lead to development of a congenial and uniform salinity profile in the region.

References

Dube SK, Rao AD, Sinha PC, Murty TS, Bahulayan N (1997) Storm surge in the Bay of Bengal and Arabian Sea: the problem and its prediction. Mausam 48:283–304

Dube SK, Chittibabu P, Sinha PC, Rao AD, Murty TS (2004) Numerical modeling of storm surge in the head Bay of Bengal using location specific model. Nat Hazards 31:437–453

Koteswaram P (1984) Climate and mangrove forests. Report of the second introductory training course on mangrove ecosystems. Sponsored by UNDP and UNESCO, Goa, pp 29–46

Liu W, Au Doris, Anderson D, Lam P, Rudolf Wu (2007). Effects of nutrients, salinity, pH and light: dark cycle on the production of reactive oxygen species in the alga *Chattonella marina.* J Exp Mar Biol Ecol 346(½):76–86., 11p

Strickland JDH, Parsons TR (1972) A practical handbook of sea-water analysis (2nd edn). J Fish Res Bd Canada 167:311

Mitra A, Choudhury A (1992) Seasonal variations in metal content in the gastropod *Nerita articulata* (*Gould*). Indian J Environ Health NEERI 35(1):31–35

Mitra A, Choudhury A (1993a) Heavy metal concentrations in oyster *Crassostrea cucullata* of Sagar Island, India. Indian J Environ Health NEERI 35(2):139–141

Mitra A, Choudhury A (1993b) Trace metals in macrobenthic molluscs of the Hooghly estuary, India. Mar Poll Bull UK 26(9):521–522

Mitra A, Banerjee K, Sengupta K, Gangopadhyay A (2009) Pulse of climate change in Indian Sundarbans: a myth or reality. Natl Acad Sci Lett 32(1 & 2):1–7

Chapter 3
Earth and Its Resources

Abstract The wheels of human prosperity run on the resource base of the region. The mineral resources of the land and the aquatic resources act as the foundation of economic profile of the nation. However, rapid industrialization and urbanization are fast depleting the resource base of the Earth, which have been critically dealt in this chapter. Many of the aquatic resources are getting deteriorated due to pollution that has been highlighted in the annexure section with real-time case study.

Keywords Land resources · Aquatic Resources · Pollution · Exploitation

3.1 Land Resources

Land is an essential natural resource, both for the survival and prosperity of humanity and for the maintenance of all terrestrial ecosystems. Over millennia, people have become progressively more expert in exploiting land resources for meeting their own needs. The limits on these resources are finite, while human demands on them are not. Increased demand, or pressure on land resources, results in the declining of crop production, degradation of land quality and quantity and competition for land.

The lithosphere, sometimes called the geosphere, refers to all of the rocks of the Earth. It is the topmost layer of the Earth whose main components are (a) firm and solid rocks and (b) the unconsolidated materials, often called soil or regolith. A rock does not possess a definite composition like that of a chemical compound; rather it is a mixture of various minerals. Thus a rock may be defined as 'an aggregate of minerals that forms a more or less definite unit of the earth's crust'. The bulk of rock is made up of silica or quartz and feldspar. Rocks are of great economic value to man. The following are some of the direct benefits of rocks:

1. Rocks through the process of disintegration by mechanical and chemical processes get converted into valuable soils. The back soil of the Deccan Trap (India) is the result of extrusive igneous rocks. All other soils are also the results of disintegrated rocks.

A. Mitra, S. Zaman, *Environmental Science - A Ground Zero Observation on the Indian Subcontinent*, https://doi.org/10.1007/978-3-030-49131-4_3

2. Rocks are the storehouses of a great variety of minerals which are used in various ways in manufacturing industries.
3. Rocks are also a source of fuel minerals, such as coal, petroleum and natural gas.
4. Rocks supply different types of stones, which are of great use in constructing houses, building, dams, roads, etc. Lime and cement, which are important building materials, are also the products of rocks.
5. Rocks are the sources of precious metals like gold, silver, platinum, etc., which are used for making ornaments.

The Indian sub-continent possesses a variety of mineral resources. Among the various mineral resources of the land, the most important are coal, manganese ore, mica, iron ore and salt. Besides these, there are several other important minerals, such as bauxite, ilmenite, monazite and zircon. However, some minerals are not found in sufficient quantity in the sub-continent of India, and these include petroleum, chromite, gypsum, mercury, tin, lead, nickel, copper and zinc.

Major minerals found in the country along with their estimated reserves are given below:

1. Bauxite Ore (Aluminium)

The total reserve is 3.076 million tonnes. About 84 percent of this reserve is of metallurgical grade. The conditional resources of bauxite are about 5, 99,780 tonnes. In addition, prospective resources are placed at 90 million tonnes.

Odisha, Andhra Pradesh, Chhattisgarh, Gujarat, Maharashtra and Jharkhand are the principal states where bauxite deposits are located. Major reserves are concentrated in the east coast bauxite deposits of Odisha and Andhra Pradesh.

2. Chromite

Total in situ reserves of chromite are estimated at 114 million tonnes. Total geological resources were estimated at 187 million tonnes, consisting of around 114 million tonnes in situ reserves and about 73 million tonnes as conditional resources.

The largest share (about 96 percent) in the total geological resources is accounted by the Cuttack district in Odisha. Deposits of economic significance occur in Odisha, Karnataka, Maharashtra, Jharkhand, Madhya Pradesh, Chhattisgarh, Tamil Nadu and Manipur. However, refractory grade reserves of chromite are very meagre.

3. Copper

The total in situ reserves of copper ore in the country are 712.5 million tonnes, equivalent to 9.4 million tonnes of metal content. The all-India conditional resources of copper are 722 million tonnes (3.15 million tonnes of copper metal), and prospective resources are 0.6 million tonnes of copper ore.

Major and important copper ore deposits are located in Singhbhum district (Jharkhand), Balaghat district (Madhya Pradesh) and Jhunjhunu and Alwar districts (Rajasthan). In addition, there are small copper ore deposits in Gujarat, Karnataka, Andhra Pradesh, Uttar Pradesh, Sikkim, Meghalaya, Maharashtra and West Bengal.

4. Gold

There are three important gold fields in the country, namely, Kolar Gold Fields, Kolar district, Hutti Gold Field in Raichur district (both in Karnataka) and Ramagiri Gold Field in Anantapur district (Andhra Pradesh). Total in situ reserves of gold ore are estimated at 22.4 million tonnes, with 116.50 tonnes of metal.

5. Iron Ore

The total in situ reserves of iron ore in the country are about 1,23,17,275 thousand tonnes of haematite and 53,95,214 thousand tonnes of magnetite. The resources of very high-grade ore are limited and are restricted mainly in the Bailadila sector of Chhattisgarh and to a lesser extent in Bellary-Hospet area of Karnataka and Barajamda sector in Jharkhand and Odisha.

Haematite resources are located in Odisha, Jharkhand, Chhattisgarh, Karnataka, Goa, Maharashtra, Andhra Pradesh and Rajasthan. Magnetite resources are located in Karnataka, Andhra Pradesh, Goa, Kerala, Jharkhand, Rajasthan and Tamil Nadu.

6. Lead–Zinc

Lead–Zinc resources are located in Rajasthan, West Bengal, Andhra Pradesh, Gujarat, Madhya Pradesh, Uttar Pradesh, Odisha, Maharashtra, Meghalaya, Tamil Nadu and Sikkim. Total in situ reserves (all grades) of lead and zinc ores are 231 million tonnes comprising metal content of 5.1 million tonnes of lead and 17.02 million tonnes of zinc metal.

7. Manganese

The total in situ resources of manganese ore are 406 million tonnes out of which 104 million tonnes are proved, 135 million tonnes are in probable and 167 million tonnes are in possible categories. Main reserves fall in Karnataka, followed by Odisha, Madhya Pradesh, Maharashtra and Goa. Minor occurrences of manganese are in Andhra Pradesh, Jharkhand, Gujarat, Rajasthan and West Bengal.

8. Tungsten

The total in situ reserves of tungsten ore have been estimated at 43.15 million tonnes or 1, 32,478 tonnes of $W0_3$ content. The main reserves are at Degana, Rajasthan. It also occurs in Maharashtra, Haryana, West Bengal and Andhra Pradesh.

9. Diamond

Diamond deposits occur in three types of geological settings such as kimberlite pipes, conglomerate beds and alluvial gravels. The main diamond bearing areas in India are the Panna belt in Madhya Pradesh, Munimadugu-Banganapalle conglomerate in Kurnool district, Wajrakarur kimberlite pipe in Anantapur district and the gravels of Krishna river basin in Andhra Pradesh.

Reserves have been estimated only in the Panna belt and Krishna Gravels in Andhra Pradesh. The total in situ reserves are about 26, 43,824 carats. There are conditional resources of 19, 36,512 carat. The new kimberlite fields are discovered recently in Raichur–Gulbarga districts of Karnataka.

10. Dolomite

Dolomite occurrences are widespread in almost all parts of the country. The total in situ reserves of all grades of dolomite are 7349 million tonnes. The major share of about 90 percent reserves is distributed in the states of Madhya Pradesh, Chhattisgarh, Odisha, Gujarat, Karnataka, West Bengal, Uttar Pradesh and Maharashtra.

11. Fluorspar

Total in situ reserves of fluorspar in India are 14.15 million tonnes. Commercial deposits of fluorspar are located in Gujarat, Rajasthan, Madhya Pradesh and Maharashtra.

12. Gypsum

The in situ reserves of gypsum are estimated at 383 million tonnes. Out of this, two millions are of surgical/plaster grade. 92 million tonnes of fertilizer/pottery grade, 76 million tonnes of cement/paint grade and 13 million tonnes of soil reclamation grade, and the rest is unclassified. The production of gypsum is confined to Rajasthan, Tamil Nadu, Jammu and Kashmir and Gujarat. Rajasthan is the main producer of gypsum followed by Jammu and Kashmir.

13. Graphite

The in situ reserves of graphite are 16 million tonnes. Odisha is the major producer of graphite. Almost the entire reserves of Tamil Nadu under the proved category are in Ramanathapuram district. Deposits of commercial importance are located in Andhra Pradesh Jharkhand, Gujarat, Karnataka, Kerala, Madhya Pradesh, Maharashtra, Odisha, Rajasthan and Tamil Nadu.

14. Limestone

The total in situ reserves of limestone of all categories and grades are placed at 1, 69,941 million tonnes. The total conditional reserves have been estimated at 3713 million tonnes.

The major share of its production comes from Madhya Pradesh, Chhattisgarh, Andhra Pradesh and Gujarat: Rajasthan, Karnataka, Tamil Nadu, Maharashtra, Himachal Pradesh, Odisha, Bihar, Uttaranchal and Uttar Pradesh. The remaining part comes from Assam, Haryana, Jammu and Kashmir, Ker and Meghalaya.

15. Mica

India is the world's leading producer of sheet mica and accounts for about 60 percent of global mica trade. Important mica-bearing pegmatite occurs in Andhra Pradesh, Jharkhand Bihar and Rajasthan.

The total in situ reserves of mica in the country are placed at 59, often tonnes. The in situ reserves of mica in Andhra Pradesh are 42,626 thousand tonnes, Bihar 12,938 tonnes, Jharkhand 1494 tonnes and in Rajasthan 2007 tonnes.

16. Magnesite

The total in situ reserves of magnesite are about 415 million tonnes of which 76 million tonnes are in the proved category. Major deposits of magnesite are found in Uttaranchal, Tamil Nadu and Rajasthan, while minor deposits are in Jammu and Kashmir, Karnataka, Himachal Pradesh and Kerala.

17. Other Minerals

Other minerals occurring in significant quantities in India are bentonite (Rajasthan, Gujarat, Jharkhand and Jammu and Kashmir), corundum (Karnataka, Andhra Pradesh, Rajasthan and Chhattisgarh), calcite (Andhra Pradesh, Rajasthan, Madhya Pradesh, Tamil Nadu, Haryana, Karnataka, Uttar Pradesh and Gujarat), fuller's earth (Rajasthan, Jharkhand, Bihar, Andhra Pradesh, Tamil Nadu, Maharashtra, West Bengal and Karnataka), garnet (Tamil Nadu, Odisha, Andhra Pradesh, Rajasthan and Kerala), pyrites (Jharkhand; Rajasthan, Karnataka, Himachal Pradesh and Andhra Pradesh), steatite (Rajasthan, Uttar Pradesh, Kerala, Maharashtra and Madhya Pradesh), wollastonite (Rajasthan and Gujarat), zircon (beach sand of Kerala, Tamil Nadu, Andhra Pradesh and Odisha) and quartz, and silica minerals are widespread and occur in nearly all states.

Besides, the country has vast marble, slate and sandstone. Granite is mainly mined in Tamil Nadu, Karnataka, Andhra Pradesh and Rajasthan; marble in Rajasthan, Gujarat and Uttar Pradesh; slate in Chhattisgarh, Madhya Pradesh, Haryana and Andhra Pradesh and sandstone in Rajasthan.

3.2 Aquatic Resources

Aquatic resources are the natural resources that are used for many purposes like aquaculture, agriculture, industry, household purpose, recreation and environmental activities. 97% of the water on the planet Earth is salt water and only 3% is freshwater. The aquatic resources are thus mainly confined in the salty medium although commercially important fish species are contributed by freshwater systems like rivers, ponds, lakes, etc. For the benefit of the readers, the entire spectrum of aquatic resources has been divided into freshwater and saline water resources.

3.2.1 Freshwater Resources

Surface water is water in a river, lake or freshwater wetland. Surface water is naturally replenished by precipitation and naturally lost through discharge to the oceans, evaporation, evapotranspiration and groundwater recharge.

Although the only natural input to any surface water system is precipitation within its watershed, the total quantity of water in that system at any given time is also dependent on many other factors. These factors include storage capacity in lakes, wetlands and artificial reservoirs, the permeability of the soil beneath these storage bodies, the runoff characteristics of the land in the watershed, the timing of the precipitation and local evaporation rates. All of these factors also affect the proportions or quantum of water loss.

Human activities can have a large and sometimes devastating impact on these factors. Humans often increase storage capacity by constructing reservoirs and decrease it by draining wetlands. Humans often increase runoff quantities and velocities by paving areas and channelizing the stream flow.

The total quantity of water available at any given time is an important consideration. Some human water users have an intermittent need for water. For example, many farms require large quantities of water in the spring, and no water at all in the winter. To supply such a farm with water, a surface water system may require a large storage capacity to collect water throughout the year and release it in a short period of time. Other users have a continuous need for water, such as a power plant that requires water for cooling. To supply such a power plant with water, a surface water system only needs enough storage capacity to fill in when average stream flow is below the power plant's need.

Nevertheless, over the long term, the average rate of precipitation within a watershed is the upper bound for average consumption of natural surface water from that watershed.

Natural surface water can be augmented by importing surface water from another watershed through a canal or pipeline. It can also be artificially augmented from any of the other sources; however, in practice the quantities are negligible. Humans can also cause surface water to be 'lost' (i.e. become unusable) through pollution.

Exploitation

The concept of water stress is relatively simple: according to the World Business Council for Sustainable Development, it applies to situations where there is not enough water for all uses, whether agricultural, industrial or domestic. Defining thresholds for stress in terms of available water per capita is more complex, however, entailing assumptions about water use and its efficiency. Nevertheless, it has been proposed that when annual per capita renewable freshwater availability is less than 1700 cubic metres, countries begin to experience periodic or regular water stress. Below 1000 cubic metres, water scarcity begins to hamper economic development and human health and well-being.

Population Growth
In 2000, the world population was 6.2 billion. The UN estimates that by 2050 there will be an additional 3.5 billion people with most of the growth in developing countries that already suffer water stress. Thus, water demand will increase unless there are corresponding increases in water conservation and recycling of this vital resource. In building on the data presented here by the UN, the World Bank goes on to explain that access to water for producing food will be one of the main challenges in the decades to come. Access to water will need to be balanced with the importance of managing water itself in a sustainable way while taking into account the impact of climate change and other environmental and social variables.

Expansion of Business Activity
Business activity ranging from industrialization to services such as tourism and entertainment continues to expand rapidly. This expansion requires increased water services including both supply and sanitation, which can lead to more pressure on water resources and natural ecosystem.

Rapid Urbanization
The trend towards urbanization is accelerating. Small private wells and septic tanks that work well in low-density communities are not feasible within high-density urban areas. Urbanization requires significant investment in water infrastructure in order to deliver water to individuals and to process the quantum of wastewater – both from individuals and from business. These polluted and contaminated waters must be treated, which may otherwise pose unacceptable public health risks.

Climate Change
Climate change could have significant impacts on water resources around the world because of the close connections between the climate and hydrological cycle. Rising temperatures will increase evaporation and lead to increases in precipitation, though there will be regional variations in rainfall. Both droughts and floods may become more frequent in different regions at different times, and dramatic changes in snowfall and snow melt are expected in mountainous areas. Higher temperatures will also affect water quality in ways that are not well understood. Possible impacts include increased eutrophication. Climate change could also mean an increase in demand for farm irrigation, garden sprinklers, and perhaps even swimming pools. There is now ample evidence that increased hydrologic variability and change in climate has and will continue to have a profound impact on the water sector through the hydrologic cycle, water availability, water demand and water allocation at the global, regional, basin and local levels. Climate change-induced sea level rise results in the intrusion of saline water in the agricultural fields and destroys crop production (particularly paddy).

Depletion of Aquifers
Due to the expanding human population, competition for water is growing such that many of the world's major aquifers are becoming depleted. This is due both for direct human consumption and agricultural irrigation by groundwater. Millions of pumps of all sizes are currently extracting groundwater throughout the world.

Irrigation in dry areas such as northern China, Nepal and India is supplied by groundwater and is being extracted at an unsustainable rate. Cities that have experienced aquifer drops between 10 and 50 metres include Mexico City, Bangkok, Beijing, Chennai and Shanghai.

Water Pollution

Water pollution is one of the main concerns of the world today. The Governments of numerous countries have striven to find solutions to reduce this problem. Many pollutants threaten water supplies, but the most widespread, especially in developing countries, is the discharge of raw sewage into natural waters; this method of sewage disposal is the most common method in underdeveloped countries but also is prevalent in quasi-developed countries such as China, India, Nepal and Iran. Sewage, sludge, garbage and even toxic pollutants are all dumped into the water. Even if sewage is treated, problems still arise. Treated sewage forms sludge, which may be placed in landfills, spread out on land, incinerated or dumped at sea. In addition to sewage, non-point source pollution such as agricultural runoff is a significant source of pollution in some parts of the world, along with urban storm water runoff and chemical wastes dumped by industries. A detailed study conducted by the present authors for three decades (1984–2014) on two major estuaries of Indian Sundarbans exhibit significant increasing trends in nitrate, phosphate and silicate levels due to intense urbanization and several anthropogenic activities in and around the study sites (vide Annexure 3A for detailed data and information).

3.2.2 Ocean Resources

The ocean is a valuable reservoir of food, minerals, natural gas and oil. The ocean basin and beaches are rich sources of sand gravel, precious metal placers, titanium and chromite placers manganese nodules, cobalt-rich crust and polymetallic sulphide. The living resources of the ocean encompass various categories of finfish and shellfish, seaweeds, phytoplankton and the macrophytes. The blue soup of the Earth, occupying about 70% of the Earth's surface, can thus act as the key of strengthening the economic backbone of the country. The major industries based on marine resources are hard minerals, fisheries and mariculture, hydrocarbons, ocean energy, recreation and tourism and shipping and shipbuilding.

The vast spectrum of resources that can be tapped from the ocean may be classified into two broad headings, namely, abiotic resources and biotic resources.

(A) **Abiotic Resources**

Abiotic resources may be of various types as listed here.

Sediment Deposits at the Continental Margin

In the continental margin, sediment deposits include sand and gravel, calcium shells, placers beaded phosphorites and metalliferous or brine rich sediments. Found in the nearshore zones, placer deposits are mechanically concentrated heavy detrital materials which originate from eroded offshore rocks, often discharged by black or dark stringers and lenses or modern beaches. In the Indian sub-continent, huge deposits are concentrated on the western coast in the form of tin placers, complex heavy minerals and precious metal placers. Besides biotite, biotite–muscovite and cassiterite, tin placer deposits that contain pyrite, ilmenite, zircon, tourmaline and monazite are found in plenty in the Indian coast. Among the complex heavy minerals, sand containing principally ilmenite, rutile, zircon, monazite, garnet and magnetite and the iron-titaniferous sands have been identified at commercially exploitable level at southern Japan, East and West Australia, California, Oregon, Atlanta, New Zealand, India, Bangladesh and Sri Lanka. Precious metal placers containing gold, platinum and related metals have been found in Alaska, Oregon, California, Nova Scotia, Lapter Sea (Russia), Southeast Australia, Bering Sea, etc.

Polymetalllic Nodules

Ferromanganese nodules are important nonliving resources of the deep oceanic environment that are usually disturbed in the regions where sediments accumulate more slowly than about 7 m per million years. Variation in the nodule composition is a function of the variations in the vertical flux of transition metals carried by settling debris (largely of biogenic origin) and in the oxic and suboxic diagenetic reactions that break down biogenic debris at the sea floor. Nodules are usually spherical in shape and range from 1 cm to about 10 cm in diameter. Discoidal or irregular shapes are also common in addition to spherical shapes. When protected from burial, the nodules can grow even up to 30 cm. Although the polymetallic nodules are rich sources of heavy metals like Mn, Fe, Ni, Cu, Co, etc., their relative proportion or percentage varies significantly from one basin to another and also in different geologic settings.

Oil and Natural Gas

The seabed is rich sources of oil and natural gas. It has been recorded that the continental shelf around India up to depth of 200 m, with an area of around 4,52,000 km^2, can have a potential oil reserve of 1 billion tonnes and a nature gas around 271 billion cubic metres approximately.

In India most of the petroleum resources are confined to the narrow belt of tertiary strata which constitute the outer margin of the extra-peninsular mountains along their whole line of contact with the peninsular block of Deccan like the Rann of Kutch, Bombay High and the Andaman and Nicobar Islands. In early 1974, the Sagar Samrat drilling platform at Bombay High, acquired from Japan, started functioning, reaching production of more than 230.95 million tonnes in 1980, and thus initiated to strengthen the economic backbone of the country.

Natural gas, whose major component is marsh gas, is usually associated with petroleum deposits. It supplies the propulsive force which carries the oil from the oil sands into the wells, and if the pressure is sufficient, the oil is brought up to surface. The natural gas has a special status not only as a fuel but also for its role in the production of oil. In India, the production of natural gas was 19.25 and 19.97 billion cubic metres in 1980 and 1981, respectively.

(B) **Biotic Resources**

The ocean sustains billions of species ranging from tiny phytoplankton to blue whale, many of which have immense ecological, economical and biomedical values. The entire spectrum of the biotic resources may be conveniently divided into some important categories.

Fish Resource

The Food and Agriculture Organization of the United Nations (FAO), which is the principal and competent international dealing with global fisheries, has in its 1995 report on *The state of World Fisheries and Agriculture* a comprehensive analysis of the trends in global fish catch, inland and marine; capture fisheries and aquaculture, for consumption as human food and as fed for animals; and fish trade.

The global fish catch, which was 20.75 million tonnes in 1950 and was estimated in 1971 to reach 100 million tonnes by the year 2000, had in fact crossed 100 million tonnes in 1989, declined thereafter until 1992 and was 101.27 million tonnes in 1993. Out of 101.27 million ones (in 1993), the marine fish catch was 84.26 million tonnes, the inland catch being 17 million tonnes.

The annual potential yield of the marine fishery resources in the India EEZ has been assessed to be between 2.3 and 5.5 million tonnes by various authors. However, a recent assessment of a combination factors and using tested methodologies is more reliable. It puts it at a modest level of 3.92 million tonnes. The potential estimated annual yield in the fishery sector, the level of exploitation and the potential available for the exploitation in a depth-wise manner are given below in Table 3.1.

The estimate of total demersal stock is 1.93 million tonnes, of which the coastal zone up to 50 m depth supports 1.28 million tonnes, while the areas beyond this depth is expected to yield 0.653 million tonnes. The principal varieties of fish available from the deep-sea sector with their estimated potential are:

Threadfin breams: 110,600 tonnes
Mackerel: 62,200 tonnes
Bulls eye: 54,800 tonnes
Scads: 23,400 tonnes
Ribbon fishes: 23,300 tonnes
Sciaenids: 22,000 tonnes

Others important demersal species identified from the deep-sea sector with significant contributions are *Caranx* spp., *Saurida* spp., (lizard fish), *Ariomma indica* (Indian draft fish), clupeids, squids and cuttlefishes, perches, etc.

Table 3.1 Depth-wise estimated yield of fishery

Resource/depth	0–50 m	50–500 m	Oceanic	Total
Demersal	1.280	0.653	–	1.933
Pelagic	1.000	0.742	–	1.752
Oceanic	–	–	0.246	0.246
Total	2.280	1.395	0.246	3.921

Cartilaginous fishes like shark are also important biotic resources. There is an increasing demand for shark meat and shark fins, the latter being considered a delicacy in Asia. Numbers of sharks have sharply and rapidly declined in regions where fishing has intensified, and many sharks are also captured incidentally in commercial fishing operations for other species. In the Indian coastal waters, the common shark species are highlighted in Table 3.2.

Seaweed Resource

Seaweeds are large visible plants of the oceans, which are found attached to rocks, corals and other submerged strata in the intertidal and sub tidal zones of the sea.

Seaweeds are similar in form with the higher vascular plants, but the structure and function of the parts significantly differ from the higher plants. Seaweeds do not have true roots, stem or leaves and whole body of the plant is called thallus that consists of the holdfast, stipe and blade. The holdfast resembles the root of the higher plants, but its function is for attachment and not for nutrient absorption. The holdfast may be discoidal, rhizoidal, bulbous or branched depending on the substratum it attaches. The stipe resembles the stem of the higher plants, but its main function is for support of the blade for photosynthesis and for absorption of nutrients from surrounding seawater. The blade may resemble leaves of the higher plants and have variable forms (smooth, perforated, segmented, dented, etc.). The important functions of the blade are photosynthesis and absorption of nutrient. The blades may be flat, ruffled, feathery or even encrusted with calcium carbonate. The most significant difference of seaweeds from the higher plants is that their sex organs and sporangia are usually one celled or if multicellular; their gametes and spores are not enclosed within a wall formed by a layer of sterile or nonreproductive cells.

Today seaweeds are used for various commercial purposes. Agar and algin extracted from the seaweeds have a variety of industrial applications. Apart from consuming the seaweed as food, they are used as fodder, manure and medicine and also as ingredients of several health products as discussed here under different headings.

1. As Food: In many countries like Malaysia, Indonesia, Korea and Australia, seaweeds are used for human consumption. These are used in the preparation of salads, soups, jellies and vinegar. The common species utilized for the preparation of food are *Porphyra* sp., *Gracilaria* sp., *Laurencia* sp., *Caulerpa* sp., *Sargassum* sp., etc. These algae are rich in vitamin B.
2. As Fodder: Some important species of seaweeds that are used in cattle feed are *Ulva* sp., *Enteromorpha* sp., *Gracilaria* sp., *Padina* sp., *Sargassum* sp., etc. Poultry feed is also made in certain areas by mixing seaweeds with trash fishes.
3. As Manure: Seaweeds have unique capacity to bioaccumulate trace elements from ambient media, and therefore they are rich sources of macro and microelements required for the growth of plants. Seaweeds like *Sargassum* sp. and *Gracilaria* sp. are used as manure for coconut plantation in India, particularly in the south Indian states.
4. As Medicine: Seaweeds have wide applications in the sphere of medicine. *Sargassum* sp. and *Turbinaria* sp. are good sources of alginates, which are known to prolong the

Table 3.2 Common shark in Indian coastal waters

S. No.	Species	Distribution	Diet	Length	Reproduction
1.	*Carcharhinus limbatus* (blacktip shark) [IUCN status: lower risk]	Cosmopolitan in distribution near the inshore regions of tropical waters. It is capable of tolerating reduced salinity but never penetrates into freshwater	Fish such as sardine, mackerel, croaker, sole along with cephalopods and crustaceans	Matured ones are 2.5 m in length. Male matures at 140–150 cm, and female matures at 150–160 cm	Produces an average of six embryos per litter. Size at births is 55–60 cm
2.	*Carcharhinus sorrah* (spot-tail shark)	Commonly found in coral reefs region	Bony fish such as mackerels, sardines, etc. and shellfish like squids and prawns	Short in length, about 2.5 m. Male matures at 115 cm, and female matures at 120 cm	Litter size is 2–6 usually delivered during March to May. The size at birth is around 40 cm
3.	*Carcharhinus dussumieri* (white-cheeked shark) [IUCN status: lower risk]	Common species in inshore waters	Small fishes, squids and crustaceans	Matured ones are 1 m. Male matures at 65 cm, and female matures at 75 cm	Produces an average of two embryos per litter. Size at birth is around 35 cm
4.	*Carcharhinus melanopterus* (blacktip reef shark) [IUCN status: lower risk]	Indo-Pacific tropical shark migrates into estuaries and brackish waters for delivering its pups	Fish like mullets, silver bellies, anchovies, hilsa, skate, prawns and squilla	Matured ones are about 2.5 m	Size at birth ranges between 45 and 50 cm
5.	*Carcharhinus macloti* (hardnose shark)	Widely distributed in the inshore region	Small fishes, squids and crustaceans	Matured ones are 1 m in length. Male matures at 60 cm, and female matures at 70 cm	Produces an average of two embryos per litter. Size at birth is 35 cm approximately
6.	*Galeocerdo cuvier* (tiger shark) [IUCN status: lower risk]	Widely distributed tropical shark. It is capable of cruising in mid ocean and shows nocturnal movement into bays and estuaries	Eels, cat fish, parrot fish, flat fish, flying fish, skates, rays, marine turtle, sea snake, sea birds, sea lion, dolphin, etc.	Matured ones are about 7.4 m in length	Development is oviparous, and the litter size is between 10 and 82. The size at birth is 50–75 cm
7.	*Scoliodon laticaudus* (Indian dog shark) [IUCN status: lower risk]	West and south coasts of India, dominant in east coast and in Indian Sundarbans	Small fishes, crustaceans and squids	Majority of males is 50–55 cm and females are about 65 cm. Males and females mature at 30 cm and 35 cm, respectively	Produces up to 20 embryos per litter. The size at birth is 14.5 cm

8.	*Rhizoprionodon acutus* (Brazilian sharpnose shark)	Abundant in the shore waters, particularly in the west coast of India during September to February and east coast during summer months	Small fishes, squids, cuttlefish, crabs and shrimps	Matured ones are about 1 m in length	Produces an average of two to six embryos per litter, which are 26–27 cm long
9.	*Sphyrna lewini* (hammerhead shark) [IUCN status: vulnerable]	Abundantly found in the Indian seas, notable for its unique migratory behaviour	Sardine, anchovies, mackerel, eel, milkfish, sole fish. It also feeds on sharks and rays	Matured ones are up to 4.2 m in length	Development is viviparous and produces an average of 15–30 embryos per litter. The size at birth is 45–55 cm
10.	*Rhincodon typus* (whale shark) [IUCN status: vulnerable]	Appears occasionally at Indian coastal waters	Filterfeeder and is believed to sieve plankton as small as 1mm diameter through the fine mesh of their gill rakers	Largest length up to 1200 cm, but mean length is around 700 cm	Ovoviviparous, embryo length varies from 48–58 cm
11.	*Atelomycterus marmoratus* (coral cat shark)	Appears occasionally at Indian coastal waters	Squids, crabs, etc.	70 cm	Ovoviviparous
12.	*Glyphis gangeticus* (river shark) [IUCN status: extremely endangered; Wildlife Protection Act: Schedule I]	Hooghly–Ganges river system of West Bengal coast	Fish	Maximum length is around 200 cm. Matured males are about 178 cm	Viviparous. Newly born individuals are 56–61 cm long
13.	*Glyphis glyphis* (sharptooth shark or speartooth shark)	Confined to turbid waters of rivers, estuaries and inshore waters of coastal West Bengal	Fish	Minimum length is around 100 cm	Livebearing probably with yolk sac placenta
14.	*Glyphis siamensis* (Irrawaddy river shark)	Confined to turbid waters of rivers, estuaries and inshore waters of coastal West Bengal	Fish	Minimum length is around 63 cm	Livebearing probably with yolk sac placenta

period of activity of certain drugs. *Hypnea* sp. and *Acanthophora* sp. are the major sources of carrageenans, which have been found to be useful in ulcer therapy.

5. As industrial raw material: Agar–agar is a gelatinous colloidal carbohydrate present in the cell wall of algae and is extracted from some members of Rhodophyceae. In India, agar has wide use in the preparation of food, ice creams, jellies, soups, bacteriological samples and cosmetics. Being a nontoxic substance, agar is also used for the preparation of wine, beer, hand lotions, etc. Algin and alginates have extensive uses in the preparation of various medicines, cosmetics, paper products, textile products, paints, milk products, etc.

The total world seaweeds production is estimated to be about 1821×10^4 tonnes (wet weight) annually. Of this, about 4.83% is being harvested from the eastern and western Indian Ocean.

Phytoplankton Resource of the Ocean

Phytoplankton are very minute organisms that drift in water and may be found in either shallow or deep water. Thus, phytoplankton are passively floating or drifting producers of the aquatic ecosystem that cannot move against the current. Phytoplankton form the primary biological component of the sea from which energy is transformed to higher trophic level. In comparison to land plants and submerged macrophytes, phytoplankton populations turn out on a much shorter time scale, blooms being of only few weeks 'duration and species' replacement taking only a matter of months or less. The fertility of the marine and estuarine system is reflected through the bioproductivity of the phytoplankton. Marine phytoplankton, as primary producers, and zooplankton, as secondary producers, are the key players in spinning the marine food web in which large fishes and marine mammals also exist. In any aquaculture farm, phytoplankton hold the primary rank in order of importance as they sustain the fishery resource either directly or through the growth of zooplankton.

It has been estimated that the total annual primary production of the world seas is around 20×10^9 tonnes of carbon which could yield 240 million tonnes of fish although our present harvest is only about 60 million tonnes.

The distribution and abundance of the commercially important finfish and shellfish and their larvae are dependent on some species of the phytoplankton by serving as their main food source. Among the species of diatoms, *Fragilaria oceanica* and *Hemidiscus hardmannianus* have been recorded to indicate the abundance of the clupeid fish, hilsa in Hooghly estuary and oil sardine, *Sardinella longiceps*, in the west coast of India. The 'white water phenomenon' due to abundance of the coccolithophores has been a good omen for the herring fisheries. Similarly the colonial diatom, *Fragilaria antarctica*, is known to indicate the abundance of the Antarctic krill, *Euphausia superb*.

The diatoms are commercially important as 'diatomaceous earth' (diatomite or kieselguhr), a congregation of dead silicon-rich diatom frustules in the seabeds like protozoan oozes. This material is employed in the filtration of fruit juice, syrup and varnish; in the boilers, electric ovens and refrigerators as insulators; as polishing agent for precious metals; in the separation of paraffin wax from petroleum; and in the manufacture of concrete. The presence of the hydrocarbon compound 'diatomin' has also been the indicator of rich petroleum grounds. Present-day scientists are of the opinion that the diatoms are the chief precursors of the petroleum-rich fields of Venezuela at Los Angeles.

Mass culture of phytoplankton may be undertaken in closed system of the space capsules. The nitrogenous wastes of the travelers could be diluted and circulated as algal nutrients and the respired air as a source of carbon dioxide. In exchange, the travelers could utilize the oxygen released by the photosynthesis of algae and benefit from the food value of cultured phytoplankton.

Although elaborated studies have been carried out on marine phytoplankton in many countries, knowledge of this group of marine flora is very scanty in India. Hence, large-scale investigations on the phytoplankton especially on their dynamics in India seas are imperative to locate new fishing grounds and their role in sustaining the blue revolution.

The phytoplankton community consists of a variety of organisms, namely, diatoms, dinoflagellates, blue-green algae, silicoflagellates, coccolithophores, etc., which ranges in terms of size from 0.001 to 0.2 mm.

Phytoplankton may be classified variously from different angles which are discussed here:

(a) On the basis of size, the phytoplankton may be grouped under five categories (Table 3.3).

(b) Phytoplankton can also be classified on the basis of the cell characteristics (Table 3.4).

Table 3.3 Classification of phytoplankton on the basis of size

Plankton category	Maximum dimension
Ultraplankton	<2 μm
Nanoplankton	2–20 μm
Microplankton	20–200 μm
Macroplankton	200–2000 μm
Megaplankton	>2000 μm

Table 3.4 Classification of phytoplankton on the basis of cell characteristics

Class	Common name	Area(s) of predominance	Common genera
Cyanophyceae (cyanobacteria)	Blue-green algae	Tropical	*Oscillatoria*, *Synechococcus*
Rhodophyceae	Red algae	Cold temperate	*Rhodella*
Cryptophyceae	Cyptomonas	Coastal	*Cryptomonas*
Chrysophyceae	Chrysomonads Silicoflagellates	Coastal Cold waters	*Aureococcus* *Dictyocha*

(continued)

Table 3.3 (continued)

Class	Common name	Area(s) of predominance	Common genera
Bacillariophyceae (Diatomophyceae)	Diatoms	All waters, *esp.* coastal	*Coscinodiscus*, *Chaetoceros*, *Rhizosolenia*
Raphidophyceae	Chloromonads	Brackish	*Heterosigma*
Xanthophyceae	Yellow-green algae	Brackish	Very rare
Eustigmatophyceae	Yellow-green algae	Estuarine	Very rare
Prymnesiophyceae	Coccolithophorids Prymnesiomonads	Oceanic Coastal	*Emiliania* *Isochrysis* *Prymnesium*
Euglenophyceae	Euglenoids	Coastal	*Eutreptiella*
Prasinophyceae	Prasinomonads	All waters	*Tetraselmis* *Micromonas*
Chlorophyceae	Green algae	Coastal	Rare
Pyrrophyceae (Dinophyceae)	Dinoflagellates	All waters, esp. warm	*Ceratium* *Gonyaulax* *Protoperidinium*

3.3 Overexploitation of Resources

The Earth's natural resources are vital to the survival and development of the human population. These natural resources are air, water, soil and minerals, along with the climate and solar energy, which are known as nonliving or abiotic part of nature. The biotic or living part of nature consists of plants, animals and microbes. These all are closely linked to each in their own habitat and require specific abiotic condition. Their products are required for food, medicines, shelter and fibre. Increasing population placed great demands on the Earth's natural resources. Large stretches of land such as forests, grasslands and wetlands have been converted into intensive agriculture. Land has been taken for industry and the urban sectors. These changes have brought about dramatic alterations in land-use patterns and rapid disappearance of valuable natural ecosystems. As the world population rises and economic growth occurs, the depletion of natural resources influenced by the unsustainable extraction of raw materials becomes an increasing concern. The need for more water, more food, more energy and more consumer goods is not only the result of a greater population but also the result of overutilization of resources by people. Industrial development is aimed at meeting growing demands of population and for these industries also generates waste in ever larger quantities. It shifts the people from their traditional, sustainable, rural way of life to urban centres that developed around industry.

For the convenience of the readers, the overexploitation of natural resources has been sub-divided into two major groups, namely, (i) overexploitation of land resources and (iii) overexploitation of aquatic resources.

3.3.1 Overexploitation of Land Resources

Land is a major resource, needed for food production, animal husbandry, industry and our growing human settlements. Land as a resource is now under serious pressure due to an increasing 'land hunger' – to produce sufficient quantities of food for an exploding human population. It is also affected by degradation due to misuse. Land and water resources are polluted by industrial waste and rural and urban sewage. They are increasingly being diverted for short-term economic gains to agriculture and industry. Natural wetlands of great value are being drained for agriculture and other purposes. Forests provide us with a variety of services. These include processes such as maintaining oxygen levels in the atmosphere, removal of carbon dioxide, control over water regimes and slowing down erosion and also produce products such as food, fuel, timber, fodder, medicinal plants, etc. Man needs land for building homes, cultivating food, maintaining pastures for domestic animals, developing industries to provide goods and supporting the industry by creating towns and cities. Equally importantly, man needs to protect wilderness area in forests, grasslands, wetlands, mountains, coasts, etc. to protect our vitally valuable biodiversity. Unsustainable land-use drives land degradation through contamination and pollution, soil erosion and nutrient depletion. In some areas there is an excess of nutrients causing eutrophication, and there can be water scarcity and salinity. The enormous increase in the production of farm and forest products has brought greater wealth and more secure livelihoods for billions but often at the cost of land degradation, biodiversity loss and disruption of biophysical cycles, such as the water and nutrient cycles. These impacts create many challenges and opportunities. When soil is used more intensively by farming, it is eroded more rapidly by wind and rain. Over-irrigating farmland leads to salinization, as evaporation of water brings the salts to the surface of the soil on which crops cannot grow. As urban centres grow and industrial expansion occurs, the agricultural land and forests shrink. This is a serious loss and has long-term ill effects on human civilization. Soil erosion is the natural process of removal of soil by water or wind. Soil erosion becomes a problem when the natural process is accelerated by inappropriate land management, such as clearance of forest and grasslands followed by cropping which results in inadequate ground cover, inappropriate tillage and overgrazing. It is also caused by activities such as mining and infrastructural and urban development without well-designed and well-maintained conservation measures. Loss of topsoil means loss of soil organic matter, nutrients, water holding capacity and biodiversity, leading to reduced production on site. The misuse of an ecosystem leads to loss of valuable soil through erosion by the monsoon rains and, to a smaller extent, by wind. The roots of the trees in the forest hold the soil.

Deforestation thus leads to rapid soil erosion. Soil is washed into streams and is transported into rivers and finally lost to the sea. Nutrient depletion is a decline in the levels of plant nutrients, such as nitrogen, phosphorous and potassium, and in soil organic matter, resulting in declining soil fertility. Deforestation causes extinction, changes to climatic conditions, desertification and displacement of populations as observed by current conditions and in the past through the fossil record. It is also accompanied by soil acidification, which increases the solubility of toxic elements, such as aluminium. The removal of the harvest and crop residues depletes the soil, unless the nutrients are replenished by manure or inorganic fertilizers. Nutrient mining refers to high levels of nutrient removal and no inputs. Deficiency of plant nutrients in the soil is the most significant biophysical factor limiting crop production across very large areas in the tropics, where soils are inherently poor. Soils, streams and groundwater in dry lands contain significant amounts of naturally occurring salt, which inhibits the absorption of water by plants and animals, breaks up roads and buildings and corrodes metal. Salinity is defined by the desired use of land and water; it is salt in the wrong place when found in farmland, drinking and irrigation water and in freshwater habitats. It is caused by inappropriate forms of land use and management. The rising, saline groundwater drives more salt into streams, and, where the water table comes close to the surface, evaporation pulls salt to the surface.

3.3.2 Overexploitation of Water Resources

Over two thirds of the Earth's surface is covered with water. But this water cannot be used for drinking purposes. This water is salty. Groundwater is used for drinking and kitchen use. Groundwater is almost pure and is the main source for our drinking water. Putting unwanted materials in rivers and lakes creates water pollution. When wastes are deposited in the water, the water is no longer good for use. Industries should reduce the dumping of wastes in water sources. It is a renewable resources; it should be used in a sustainable manner. Farmers should adopt water harvesting techniques. Overexploitation anywhere is often accompanied by detrimental environmental side effects, such as poor water quality and reduced groundwater flowing to springs, streams and wetlands. Some causes and effects of groundwater depletion are not obvious. With the growth of human population, there is an increasing need for larger amounts of water to fulfil a variety of basic needs. Today in many areas, this requirement cannot be met. Overutilization of water occurs at various levels. Most people use more water than they really need. Most of us waste water during a bath by using a shower or during washing of clothes. Many agriculturists use more water than necessary to grow crops. There are many ways in which farmers can use less water without reducing yields such as the use of drip irrigation systems. Agriculture also pollutes surface water and underground water stores by the excessive use of chemical fertilizers and pesticides. Methods such as the use of biomass as fertilizer and nontoxic pesticides such as neem products and using integrated pest management systems reduce the agricultural pollution of surface and groundwater. Industry tends to maximize short-term economic gains by not bothering about its liquid waste and releasing it into streams, rivers and the sea. In the longer term, as people become more conscious of using 'green products' made by eco-sensitive industries, the polluter's products may not be used. The polluting industry that does not care for the environment and pays off bribes to get away from the cost needed to use effluent treatment plants may eventually be caught, punished and even closed down. Public awareness may increasingly put pressures on industry to produce only eco-friendly products which are already gaining in popularity. As people begin to learn about the serious health hazards caused by pesticides in their food, public awareness can begin putting pressures on farmers to reduce the use of chemicals that are injurious to health. Everywhere the 'greenhouse effect' due to atmospheric pollution is leading to increasingly erratic and unpredictable climatic effects. This has seriously affected regional hydrological conditions. Floods have been a serious environmental hazard for centuries. However, the havoc raised by rivers overflowing their banks has become progressively more damaging, as people have deforested catchments and intensified use of river flood plains that once acted as safety valves. Wetlands in flood plains are nature's flood control systems into which overfilled rivers could spill and act like a temporary sponge holding the water, preventing fast-flowing water from damaging surrounding land. In most arid regions of the world, the rains are unpredictable. This leads to periods when there is a serious scarcity of water to drink, use in farms or provide for urban and industrial use. Irrigation to support cash crops like sugarcane produces an unequal distribution of water. Large landholders on the canals get the lion's share of water, while poor, small farmers get less and are seriously affected.

3.4 Take Home Messages

(A) The lithosphere is the topmost layer of the Earth whose main components are (a) firm and solid rocks and (b) the unconsolidated materials, often called soil or regolith. A rock does not possess a definite composition like that of a chemical compound; rather it is a mixture of various minerals. Thus a rock may be defined as 'an aggregate of minerals that forms a more or less definite unit of the earth's crust'. The bulk of rock is made up of silica or quartz and feldspar. Rocks are of great economic value to man.

(B) The ocean is a valuable reservoir of food, minerals, natural gas and oil. The ocean basin and beaches are rich sources of sand gravel, precious metal placers, titanium and chromite placers manganese nodules, cobalt-rich crust and polymetallic sulphide. The living resources of the ocean encompass various categories of finfish, and shellfish, seaweeds, phytoplankton and the macrophytes. The blue soup of the Earth, occupying about 70% of the Earth's surface, can thus act as the key of strengthening the economic backbone of the country. The major industries based on marine resources are hard minerals, fisheries and mariculture, hydrocarbons, ocean energy, recreation and tourism and shipping and shipbuilding.

(C) Land resource is now under serious stress due to an increasing 'land hunger' – to produce sufficient quantities of food for an exploding human population. It is also affected by degradation due to misuse. The exponential increase of industries coupled with rapid urbanization has posed an adverse impact on the forest land.

(D) Water resources are polluted by industrial waste, rural and urban sewage. They are increasingly being diverted for short-term economic gains to agriculture and industry. Natural wetlands of great value are being drained for agriculture and other purposes. In addition coastal and estuarine waters are used to accelerate aquaculture (preferably shrimp culture), and the waste water from these aquaculture farms are finally drained in to the source water (coastal water and estuaries).

3.5 Brain Churners

This section is an evaluation of the concept gained by the readers after scanning the resources of the Earth. It is expected that the readers will find interest in answering the questions asked on the resources of the land and water and their exploitation due to rapid industrialization and urbanization. The section will also help the readers to face various competitive examinations as environment has been given thrust in recent times. The readers can consider this section as a sort of *self-appraisal approach* as correct answers to all the questions are filled in black in the Answer Box at the end of this section.

1. The lithosphere is also known as

   (a) Cryosphere
   (b) Geosphere
   (c) Hydrosphere
   (d) None of these

2. The ore of aluminium is called

   (a) Bauxite
   (b) Chromite
   (c) Haematite
   (d) Magnetite

3. Polymetallic nodules are found in the

   (a) Cave
   (b) Pond
   (c) Lake
   (d) Ocean

4. The biotic resource of the ocean encompasses
   (a) Petroleum
   (b) Rocks
   (c) Minerals
   (d) Fishes
5. Seaweeds are characterized by the presence of
   (a) Root, stem and flower
   (b) Root, stem and leaves
   (c) Holdfast, stipe and blade
   (d) Holdfast, root and leaves
6. *Fragilaria antarctica* is the indicator of
   (a) Shad
   (b) Salmon
   (c) Antarctic krill
   (d) None of the above
7. Ultra plankton has the dimension
   (a) Less the < 2μm
   (b) 5mm
   (c) 2–5 mm
   (d) 20 μm
8. *Oscillatoria* sp. is a
   (a) Red alga
   (b) Blue-green alga
   (c) Green alga
   (d) Yellow-green alga
9. Atmospheric humidity increases due to
   (a) Transpiration of plants
   (b) Respiration of animals
   (c) Respiration of animals
   (d) Photosynthesis
10. *Glyphis gangeticus* is
    (a) Extremely endangered
    (b) Not at all threatened
    (c) Threatened
    (d) None of the above
11. *Ulva* sp. is a
    (a) Red seaweed
    (b) Green seaweed
    (c) Brown seaweed
    (d) None of the above

12. Deforestation leads to
    (a) Volcanic activity
    (b) Earthquake
    (c) Rapid soil erosion
    (d) None of the above
13. *Fragilaria* sp. is the indicator of abundance of
    (a) Carp
    (b) Hilsa
    (c) Catfish
    (d) Mackerel
14. *Enteromorpha* sp. is a thread-like
    (a) Brown seaweed
    (b) Red seaweed
    (c) Green seaweed
    (d) None of the above
15. The polymetallic nodules are the sources of
    (a) Mn, Fe, Ni, Cu
    (b) Al, C, H, N
    (c) Sn, Cr, Hg, Pb
    (d) Zn, Hg, Cu, Cr
16. An important non-point source of pollution is
    (a) Industrial outlets
    (b) Agricultural runoff
    (c) Factory chimney
    (d) None of the above
17. Excessive increase of algae leads to
    (a) Eutrophication
    (b) Red tide
    (c) El Niño
    (d) Salinization

18. Terrestrial forest are

(a) Sink of Hg
(b) Sink of C
(c) Sink of O^2
(d) None of the above

19. Forest soil is the store house of

(a) Uranium
(b) Polonium
(c) Carbon
(d) Chromium

20. The primary precursors of petroleum are

(a) Diatom
(b) Fungus
(c) Seaweeds
(d) Sea grass

Answer Sheet

S. No.	a	b	c	d
1.		■		
2.	■			
3.				■
4.				■
5.			■	
6.			■	
7.	■			
8.		■		
9.	■			
10.	■			
11.		■		
12.			■	
13.		■		
14.			■	
15.	■			
16.		■		
17.	■			
18.		■		
19.			■	
20.	■			

Annexure 3A: Rising Nutrient Levels in Two Major Estuaries in the Indian Sundarbans: A Time Series Analysis

Introduction

Coastal waters and estuaries are facing a variety of adverse impact affecting both the ecosystem and human health through sewage–wastewater discharge and disposal practices that often lead to introduction of high nutrient loads, hazardous chemicals and pathogens causing diseases. In countries like India, shrimp culture practices and aquaculture waste have magnified the adverse situation (Mitra 1998; Mitra and Zaman 2015). In regional level like Sundarbans deltaic ecosystem (located at the tail end of the mighty River Ganga), phenomenon like erosion and subsequent washing of the topsoil from the intertidal mudflats of mangroves also contributes considerable amount of nutrients in the adjacent aquatic ecosystem. The adverse public health, environmental, socio-economic, food quality, security and aesthetic impact from sewage contamination in coastal areas are well documented (Luger and Brown 1999; Tyrrel 1999; Danulat et al. 2002; WHO 2003). Apart from these causes, erosion of river banks due to tidal surges also conveys nitrate, phosphate and silicate in the estuarine water (Mitra et al. 2009), which may have a far-reaching effect on the environment. However, very few researches have focused on the trend of nutrient load in the estuarine waters based on past long-term data bank. The present paper is a road map towards this direction in the frame work of Indian Sundarbans.

Materials and Methods

Study Area

The mangrove-dominated Indian Sundarbans in the lower Gangetic delta region at the apex of Bay of Bengal is inhabited by some 4.2 million populations. Two major estuaries in this delta complex are Hooghly (in the western sector) and Matla (in the central sector). There is a multitude of industries located on the western bank of the Hooghly estuary. Apart from this, large number of tourism units and shrimp culture farms is also located adjacent to the Hooghly estuary. A number of literatures are available on the salinity profile, surface water temperature, and pH of the area (Mitra 1998; Mitra and Bhattacharyya 1999; Mitra 2000; Majumder et al. 2002; Panja et al. 2003; Banerjee et al. 2005; Mitra and Banerjee 2005; Mitra et al. 2011; Zaman

Table 3A.1 Location of sampling stations

Stations	Geographical locations	
	Latitude	Longitude
Diamond Harbour	22°11′4.2″N	88°11′22.2″E
Namkhana	21°45′53.7″N	88°13′51.5″E
Ajmalmari	21°49′42.9″N	88°37′13.7″E

and Mitra 2014; Mitra and Zaman 2015; Ray Choudhuri et al. 2015; Trivedi et al. 2015a, b; Chaudhuri et al. 1994). However, very few researches have addressed the long-term variation of nutrient level in the estuarine waters, although the area is presently experiencing population growth, industrial activities, mushrooming of shrimp farm, growth of tourism units and establishment of fish landing stations (Mitra 2013). Apart from these primary sources of nutrients, churning action of bed substratum due to wave action and currents conveys silica to the overlying aquatic phase.

Sample Collection

Sampling of surface water was done during high tide from three stations, namely, Diamond Harbour, Namkhana (in the Hooghly estuary) and Ajmalmari (in the Matla estuary) (Table 3A.1). Sample collection was carried out during May (premonsoon), September (monsoon) and December (postmonsoon) for a period of 31 years (1984–2014).

Nutrient Analysis

Surface water samples collected for nutrient analysis were filtered through a 0.45 μm polycarbonate filters (Millipore 47 mm diameter) and then deep frozen for further analysis in the laboratory. The standard spectrophotometric method of Strickland and Parsons (1972) was adopted to determine the nutrient concentrations in surface water. Nitrate was analyzed by reducing it to nitrite by means of passing the sample with ammonium chloride buffer through a glass column packed with amalgamated cadmium filings and finally treating the solution with sulphanilamide. The resultant diazonium ion was coupled with N-(1-napthyl)ethylenediamine to give an intensely pink azo dye. Estimation of the phosphate was carried out by treatment of an aliquot of the sample with an acidic molybdate reagent containing ascorbic acid and a small proportion of potassium antimony tartrate. Dissolved silicate was analyzed by treating the sample with acidic molybdate reagent. The resultant silicomolybdic acid was reduced to molybdenum blue complex by ascorbic acid and incorporating oxalic acid prevented formation of similar blue complex by phosphate.

Statistical Analysis

ANOVA was performed in order to analyze whether the selected nutrients vary significantly between years and stations ($p < 0.01$).

Results

We note the significant spatiotemporal variations of nitrate, phosphate and silicate in the study region. Also sudden rise of the nutrient level during premonsoon, 2009 is attributed to supercyclone, AILA that contributes nutrients through massive erosion of river banks, washing of topsoil of intertidal mudflats along the estuaries and churning of the river bed.

Dissolved Nitrate

At Diamond Harbour, dissolved nitrate concentration ranged from 20.84 ppm (during premonsoon in 1984) to 48.15 ppm (during premonsoon in 2009). At Namkhana, dissolved nitrate concentration ranged from 18.90 ppm (during premonsoon in 1984) to 43.44 ppm (during premonsoon in 2009). In case of Ajmalmari, the values ranged between 9.44 ppm (during premonsoon in 1984) and 24.89 ppm (during premonsoon in 2009).

In both Diamond Harbour and Namkhana of the Hooghly estuarine system, nitrate has increased 9.49 μg at l^{-1}/decade and 10.57 μg at l^{-1}/decade, respectively. In Ajmalmari, located in the Matla estuarine complex, the increase is relatively low (5.98 μg at l^{-1}/decade).

Dissolved Phosphate

At Diamond Harbour, dissolved phosphate concentration ranged from 1.54 ppm (during premonsoon in 1984) to 6.99 ppm (during premonsoon in 2009). At Namkhana, dissolved phosphate concentration ranged from 0.98 ppm (during premonsoon in 1984) to 6.05 ppm (during premonsoon in 2009). Ajmalmari, the station in the Matla estuary exhibited a phosphate value from 0.56 ppm (during premonsoon in 1984) to 2.96 ppm (during premonsoon in 2009). In both Diamond Harbour and Namkhana of the Hooghly estuarine system, phosphate has increased 1.96 μg at l^{-1}/decade and 2.11 μg at l^{-1}/decade, respectively. In Ajmalmari, located in the Matla estuarine complex, the rate of increase is 0.92 μg at l^{-1}/decade.

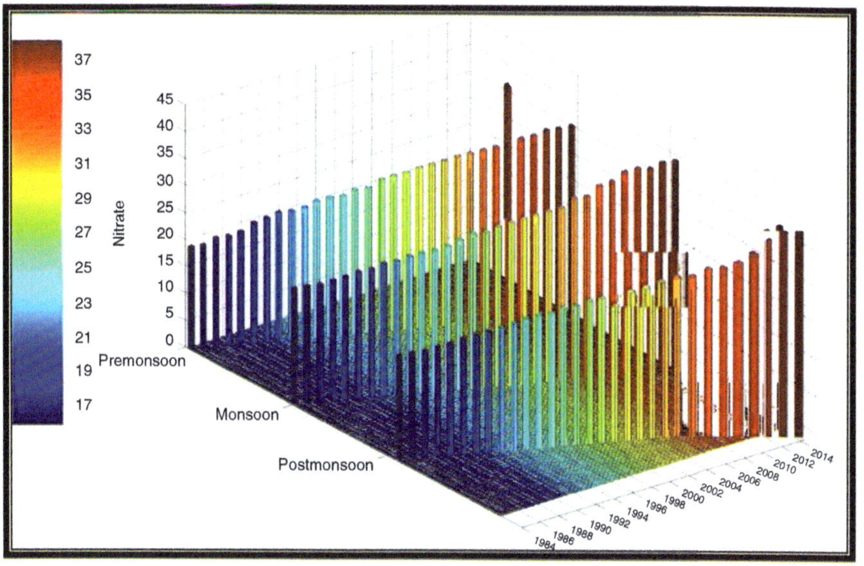

Fig. 3A.1 Dissolved nitrate (in μg at l^{-1}) in Diamond Harbour during 1984–2014

Dissolved Silicate

At Diamond Harbour, dissolved silicate concentration ranged from 101.83 ppm (during premonsoon in 1984) to 242.78 ppm (during premonsoon in 2009). At Namkhana, it ranged from 44.68 ppm (during premonsoon in 1984) to 211.49 ppm (during premonsoon in 2009). In case of Ajmalmari, the silicate ranged from 31.43 ppm (during premonsoon in 1984) to 111.99 ppm (during premonsoon in 2009). In both Diamond Harbour and Namkhana of the Hooghly estuarine system, silicate has increased 64.57 μg at l^{-1}/decade and 54.56 μg at l^{-1}/decade, respectively. The rate of increase is 31.18 μg at l^{-1}/decade in Ajmalmari, which is relatively low compared to stations selected in the Hooghly estuary.

The temporal variations of the selected nutrients during 1984–2014 are shown in Figs. 3A.1, 3A.2, 3A.3, 3A.4, 3A.5, 3A.6, 3A.7, 3A.8 and 3A.9.

Discussion

The enrichment of the aquatic system by nutrients has both natural and anthropogenic origin. The main sources of nutrient input in the present study area are runoff from the adjacent landmasses (Mitra 2013), erosion and leaching (Mitra et al. 2009), sewage from cities and industrial wastewater (Mitra and Choudhury 1993; Mitra 1998; Bhattacharyya et al. 2013), untreated sewage disposal from shrimp farms and tourism units (Mitra 2013; Zaman and Mitra 2014; Mitra and Zaman 2015). Atmospheric deposition of nitrogen (from combustion gases) can also be important

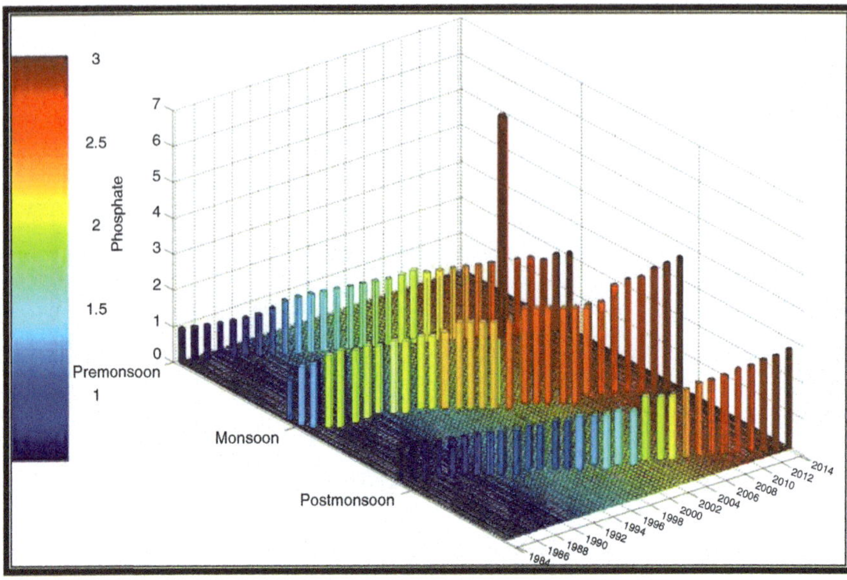

Fig. 3A.2 Dissolved phosphate (in μg at l^{-1}) in Diamond Harbour during 1984–2014

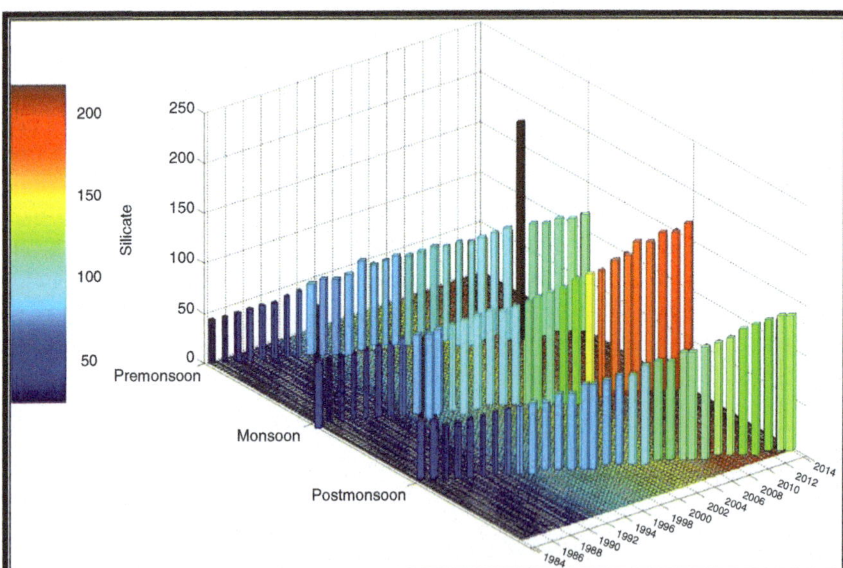

Fig. 3A.3 Dissolved silicate (in μg at l^{-1}) in Diamond Harbour during 1984–2014

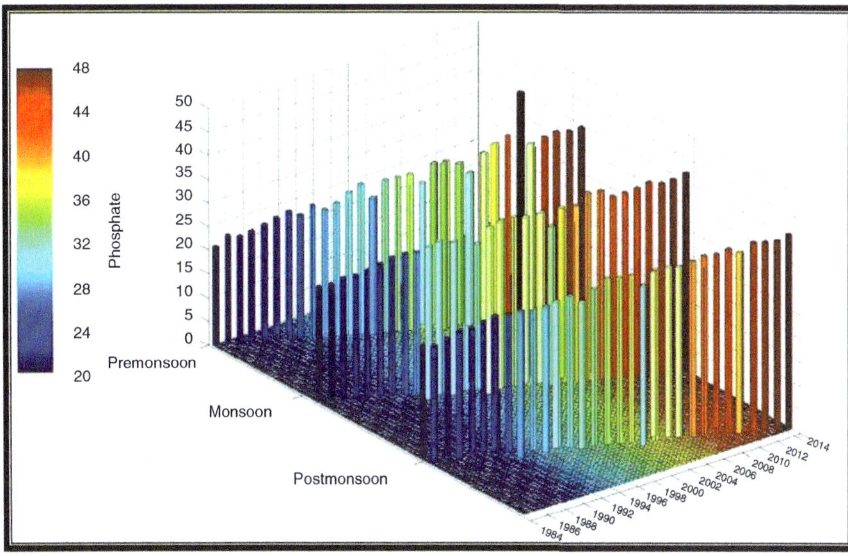

Fig. 3A.4 Dissolved nitrate (in μg at l^{-1}) in Namkhana during 1984–2014

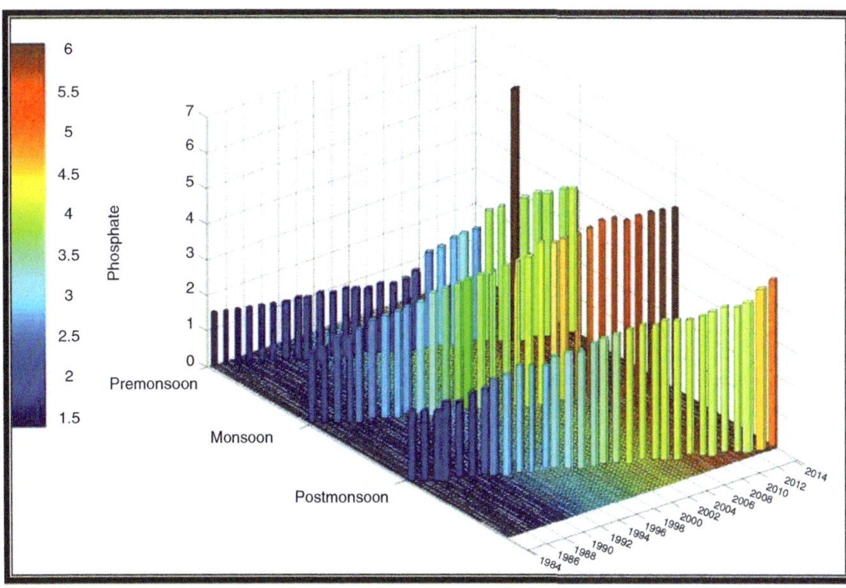

Fig. 3A.5 Dissolved phosphate (in μg at l^{-1}) in Namkhana during 1984–2014

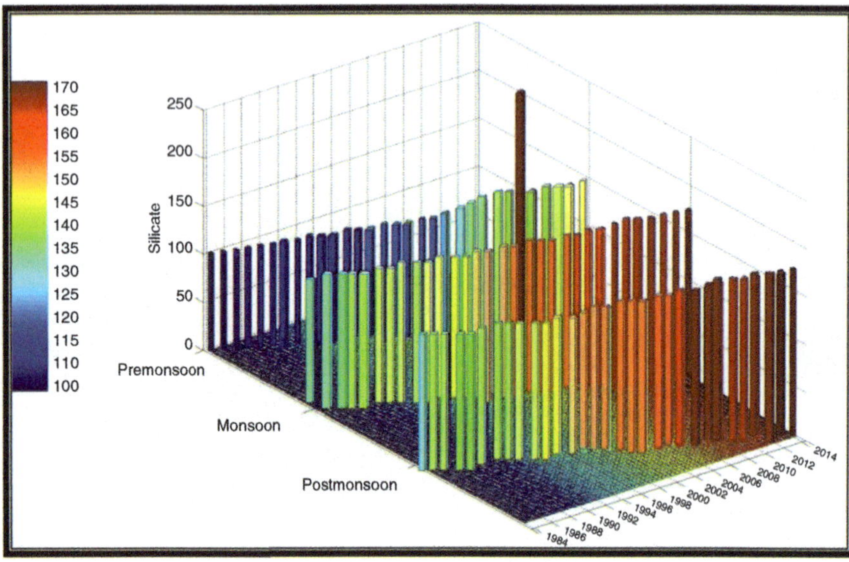

Fig. 3A.6 Dissolved silicate (in μg at l^{-1}) in Namkhana during 1984–2014

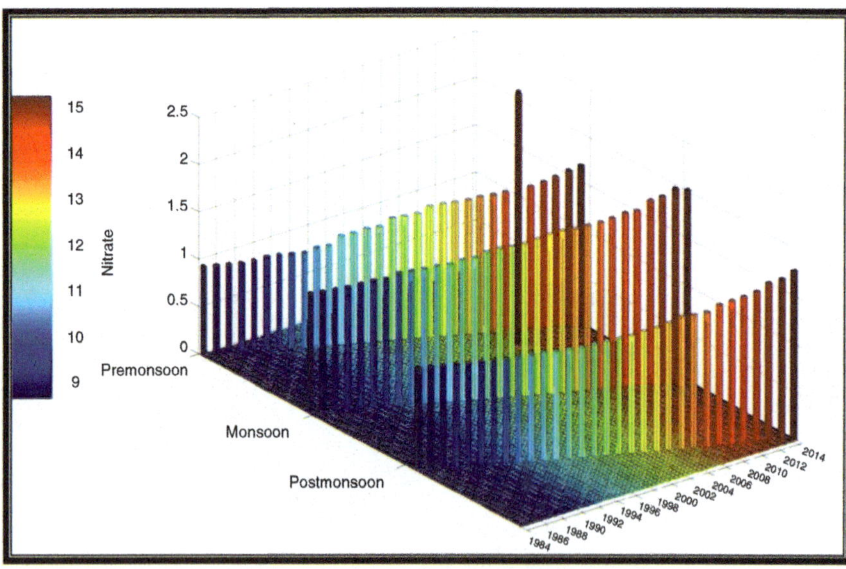

Fig. 3A.7 Dissolved nitrate (in μg at l^{-1}) in Ajmalmari during 1984–2014

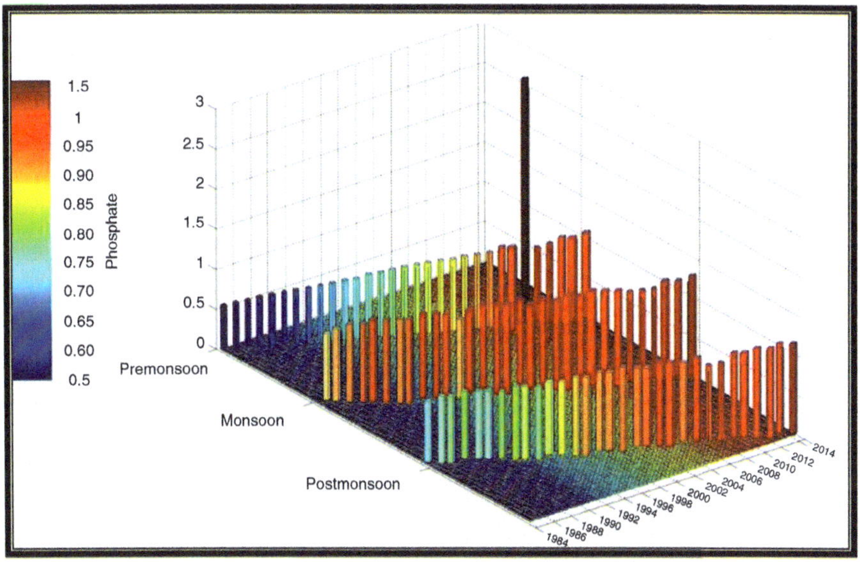

Fig. 3A.8 Dissolved phosphate (in μg at l^{-1}) in Ajmalmari during 1984–2014

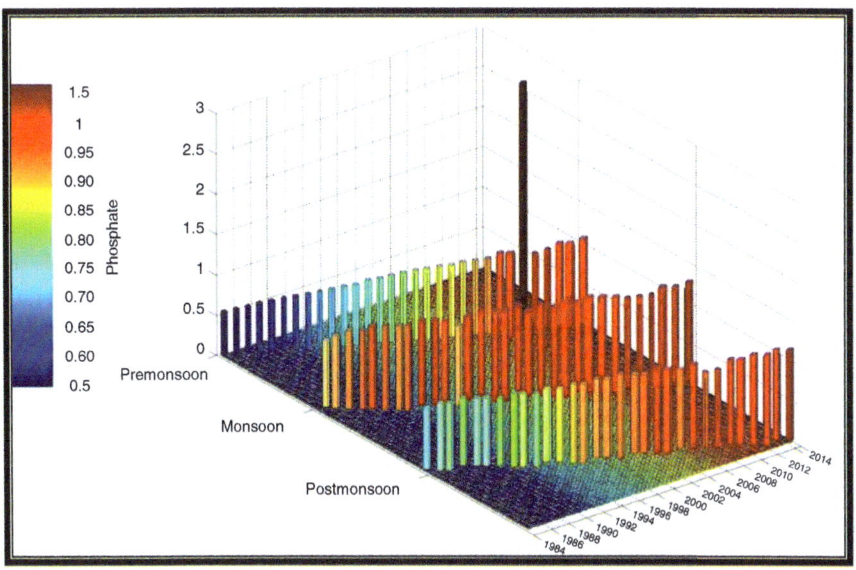

Fig. 3A.9 Dissolved silicate (in μg at l^{-1}) in Ajmalmari during 1984–2014

for Hooghly estuarine system and its surrounding area as multifarious industries are concentrated in this estuarine region. The effects of nitrogen on marine and estuarine systems, the pathways for nitrogen transport between land and aquatic habitats and the positive correlation between nitrogen and primary production and often secondary production (i.e. fishery yields) have been widely reviewed (Hecky and Kilham 1988; Howarth 1988; Rabalais 2002).

Most of the phosphorus pollution in the present study area comes from the households and industries including phosphorus-based detergents which are widely used in thickly populated cities of Kolkata, Howrah and Haldia industrial complex adjacent to the Hooghly estuary.

Silicate, being an important ingredient of bed material and soil substratum, originated due to erosion and churning action of the water that amplifies during tropical cyclone, which is common in the present geographical locale.

The ratio of nitrogen and phosphorus plays a regulatory role in the phytoplankton diversity spectrum of brackish water. It determines which of the two elements will be the limiting factor and consequently which one has to be controlled in order to arrest the bloom condition (Table 3A.2).

Large marine areas frequently have nitrogen as the limiting nutrient, especially in summer. Intermediate areas such as river plumes are often phosphorus-limited during spring but may turn to silica or nitrogen limitation in summer. When phosphorus is the limiting factor, a phosphate concentration of 0.01 mg l^{-1} is enough to support plankton, and concentrations from 0.03 to 0.1 mg l^{-1} or higher will be likely to promote blooms.

In coastal areas, the growth and proliferation of diatoms are promoted by the presence of silica. When the silica concentration is low, diatoms cannot develop. Then other opportunistic toxic algal species, which are no longer submitted to competition, can grow and form blooms. Species from the genus *Phaeocystis* and several dinoflagellates (*Prorocentrum*, *Dinophysis*, *Gymnodinium*) are known to proliferate under such conditions.

In this study, the N/P ratios in all the selected stations and seasons are greater than 10 (except monsoon season in Namkhana) (Table 3A.3). This implies that the aquatic phase of the present study area is P-limiting (WHO 2003). A high N/P ratio normally increases the standing stock of dinoflagellates and diatoms.

Table 3A.2 Nitrogen/phosphorus ratios (expressed in weight) for various limiting conditions in freshwater and estuarine/coastal water

Type of waterbody	N-limiting Ratio N/P	Intermediate Ratio N/P	P-limiting Ratio N/P
Freshwater	≤ 4.5	4.5–6	> 6
Estuarine/coastal water	≤ 5	5–10	≥ 10

Source: WHO (2003)

Table 3A.3 Average nitrogen/phosphorus ratios in different seasons in the three sampling stations

N/P ratio			
	Seasons		
Stations	Premonsoon	Monsoon	Postmonsoon
Diamond Harbour	15.53	12.60	17.04
Namkhana	12.26	8.85	10.18
Ajmalmari	14.55	15.39	14.18

Table 3A.4 ANOVA result showing temporal and spatial variations of dissolved nutrients (nitrate, phosphate and silicate)

Variables			F_{cal}	F_{crit}
Dissolved nitrate	Premonsoon	Between stations	876.8915	3.150411
		Between years	21.43769	1.649141
	Monsoon	Between stations	709.1798	3.150411
		Between years	15.52743	1.649141
	Postmonsoon	Between stations	846.2506	3.150411
		Between years	13.00163	1.649141
Dissolved phosphate	Premonsoon	Between stations	92.25571	3.150411
		Between years	8.000291	1.649141
	Monsoon	Between stations	268.6471	3.150411
		Between years	4.05633	1.649141
	Postmonsoon	Between stations	337.6358	3.150411
		Between years	4.725808	1.649141
Dissolved silicate	Premonsoon	Between stations	517.3935	3.150411
		Between years	21.15371	1.649141
	Monsoon	Between stations	426.0496	3.150411
		Between years	7.435455	1.649141
	Postmonsoon	Between stations	1061.878	3.150411
		Between years	11.78569	1.649141

Significant variations in the level of dissolved nitrate, phosphate and silicate between years and between stations were observed ($p < 0.01$) which reveal the impact of season and anthropogenic pressure in the present study area (Table 3A.4).

Conclusion

The core findings of the present study are listed here:

1. The main sources of nutrients in the present study area are primarily anthropogenic in nature, although natural disaster (like super cyclone AILA) resulted in sudden rise in nutrient level during premonsoon, 2009.

2. The concentrations of nutrients exhibit a gradual increasing trend which is an evidence of rising anthropogenic pressure in and around the mangrove-dominated Indian Sundarbans.
3. The situation seems to be alarming in terms of nutrient enrichment if proper management/control measure is not adopted. The policymakers must foster appropriate action with true partnership with private sectors. The regulation should also be put in place to enforce the polluters (from urban areas, industries, shrimp farms and tourism units) to pay the principal cost and also foster a willingness to pay among polluters through provision of better and efficient services. This should ensure operational sustainability of the services.

References

Banerjee K, Mukherjee D, Das S, Banerjee S, Mitra A (2005) Physico-chemical characteristics of coastal waters of West Bengal. J Ecotoxicol Environ Monit 7:37–46

Bhattacharyya SB, Roychowdhury G, Zaman S, Raha AK, Chakraborty S, Bhattacharjee AK, Mitra A (2013) Bioaccumulation of heavy metals in Indian white shrimp (*Fenneropenaeus indicus:* a time series analysis). Int J Life Sci Biot Pharm Res 2(2):97–113

Chaudhuri A, Mitra A, Trivedi S, Gupta A, Choudhury A (1994) Phosphate and nitrate status in the east coast of Indian sub-continent. In: Seminar on our environment: its challenges to development projects. Kolkata, American Society of Civil Engineers - India Section

Danulat E, Muniz P, García-Alonso J, Yannicelli B (2002) Fist assessment of the highly contaminated harbour of Montevideo, Uruguay. Mar Pollut Bull 44:554–565

Hecky RE, Kilham P (1988) Nutrient limitation of phytoplankton in freshwater and marine environments: a review of recent evidence on the effects of enrichment. Limnol Oceanogr 33:796–782

Howarth RW (1988) Nutrient limitation of net primary production in marine ecosystems. Annu Rev Ecol Syst 19:898–910

Luger M, Brown C (1999) The impact of treated sewage effluent on Urban Rivers. An ecological, social and economic perspective

Majumder S, Mitra A, Panda UC, Choudhury A Bhattacharyya DP (2002) Baseline studies of hydrological parameters in and around Digha, Northeast coast of India. Proceedings of national seminar on pollution in Bay of Bengal

Mitra A (1998) Status of coastal pollution in West Bengal with special reference to heavy metals. J Indian Ocean Stud 5(2):135–138

Mitra A (2000) The Northeast coast of the Bay of Bengal and deltaic Sundarbans (Chapter 62). In: Sheppard C (ed) Seas at the Millennium – an environmental evaluation. University of Warwick, Coventry, pp 143–157

Mitra A (2013) Blue carbon: a hidden treasure in the climate change science. J Mar Sci Res Dev 3(2):1–2

Mitra A, Banerjee K (2005) Living resources of the sea: focus Indian Sundarbans. WWF India, Sundarbans Landscape Project, Canning Field Office, West Bengal, India

Mitra A, Bhattacharyya DP (1999) An ecological profile of Sundarbans ecosystem. Indian J Biodivers II(2):63–71

Mitra A, Choudhury A (1993) Seasonal variations in metal content in the gastropod *Nerita articulata* (Gould). Indian Environ Health 35(1):31–35

Mitra A, Halder P, Banerjee K (2011) Changes of selected hydrological parameters in Hooghly estuary in response to a severe tropical cyclone (Aila). Ind J Geo Mar Sci 40(1):32–36

Mitra A, Sengupta K, Banerjee K (2009) AILA and its impact on Gangetic delta. Environment watch - a newsletter of Indian Chamber of Commerce, pp 5–6

Mitra A Zaman S (2015) Blue carbon reservoir of the blue planet, published by Springer, ISBN 978-81-322-2106-7 (Springer, https://doi.org/10.1007/978-81-322-2107-4)
Panja U, Majumdar S, Mitra A (2003) An ecological profile of Sundarbans ecosystem. Sea Explor 6:21–27
Rabalais N (2002) Nitrogen in aquatic ecosystems. Ambio 31:102–112
Ray Choudhuri T, Banerjee K, Zaman S, Chakraborty S, Rudra T, Guha A, Karmakar D, Pramanick P, Fazli P, Mitra A (2015) Diluted proxies of climate change in the inshore estuarine complex of Bay of Bengal: a case of overlapping noise and facts. J Environ Sci Comp Sci Eng Technol 4(2):1–15
Strickland JDH, Parsons TR (1972) A practical handbook of seawater analysis. Fish Res Board Canada 167(1977):45–103
Trivedi S, Chakraborty S, Zaman S, Pramanick P, Fazli P, Amin G, Mitra A (2015a) Impact of salinity on the condition factor on the commercially important fin fish in the lower Gangetic Delta. J Environ Sci Comp Sci Eng Technol 4(2):473–480
Trivedi S, Zaman S, Roy Choudhury T, Pramanick P, Fazli P, Amin G Mitra A (2015b) Inter-annual variation of salinity in Indian Sundarbans. Ind J of Geo Mar Sci (in press)
Tyrrel SF (1999) The microbiological quality of water used for irrigation. Irrig News 27:39–42
WHO (World Health Organization) (2003) Looking back: looking ahead: five decades of challenges and achievements in environmental sanitation. WHO, Geneva
Zaman S, Mitra A (2014) Warning bell of climate change in the lower Gangetic Delta. Res Rev J Ecol 3(1):41–45

Chapter 4
Basics of Ecosystem

Abstract The concept and types of ecosystem are the basic pillars of the chapter along with energy flow through major types of ecosystem. The sustainability of an ecosystem is maintained if the energy flows through different tiers (*viz.* producer, primary consumer, secondary consumer, tertiary consumer and decomposer) in a perfect fashion. However, damage to a tier often results in the disruption of the food chain and breaks the flow of energy. The wild harvest of tiger prawn seeds in Indian Sundarban estuaries is a relevant example in this context, where huge quantum of fish juveniles (primary consumers of the ecosystem) are destroyed during tiger prawn (*Penaeus monodon*) seed collection and thus poses a negative impact on the demersal fish stock of the aquatic ecosystem.

Keywords Food web · Food pyramid · Energy flow · Sundarban estuary · Prawn seed

4.1 Definition and Overview

The word ecosystem was coined by A.G. Tansley (1935) in an article that appeared in 'Ecology'. The prefix *eco* means environment. A system, according to Webster, is 'an aggregation or assemblage of objects joined in a regular interaction or interdependence.' A system is more simply defined by Churchman (1968) as 'a set of parts coordinated to accomplish a set of goals.' The goal of a living system such as the tissues and organs that make up the animal body is coordinated effort to cause a flow of energy through the system and thus maintain life.

The concept of ecosystem is essential nowadays for two reasons. Firstly, it helps to understand the modification of environment caused by various types of organisms (including human beings) which has profound influence on other types of organisms. Secondly, it helps to understand the global biogeochemical cycles that support the foundation of life on Earth and the various human activities that are strongly influencing these cycles. Today the concept of ecosystem finds wide application in

A. Mitra, S. Zaman, *Environmental Science - A Ground Zero Observation on the Indian Subcontinent*, https://doi.org/10.1007/978-3-030-49131-4_4

different spheres of life like town planning, resource conservation, pollution control, fisheries and even in feasibility study for constructing dams, hotels, bridges, industries, national highways, power plants, etc.

Ecosystems are dynamic, not static. The populations of plants and animals they contain are always changing and adapting in response to major and minor changes in environmental conditions. Today human beings are creating artificial ecosystems and are manipulating their biotic components by changing the environment. It has become clear now that the price we are paying for simplifying, maintaining and protecting such stripped-down ecosystems is high. It includes time, money, increased use of energy and resources, loss of genetic, diversity and loss of natural landscape (Table 4.1).

The biosphere is composed of diverse types of ecosystems and all ecosystems by default consist of biotic (living) and abiotic (nonliving) components.

The different types of flora and fauna that constitute the biotic component of an ecosystem are classified as producers, consumers and decomposers depending on the basis of their sources of energy and matter.

The abiotic component of an ecosystem includes various physical factors (like sunlight, humidity, atmospheric temperature, atmospheric pressure, precipitation, water currents, waves, wind, etc.) and chemical factors (like the various elements or compounds in the atmosphere, hydrosphere or lithosphere) that are essential for living organisms.

The biotic tier of the ecosystem is totally dependent on the abiotic compartment as the latter is the source of all matter and energy that is transferred to the biotic tier through producers, which are like the antenna of the biotic sphere.

Producers are organisms that can prepare their own food and include all the green plants of the Earth. They can manufacture carbohydrates and glucose ($C_6H_{12}O_6$) through the process of photosynthesis from carbon dioxide (CO_2) and water (H_2O) utilizing sunlight as the source of energy. Since producers are the only

Table 4.1 Comparison of a natural ecosystem and a simplified human system

Natural ecosystem (marsh, grassland, forest)	Simplified human system (cornfield, factory, house)
Captures, converts and stores energy from the Sun	Consumes energy from fossil or nuclear fuels
Produces oxygen and consumes carbon dioxide	Consumes mostly oxygen and produces carbon dioxide from the burning of fossil fuels
Creates fertile soil	Depletes or deteriorates fertile soil
Stores, purifies and releases water gradually	Often uses and contaminates water and releases it rapidly
Provides wildlife habitats	Destroys some wildlife habitats during mining operations, setting industries, urban sectors, etc.
Filters and detoxifies pollutants and waste products free of charge	Produces pollutants and wastes, which are cleaned up with high expenses
Usually capable of self-maintenance and self-renewal	Requires continual maintenance and renewal at great cost

biotic component who can manufacture food, therefore, all animals depend on them for their nourishment. Consumers are organisms, mostly animals that cannot manufacture or prepare their own food and depend directly or indirectly on producers for their nourishment.

Depending on their food sources, consumers may be:

1. Primary consumers, which directly feed on plants like herbivores
2. Secondary consumers and higher levels consumers, which feed on plant eating animals like the carnivores

Decomposers constitute another important group of biotic component, which obtain nutrient breaking down complex molecules in the wastes and dead bodies of other organisms into simpler chemicals, some of which are returned to the soil and water for reuse by producers.

The interrelationship between the components of ecosystem can be visualized through food chains and food webs existing in all ecosystems of the Earth. The food chain represents the sequence of transfer of energy in the form of food from organisms of one trophic level to organisms of another trophic level when one organism eats or decomposes another. In the language of Odum (1971) '… food chain is the transfer of energy from the source in plants through a series of organisms with repeated stages of eating and being eaten….'

Food chains are not isolated sequences but are interconnected with one another in all types of ecosystem. The interlocking pattern is usually referred to as food web. Thus food webs are complex network of many interconnected food chains and feeding interactions. The food webs in the grassland and marine ecosystems existing since long evolutionary period of time are presented in Figs. 4.1 and 4.2, respectively.

To understand the interplay of structure and function of biotic and abiotic components of an ecosystem, the entire phenomenon is considered from different angles. According to Evans (1956), Whittaker (1960) and Odum (1971), the producer–consumer relationship highlights one kind of structure called the trophic structure. Trophic levels in this structure are nothing but the steps or tiers through which food energy passes into an ecosystem.

Trophic structure and also trophic function may be represented by means of ecological pyramids in which the first trophic level or producer forms the base and

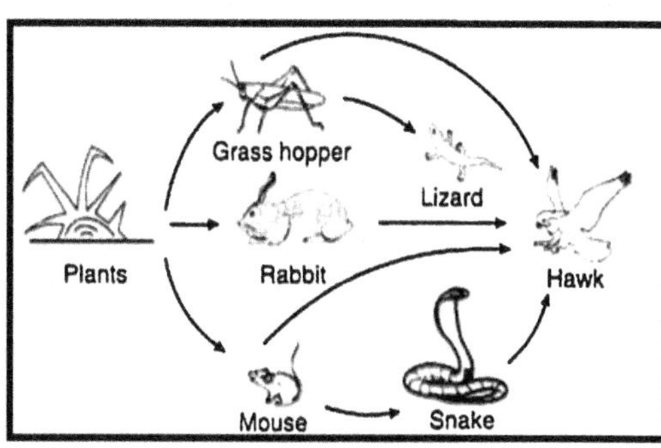

Fig. 4.1 Food web in the grassland ecosystem

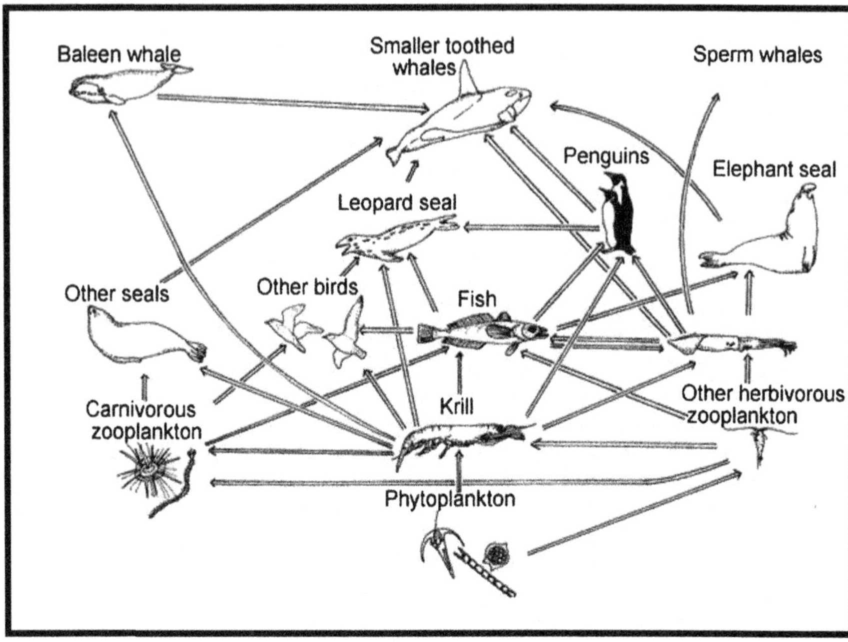

Fig. 4.2 Food web in the marine ecosystem

successive levels, the tiers, make up the apex. Ecological pyramids may be broadly classified under three main headings, namely:

1. The pyramid of numbers in which the number of individual organisms is depicted
2. The pyramid of biomass, which is based on the total dry weight, caloric value or any other measure reflecting the total amount of living materials
3. The pyramid of energy, which reflects the rate of energy flow through successive trophic levels considering the wastage of energy at each trophic level due to life processes like respiration, excretion and reproduction etc.

Out of three types of ecological pyramids, the pyramid of energy gives a unique overall picture of the functional nature of communities, since the number and weight of organisms that can be supported at any level in any situation depend not only the amount of fixed energy present at any instant of time in the level just below but also on the rate at which food is being produced. In contrast to the pyramids of numbers and biomass, which highlights pictures of standing states, i.e. organisms present at any particular instant of time, the pyramid of energy reflects the rates of passage of energy through food mass via the food chain. The pyramid of energy takes the second law of thermodynamics into consideration by depicting the wastage of energy through heat during the process of transference of energy from one trophic level to another.

4.2 Types of Ecosystem

4.2.1 Terrestrial Ecosystem

A terrestrial ecosystem is a type of ecosystem found only on landforms. Seven primary terrestrial ecosystems exist in the planet Earth, namely, tundra, taiga, temperate deciduous forest, tropical rain forest, grassland and deserts. The salient features of each of these ecosystems are given here in points.

(a) **The Tundra**

- The tundra is the northern most biome, forming a circle around the Earth above the taiga.
- Plants are small and ground touching since water is scarce and frozen through most of the year. Roots cannot penetrate because of permafrost.
- Light also limits plant growth, and the optimum intensity is available only during the brief summer season.
- Lichens and mosses are common ground cover, along with dwarf trees, grasses and sedges.
- The animal life includes some large herbivores along with numerous hares and lemmings. In summer the diversity of insects and also migratory birds flourishes.

(b) **The Taiga**

- The taiga is confined to the Northern Hemisphere. Rainfall varies but usually less than the deciduous forests. Bogs are very common.
- Light can be limiting factor to plant growth.
- The dominant plants are conifers with a few species of broad-leaved deciduous trees. Conifers have fine, waxy leaves adapted for resisting water loss. Their decomposition in the litter produces an acidic soil condition.
- Large herbivores and carnivores still survive in the taiga. The insect diversity in the taiga is also unique but often seasonal. The taiga is important to people for its timber.

(c) **The Temperate Deciduous Forest**

- This biome is found in all continents and like the grassland is greatly disturbed by the human activities.
- Seasonal change is a unique characteristic of this biome, and deciduous plants are common. Rainfall average about 40 inches distributed throughout the year. The forest floor shows heavy growth.
- The forest soil is rich with diverse microbial population.
- Plants adapt to winter by losing their leaves and becoming dormant. Animals migrate or hibernate as protective adaptation.

(d) **Tropical Rain Forest**

- This category of forest is supported by heavy rainfall and a warm climate. These forests are found in South America, Africa, India, Burma, Central America, Indonesia and the Philippines.
- There are no dominant species of plants, but there are dominant types. The canopy is composed of very tall hardwoods. Vines abound as do epiphytes, which absorb water from the air.
- The forest floor is dark and devoid of foliage. Scavengers and decomposers are widely distributed in this biome. Arboreal animals are also found in plenty.
- The jungle refers to thick growth along rivers or where the sunlight reaches the forest floor after disturbances.
- Tropical soils are relatively infertile and marginally useful because of the rapid cycling of nutrients. Agricultural interests and foresting have depleted much of this biome.

(e) **The Grassland**

- Grasslands exist as huge island plains in North America, Asia, Australia, South America and Africa. Grasslands are known by different names like the North American tall grass prairies, Eurasian steppes, South African belts and Argentinean pampas. Savannahs are tropical grasslands with scattered trees that occur either where the climate is too dry for dense forests to grow or where soil conditions and periodic fires maintain a grassland system.
- Rainfall is seasonal and limited and grasses dominate this biome.
- The animals of grasslands produce huge populations and include the largest herbivores.
- The food webs of grasslands are complex and relatively stable. Occasionally insect populations explode out of control and devastate this biome. Normally grasslands support the largest populations of animals on Earth.

(f) **The Desert Biome**

- Deserts receive less than 25 cm (10 inch) of rain per year. Temperatures in deserts vary according to latitude, and they are not necessarily hot places.
- Some deserts are produced and maintained by high mountain ranges that block coastal precipitation. Mountains form desert conditions by forming rain shadows as seen in the American desert.
- Large deserts are scattered in all continents except Europe and Antarctica. All deserts experience dramatic daylight fluctuations in temperature.
- Deserts are created along the 30° north and south latitude lines. They are concentrated at those pockets where air cells originating at the equator have lost their moisture.
- Plants adapted to deserts are termed as xerophytes. Some become dormant, while others store water. Rodents survive by spending daytime in burrows and venturing out at night. Many animals survive without water through behavioural and physiological strategies. The kangaroo rat rarely drinks but

lives off metabolic water conserved by a highly efficient kidney. The camel drinks huge quantities of water, survives dehydration and withstands unusual changes in body temperature.

4.2.2 Aquatic Ecosystem

4.2.2.1 Freshwater Ecosystem

Ecologists divide freshwater ecosystems into Lotic system like rivers, springs streams, etc. which are running bodies of water and lentic system like lakes, ponds, etc. which are standing water bodies.

(a) **Lotic System**

The lotic systems encompass running water bodies with high dissolved oxygen concentrations in the aquatic phase. Rivers and streams, which are unique examples of lotic system, depend on the surrounding land mass for their water supply and nutrient level. Dead leaves and sewage from the adjacent towns and cities increase the nutrient level of the water bodies and sustain the existing food webs of the ecosystem. Dissolved oxygen in the flowing water is usually abundant because of maximum diffusion from the adjacent layer of air. This creates a healthy environment for the growth and survival of the aquatic organisms. Dumping organic materials, in excess of the carrying capacity of the ecosystem, can deteriorate the water quality as their decomposition process uses the dissolved oxygen of the aquatic phase.

The producers of the lotic system include phytoplankton and various macro algae. Zooplankton are included in the tier of primary consumers. Snails, bass, cat, fish, crabs and various worms constitute the consumers of different levels.

(b) **Lentic System**

The lentic systems include standing water bodies like lakes, ponds, bogs, etc. Light penetrates to different degrees in the regions of a lake or a pond, and these differences determine the types of plants that live in a particular area. The littoral zone is the shallow region along the shore where light can penetrate to the bottom with sufficient intensity, so as to initiate the process of photosynthesis. The photosynthetic organisms in this region are free-floating plants, although rooted plants are also found. Typical producers include cyanobacteria, green algae and diatoms as well as some macrophytes like water hyacinth, water lily, etc. Animal life is also diverse, including damselflies, dragonfly nymphs, crayfish, rotifers, flatworms, hydra, snails, snakes, frogs and fishes.

The limnetic zone is the layer of open water where light can penetrate. It is inhabited by phytoplankton, zooplankton, and fish. The profoundal zone is the deep region beneath the limnetic zone where light does not penetrate. The biotic community in this zone depends on the rain of organic material from the surface layer and includes mostly scavengers and decomposers such as bacteria, fungi and insect larvae.

Oxygen and mineral nutrients in the lakes are distributed unevenly. Concentration of dissolved oxygen is usually greater in the upper layers, where it comes from the atmosphere through the process of diffusion. The process of photosynthesis, being carried on in the upper layers of lakes and ponds by the green algae and macrophytes, also increases the oxygen concentration. Deeper lakes in the temperate regions that have seasons often develop layers with very different water temperatures. The layers prevent complete circulation in the lake. As a result of this thermal stratification, the redistribution of oxygen and nutrients occurs seasonally.

In the summer, the upper layer of a lake is heated by the Sun, but the lower layers remain cold. Between these two layers is a third region, called thermocline, where water temperature drops quickly. In the fall, the water temperature in the upper layer of the lake drops with the air temperature. Gradually the water temperature becomes the same throughout the lake. Wind then mixes the upper and the lower layers, creating a fall turnover of nutrients and oxygen. During the winter, surface water becomes colder. When water cools to 4 °C, the temperature at which it is most dense, it sinks. Water, colder than this, floats above the 4 °C layer. With the warmth of the spring Sun, when the temperature of the upper layer rises to 4 °C, the water starts sinking, bringing oxygen with it. In turn, the nutrient-rich lower layers circulate to the top. This spring turnover often results in great algal growth due to high nutrient concentrations at the upper layer, which is called blooms.

On the basis of fertility and productivity, lakes are often classified into two types:

1. **Oligotrophic**

   These are deep lakes that are poor in nutrient level and productivity.

2. **Eutrophic**

   These are shallow lakes, rich in nutrients and productivity. The nutrients are primarily nitrate and phosphate due to which phytoplankton bloom occurs.

4.2.2.2 Estuarine Ecosystem

An estuary may be defined as the zone of intersection of fresh and marine waters with unique physico-chemical and biological characteristics. Estuaries generally occupy those areas of the coasts, which are least subject to marine features/activities and wave actions and thus are major sites for development of harbours, recreational activities and aquacultural farms. In estuaries, freshwater collected over vast regions of the land pours into the ocean, which sends salt water upstream far beyond the river mouth. According to Odum (1971), estuaries belong to different class of 'fluctuating water level ecosystems'.

Each estuary has its own physical features that influence its ecology. These features primarily encompass the amount of river discharge, depth and general topography, specific circulation pattern, climatic regime and vertical tide range. Although the physico-chemical condition in estuaries is extremely stressful (particularly in terms of salinity and turbidity), yet, the food availability is so favourable that the ecosystem is packed with life. Organisms living in this habitat exhibit wide range of tolerances due to tidal and seasonal variations of hydrological parameters (i.e. they are mainly euryhaline). Estuaries rank first among the most productive regions of marine ecosystems as they typically contain diverse group and high biomass of benthic algae, seagrasses, saltmarsh grasses and phytoplankton that support large number of fishes and birds. This is mainly because estuaries are enriched by nutrients that are contributed through land drainage. Also rivers contribute substantial load of nutrients in the estuarine system. The estuarine system possesses considerable nutrient retention capacity due to which the flora and fauna thriving within such system never faces the scarcity of nutrients.

Estuaries provide unique ecosystem services. Only a few type of fish species leave permanently in an estuary, but many species develop there; thus both larval and juvenile fishes are present in large numbers. It has been estimated that well over half of all marine fishes develop in the protective environment of an estuary, and because of this reason estuaries, are often referred to as the nurseries of the sea. Apart from commercially important finfish species, shellfishes like shrimps, molluscs and crabs also use the estuary as their nursery ground. In some maritime states of India like West Bengal, few dark environmental issues have been opened up related to the fish juvenile diversity of the estuaries. The authors have a made a ground zero survey on this aspect and observed that a variety of fish juveniles are wasted during the wild harvest of tiger prawn seeds in the Indian Sundarban estuaries and this occupation has deep rooted in the socio-economic profile of the region because of the absence of tiger prawn hatchery in the state. Because of the absence of congenial salinity, hatcheries could not be established in and around Indian Sundarbans due to which the poverty-stricken island dwellers (preferably the women) screen the estuarine water with nets of fine mesh size to trap the tiger prawn seeds. During this operation several fish juveniles are also caught in the net, which are thrown away and destroyed as they are nonremunerative unlike the shrimp seeds, which are bought by the shrimp farm owners with ready cash. Thus this case study is a relevant example of biodiversity damage. Also during dragging the nets along the shore for this operation, the mangrove seedlings are uprooted (blue carbon loss), which otherwise could be a potential sink of carbon and provide ecosystem services like erosion control, bioremediation, etc. A detailed picture of these environmental issues is presented as Annexure 4A.

4.2.2.3 Marine Ecosystem

The ocean occupies about 70% of the Earth's surface and manipulates efficiently the global hydrological cycle. It is also a unique reservoir of biological resources, natural gas, oil and minerals.

The marine ecosystem encompasses the seashores as well as the ocean basins. This ecosystem has been classified by the ecologists from various angles like depth, type of bottom, light level, etc. The various zones of marine ecosystem on the basis of depth are continental shelf, continental slope, bathyal, abyssal and hadal zone.

The continental margin includes the continental shelf and slope. The continental shelves fringe the continents and slope gently towards the ocean basins. They are relatively flat, generally less than 130 m deep. However, the range of the depth for the continental shelf is 30 m to about 275 m. The criterion for defining the seaward edge of the continental shelf is a marked change in slope. The continental shelf has a gradient of about 1:1000. Seaward from the continental shelf is the continental slope which extends to a depth of 1.6–3.2 km. The continental slope has a gradient between 1:2 and 1:40. The slope is scarred by many submarine canyons which may be as deep as 1200 m. It is believed that these canyons have been cut by turbidity current of dense sediment-laden waters flowing down the surface of the continental slope. Sediments from the continental shelf are carried into the deep ocean through the canyons present in the continental slope. At the foot of the continental slope, the suspended materials and sediments settle out that slope gently towards the ocean floor. This region is called the continental rise and has a gradient between 1:50 and 1:800. The continental rise is fairly well developed in the Atlantic and India Oceans but is narrow or almost non-existent in the Pacific Ocean.

The ocean basin floor extends seaward from the continental rise or the continental margins and includes the abyssal plains, oceanic rises, sea mounts and trenches. Abyssal plains are found at the base of the continental rise and are flat plains having a gradient less than 1:1000. They are formed due to even deposition of sediments from the continental rise carried down by the turbidity currents. Oceanic rises refer to rather isolated areas which are elevated above the abyssal floor and are distributed sporadically on the ocean basin floor. They may vary from low hills to mountains as high as 1525 m. The example of Bermuda rise is very prominent in this case on which the Bermuda Islands are formed. Seamounts are isolated peaks that rise several thousand meters above the sea floor. They are volcanic in origin and are found principally in the vicinity of the fault zones. Trenches are long narrow depressions in the oceanic floor which are over 6100 m deep. The deepest known trench in the ocean compartment is the Mariana trench of the western North Pacific which is 11,000 m deep. Trenches are invariably associated with the systems of active volcanoes and are believed to be caused down wrapping of the oceanic crust beneath the continental crust.

The entire water mass above the ocean floor (benthic substratum) with all the diverse types of organisms is referred to as the pelagic zone and may be differentiated into neritic and oceanic zone. Neritic zone is the shallow water zone that extends from the intertidal zone to the edge of the continental shelf and includes the coastal waters having an average depth of 200 m. The water masses beyond the continental shelf and covering the deep abyssal plain constitute the oceanic zone. The depth of this zone is more than 3 km.

On the basis of distribution of light in the oceanic compartment, it is possible to distinguish two district zones, namely, (1) photic zone and (2) aphotic zone.

The photic zone is restricted up to that layer where light is available and usually ranges up to 200 m from the surface. The productivity is more in this zone, and the producers include many species of large algae thriving in extensive beds, as well as huge population of phytoplankton that serve as a nutrient base for the rest of the

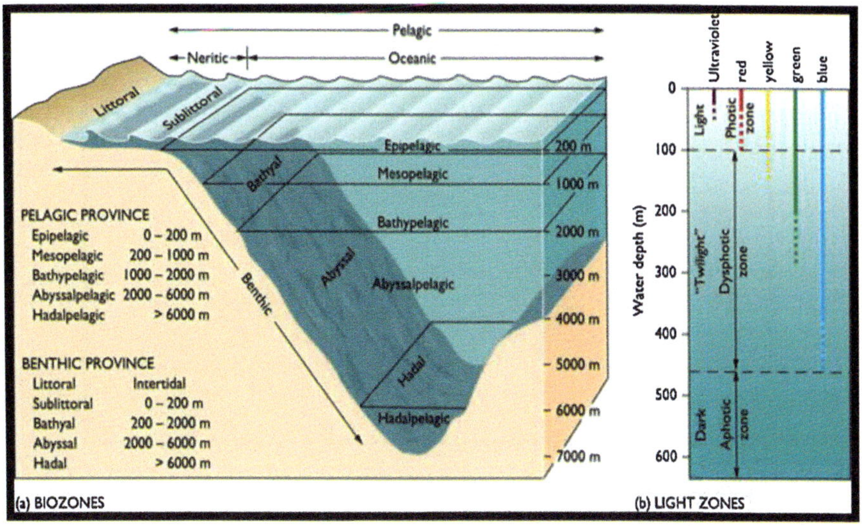

Fig. 4.3 Zonation of the marine ecosystem

consumers. The aphotic zone is the deeper lightless zone of the ocean usually below 200 m.

The divisions and subdivisions of marine environment are represented in Fig. 4.3.

4.3 Energy Flow in the Ecosystem

Energy sources on Earth include tidal energy, nuclear energy, the heat of the Earth's core and so on. But for living systems, the unique energy source is solar radiation. Biological systems depend on the flow of this energy from the Sun, through the system and back into outer space in the form of heat. The Earth is in a steady state, so over any appreciably long period, the amount of energy that enters from the source (the Sun) is equal to the amount lost as heat to the 'sink' (outer space). The existence of both source and sink is essential, because if the Earth did no maintain this energy balance, it would heat up or cool down, and in either state, biological systems would not persist. In the first stage of energy flow, the producers trap the energy due to their autotrophic nature. It is then transferred to different levels of consumers, and finally the decomposers act on the dead organic matter (of both producers and consumers) and form elements, which ultimately return to the ambient environment. A simplified diagrammatic representation of energy flow through ecosystems is presented in Fig. 4.4.

Energy flows and in doing so produces material cycles, the water cycle being an example. Part of the solar energy eventually absorbed by the atmosphere and oceans produces temperature differences; these cause convection currents in the air and

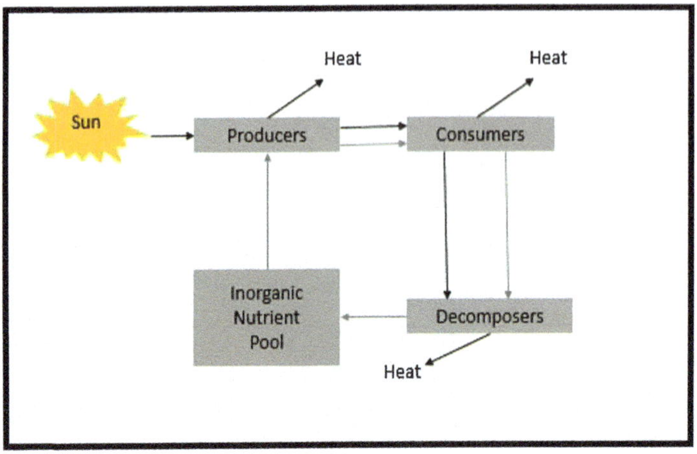

Fig. 4.4 Energy flow through forest ecosystems

produce winds. The heat energy absorbed by water also produces evaporation, and water vapour is then convected high in the atmosphere. When the water vapour precipitates over land, the heat energy absorbed during evaporation is released. The cycle is completed when the water returns to sea level via runoff, rivers and lakes, the potential energy of the water being dissipated as heat. Man concentrates the potential energy by building dams and converting the energy to hydroelectric power.

Other energy sources that play apart in cycles include the thermal energy of the Earth, which produces volcanoes and mountain lifting, and the solar energy trapped by plants, which drives biological cycles. The latter energy is transferred as food from one part of a biological system to another. Each transfer involves biochemical rearrangements during which heat is produced, and eventually all the energy first trapped by photosynthesizing plants leaves the system as heat.

4.3.1 Energy Flow Through Forest Ecosystem

The entire process of energy flow through an ideal forest ecosystem may be discussed under three broad headings.

Acquisition of Energy by the Producers

The Sun is the main source of energy in the forest ecosystem. Extraterrestrial sunlight reaches the biosphere at a rate of 2 gcal per cm^2 per min but is attenuated as it passes through different layers of atmosphere. According to Gates (1965) about 67% (1.34 gcal per cm^2 per min) may reach the Earth's surface at noon on a clear summer day. The magnitude of radiation penetrating the atmosphere is a function of several factors like atmospheric gases, dust, wavelength of the radiation, etc. Shortwave ultraviolet radiation below 0.3 μ is abruptly terminated by the ozone layer in the outer atmosphere (about 18 miles or 25 km altitude). Solar radiation is further attenuated, and the spectral distribution of its energy is greatly altered as it passes through the cloud cover, water and vegetation. Radiant energy reaching the surface of the Earth on a clear day consists of 10% ultraviolet, 45% visible and 45% infrared. Visible radiation is least attenuated as it passes through dense cloud cover and water, which means that photosynthesis can occur even on cloudy day at some depth in clear water. This concept is very important for understanding the acquisition of energy by aquatic plants which are present in the water bodies of moist deciduous forests in the tropics. The daily input of sunlight to the autotrophic layer of an ecosystem varies mostly between 100 and 800 gcal per cm^2 averaging about 300–400 gcal per cm^2 for an area in the temperate zone such as the United States. It has been observed by researchers that vegetation of the Earth strongly absorbs the blue and red visible wave lengths, less strongly the green wave length and very weakly the near infrared. The blue and red light (0.4–0.5 μ and 0.6–0.7 μ bands, respectively) are especially absorbed by chlorophyll, and the far-infrared heat energy is absorbed by the water in the leaves and the water vapour that surrounds them. Green plants thus efficiently absorb the blue and red light that are most useful in photosynthesis.

Fixing of Energy by the Producers

The flow of energy through the forest community starts with the fixation of sunlight by plants, a process that in itself demands the expenditure of energy. Initially plants rely on the food stored in the seed for energy until their own production machinery is working. Once mobilized, the green plant begins to accumulate energy. Energy accumulated by plants is called production or, more specifically, primary production, because it is the first and basic form of energy storage in an ecosystem. The rate at which energy is stored by photosynthesis activity is known as primary productivity. All of the Sun's energy that is assimilated (i.e. total photosynthesis) is gross primary production. Like other organisms, plants require energy for reproduction and maintenance. The energy required for these needs is provided by a reverse of the photosynthesis process, respiration. The energy remaining after respiration and stored as organic matter is net primary production or plant growth. Thus net production for a single plant or a plant community can be described by the formula:

$$\text{Net primary production} = \text{Gross production} - \text{Autotrophic Respiration}$$

The basic unit of energy used in ecological work is the calorie; production is usually expressed in kilocalories per square meter per year ($Kcal/m^2/year$). However, production may also be expressed as dry organic matter in grams per square meter per year ($g/m^2/year$).

Net primary production accumulates over a periods of time as plant biomass. Part of this accumulation is turned over seasonally through decomposition. Part is retained over a longer period as living material. The amount of this accumulated living organic matter found on a given area at a given time is the standing crop biomass. Biomass is usually expressed as grams dry weight of organic matter per square meter (g/m^2), kilocalories per square meter ($Kcal/m^2$) or some other appropriate measure per unit area.

Different types of forests ecosystems have characteristically different productivities depending on the quality and quantity of producers. The net primary productivity (NPP) of tropical rain forest ranges between 1000 and 3500 gm cm^{-2} $year^{-1}$. The mean values of NPP for tropical seasonal forest, temperate evergreen forest, temperate deciduous forest and boreal forest are 1600, 1300, 1200 and 800 gm cm^{-2} $year^{-1}$, respectively.

In the Indian sub-continent, the forest primary productivity accounts to a considerable value owing to the presence of diverse type of plant species in diverse category of forest.

In tropical rain forest (which is influenced by the southwest monsoon) of the West Coast extending from the North Kanara to the South and in the eastern sub-Himalayan tract in Assam, the producers include many families among which Dipterocarpaceae, Guttiferae, Lauraceae, Magnoliaceae, Euphorbiaceae and Palmae are dominant. Climbers and epiphytes are plentiful among the producers of the ecosystem. The producers of the moist deciduous forests in India mainly consist of bamboo, teak (*Tectona grandis*), sal (*Shorea robusta*) and sandalwood (*Santalum album*) of Karnataka. In India such forests are concentrated along the western side of the Deccan Plateau, i.e. Bombay, Northeast Andhra, Madhya Pradesh, Gangetic Plain and some area of the sub-Himalayan tract, extending from the Punjab in the west to the Assam valley in the east. Dry or temperate deciduous forests, in India, occur in the northern Punjab where the rainfall usually ranges between 50 and 75 cm. The important producers in this ecosystem are *Acacia arabica*, *Acacia modesta*, *Prosopis spicigera*, *Albizia lebbeck*, etc. The dry deciduous forests in Tamil Nadu are dominated by *Soymida febrifuga*, *Terminalia* spp., *Tectona grandis*, etc.

The net primary production of any forest ecosystem is not only accounted by the vegetative biomass above the ground, but the root systems below the soil surface also have contribution to the productivity.

Transference of Energy Through Trophic Level

The transfer of food energy from the source in the plants through some tiers of organisms (each tier is referred to as the trophic level) constitutes the food chain of an ecosystem. In forest ecosystem many food chains, being intricately related to each other, form the food webs. All the consumers of any type of forest ecosystem (starting from squirrel to lion) depend on the value of net primary productivity for their supply of energy. However, ecosystems differ in the efficiency with which consumers at different trophic levels are able to acquire energy from the level below them and to incorporate that energy into biomass, the weight of the living organisms. Many kinds of ecological efficiencies have been given diverse names.

Lindeman's efficiency, also called ecological efficiency, is the ratio of energy acquired by one trophic level (N + 1) to that acquired by the previous (N). Assimilation efficiency is the fraction of light striking plants that is actually used in photosynthesis, or it is the fraction of energy consumed by animals which is absorbed to form the food. Growth efficiency pertains to the splitting of the assimilated energy into the portion degraded through the metabolic activities of organisms or lost through excretion and the portion incorporated into new tissues. Exploitation efficiency pertains to the fractions of those tissues that are consumed by the next trophic level. Lindeman's efficiency is the product of assimilation efficiency, growth efficiency and exploitation efficiency.

Assimilation efficiency is a unique function of diet spectrum of an organism. For example, a millipede feeding on decaying wood of a forest tree may have an assimilation efficiency of 15%. An elephant depending on vegetative parts of the forest plants may have about 30%. In general carnivores are more efficient than herbivores due to the presence of much more indigestible material (like cellulose) in the plant tissues. Growth efficiency at the producer level can be quite high, in contrast to assimilation efficiency. Plants do not spend a lot of energy in locomotion and thus are able to incorporate a high percentage of material in their tissues. An efficiency of 30% has been estimated for a tropical rain forest. Efficiencies above 60–70% are common, especially outside of the tropics.

Generally ectotherms have higher growth efficiencies than endotherms as the former type does not divert energy towards maintaining their body temperatures. Insects tend to use about 33% of their assimilated energy in tissue building and reproduction, birds and mammals less than 1–5%. Growth efficiencies tend to be higher in young than in old animal.

The generalized pattern of energy flow often becomes complicated when some forest animals like leopards are considered for case study purpose. This is because their diet chart includes a wide spectrum of animals and hence the energy from different consumer compartments converges to a single compartment resulting in an amplification of energy in the animal body. The leopard scat analysed from North Bengal area at the base of the Himalayan range during 1996–1997 revealed the presence of 18 different species (viz. *Suncus maximus*, *Macaca mulatta*, *Macaca assamensis*, *Presbytis entellus*, *Muntiacus muntjak*, *Axis porcinus*, *Tragulus meminna*, Domestic goat *Sus scrofa*, *Paradoxurus hermaphroditus*, *Bandicota indica*, *Rattus rattus*, *Callosciurus pygerythrus*, *Cannomys badius*, *Petaurista petaurista*, *Ratufa bicolor* and *Gallus gallus*). This animal thus presents another model of energy flow in which energy of the consumers does not decrease to that extent (in comparison to the animal that feed on single species) as the energy paths from a wide range of prey species find their way into the predator. However, during all cases energy transference through trophic level in forest ecosystem, Lindeman's 10% rule is conveniently applied.

Lindeman's efficiencies are often said to be about 10% at levels above the producer level. This 10% rule of thumb is convenient for peeping into various ecological problems. We can visualize each successive trophic level as having roughly one tenth of the energy of the preceding level available to it. On this basis it can be stated that with the same primary production, about ten times more biomass of deer that eat grass could be supported than biomass of tiger that depend on deer in the forest ecosystem. This progressive decrease of available energy as it passes along food chains, following the dictates of the second law of thermodynamics, results in a pyramidal structure of ecosystems. An ideal model applicable to energy flow through terrestrial ecosystem (preferably forest) is presented in Fig. 4.5.

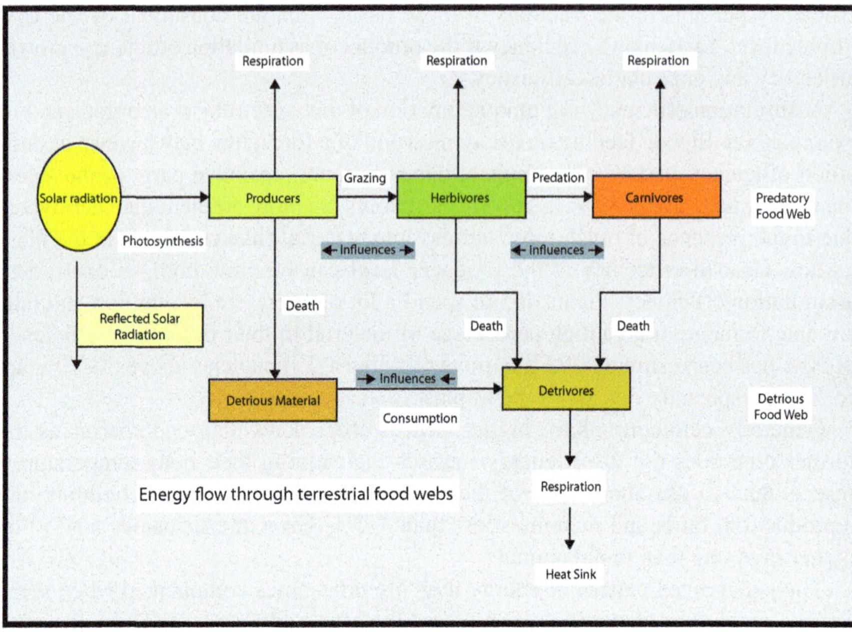

Fig. 4.5 Energy flow through terrestrial ecosystem

4.3.2 Energy Flow Through Aquatic Ecosystem

The main factors affecting the biological diversity of aquatic ecosystems are salinity, transparency of the aquatic phase (or the depth up to which sunlight can penetrate), dissolved oxygen and water temperature. Salinity, the concentrations of dissolved salts (especially sodium chloride) in aquatic phase are a limiting factor because different species of aquatic plants and animals have different salinity tolerances. Salinity levels are used to divide aquatic ecosystems into two major classes.

1. **Freshwater ecosystems** consisting of inland bodies of standing water (like lakes, reservoirs, ponds, etc.) and flowing water (like streams and rivers)
2. **Marine or salt water ecosystems** such as oceans, estuaries, coastal wetlands and coral reefs

The basic mechanism of energy flow starting from the producer level through different trophic tiers is same in all types of ecosystem, but due to some factors like depth, salinity, upwelling and nutrient load, etc., the quantum and nature of primary producing greatly differ among aquatic ecosystems.

Freshwater Ecosystem

Ponds are small, shallow usually human created impoundments of freshwater where photosynthesis can take place at all depths during daylight unless the surface water becomes clogged with masses of algae during warm weather. In this ecosystem, the energy flow is triggered by various algae like *Chlamydomonas*, *Volvox*, *Spirogyra* sp., etc. The primary consumer includes variety of zooplankton and herbivorous fishes, and they transfer the energy further to higher trophic levels consisting of large fishes, crabs, birds (like kingfishers), etc.

Streams are relatively small, flowing water bodies that empty into rivers. These water bodies receive a considerable contribution from the adjacent land in the form of runoff, and nutrient depletion is not witnessed throughout the year. Excessive addition of nutrient and organic materials to streams and other water bodies can cause undesirable effects on secondary production, but up to a certain level, the addition of small amount of nutrients to aquatic ecosystems can stimulate desirable secondary production. Researchers showed this by the experimental addition of sucrose to Berry Creek in Oregon. It was found that trout productions were very high in the sucrose-enriched section in the creek rather than the control section. The abundance of benthic insects in the sucrose-enriched sectioned also increased due to the increased availability of the bacterial food in the form of a slime bacterium, *Sphaerotilus natans*. Thus not all energy pathways through this stream community lead to trout production. Trout depend only on a few selective species of benthic invertebrates whose abundance increased due to addition of small amount of sucrose. However, if sucrose addition were increased greatly, the population of benthic invertebrates would be under severe stress as the colony of slime bacteria would smother the bottom. In such case, the production of trout would not exhibit an increasing trend. The principal of employing critical or optimum level of nutrients to initiate the process of energy flow is also practiced since a long period for farming pisciculture in and around the East Kolkata Wetlands. The sewage discharged from the highly urbanized and industrialized city of Kolkata, being packed with nutrients like nitrate and phosphate, is diverted into the water bodies of East Kolkata Wetlands where these nutrients initiate the growth of phytoplankton. These planktonic communities, being the main players behind the primary production of the aquatic ecosystems, play a great role in the production of fishery. The important fishes constituting the consumer level of these wetlands mainly include *Channa punctatus*, *Channa striatus*, *Channa gachua*, *Labeo rohita*, *Catla catla*, *Cirrhinus mrigala*, *Hypophthalmichthys molitrix*, *Cyprinus carpio*, *Tilapia mossambica*, *Liza parsia*, *Liza tade*, *Lates calcarifer*, *Oreochromis niloticus*, *Clarias batrachus*, *Heteropneustes fossilis*, *Mystus gulio*, *Puntius japonica* and *Anabas testudineus*. A survey done to investigate the inter-relationship between NPP and fish yield at 15 water bodies in the East Kolkata Wetlands during 1995–1996 revealed that the value of NPP of the water bodies receiving sewage ranges between 21.7 and 50.1 mg C m^{-3} h^{-1} and that of fish yield between 1.5 and 7.0 tonnes/ha. A unique positive correlation between NPP and secondary production ($p < 0.01$) was also revealed from the study. An ideal model of energy flow through freshwater system is highlighted in Fig. 4.6.

Marine Ecosystem

The mechanism of energy flow through the marine ecosystem is a bit complicated as life in the oceans is unevenly distributed. The neritic zone, representing 10% of

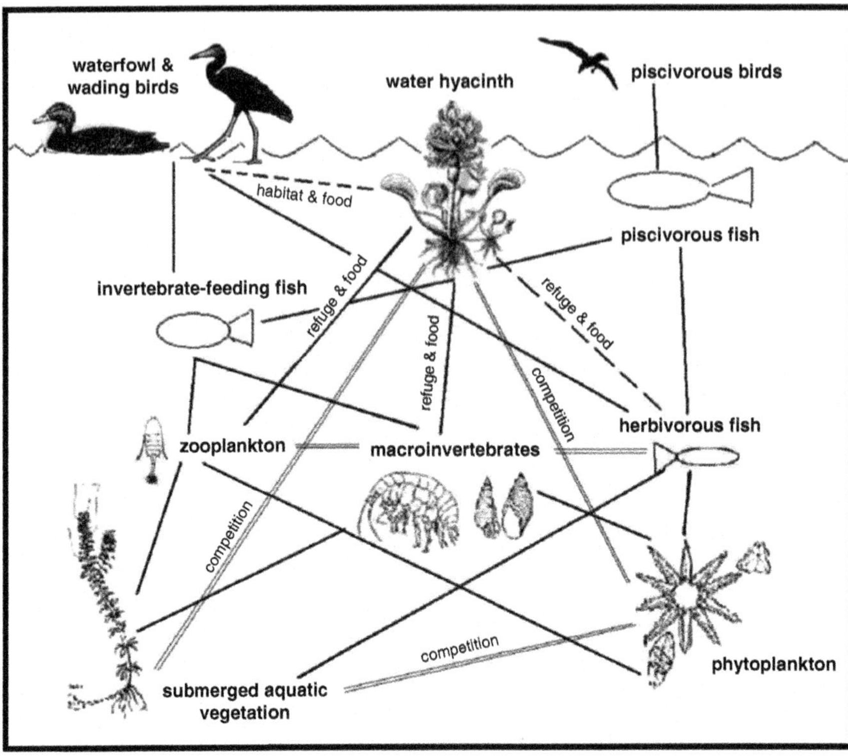

Fig. 4.6 Energy flow through freshwater system

the total ocean area, contains 90% of all ocean plant and animal life and is the site of the commercial marine fisheries. The primary productivity due to solar energy in the marine ecosystem is limited in the photic zone (approximately up to 200 m deep from the surface) as the Sun rays fail to penetrate beyond this zone. This is one of the main reasons why a huge open sea has a relatively low average net primary productivity of plant life.

The nature of energy flow in marine ecosystem differs in different zones of compartments of ocean. The neritic or coastal zone is relatively warm, nutrient-rich shallow water zone that extends from the high tide mark on land to the edge of shelf like extension of continental landmasses known as the continental shelf. Estuaries are distributed along the coastlines, where there is a continuous overlapping of fresh and saline water due to tidal actions. All these zones are the sites of high primary production owing to the presence of a variety of producers as listed here.

1. Sea grasses (*Zostera* sp.)
2. Salt marsh grasses (*Porteresia coarctata*)
3. Mangroves (*Avicennia* spp. *Rhizophora* sp., *Heritiera fomes*, etc.)
4. Sea weeds (*Ulva* sp., *Enteromorpha* spp., etc.)
5. Phytoplankton (*Coscinodiscus* spp., *Biddulphia* sp., etc.)

The estimated average net primary productivity by producers in estuaries is about 9200 Kcal/m^2/year, but when the average net primary productivity is compared among the major types of ecosystem of the world, it becomes extremely low (about 18 billion Kcal/year), whereas for open ocean the value becomes 200 billion Kcal/year. This data often becomes misleading as the productivity in open ocean is very much less in comparison to estuaries and coastal wetlands. However, as the oceans cover so much of the globe in terms of area, they head the list of NPP among the major types of ecosystem in the world although the productivity value per unit area per unit time is extremely poor. The energy from these producers passes to the zooplankton, which mostly graze over the phytoplankton for building their own biomass. The various finfishes, crabs, oysters, squids and birds like gulls are the top level consumers of the marine ecosystem. The coastal region exhibits a unique reservoir of nutrients not only because of their derivation from the adjacent landmass through runoff but also due to some processes like upwelling. Along some steep coastal area, such as those of Peru, winds blow the surface water away from the shore, allowing cold, nutrient-rich bottoms waters of upwelling to rise to the surface and support large population of plankton. In areas of upwelling, food chains are usually short. This is primary because of the large size of the phytoplankton, many of which form colonies several millimetres in diameter. These large colonies are directly consumed by herbivorous fishes like sardines, anchovies, etc. making the food chains of the area relatively shorter. Thus the process of energy flow in neritic or coastal zone is governed greatly by wind driven currents.

The sharp increase in the depth of the water at the edge of the continental shelf marks the separation of the neritic zone from the open ocean. This compartment snatches about 90% of the total surface area of the ocean but sustains only about 10% of its plant and animal life. This is mainly because light is available only at its surface layer and most nutrients are found on the bottom far below.

The open sea can be broadly divided into three distinct vertical zones on the basis of magnitude of NPP. The surface layer through which enough sunlight can penetrate for photosynthesis is called euphotic zone. Phytoplankton are the main primary producers of this zone. The phytoplankton comprise a variety of organisms, namely, diatoms, dinoflagellates, blue green algae, silicoflagellates, coccolithophores, etc. which have a size range of 0.001–0.2 mm and contribute to about 95% of the primary production in the open sea euphotic compartment. The diatoms commonly known as the pasture of the sea or meadows of the sea are unicellular autotrophic organisms and form about 90% of the phytoplankton biomass. The zooplankton community of the euphotic zone, being sustained by the phytoplankton biomass, support commercially important fishes like herrings, sardines, anchovies and other small fish that feed at the surface and larger predators such as tuna, mackerel, sword fish, etc.

The quantum of input energy, however, decreases in the bathyal zone, which is a colder, darker layer below the euphotic zone. As the light penetration is very scanty in this layer, therefore the photosynthetic activity is greatly hindered. The pitch dark, colder abyssal zone is witnessed as one goes deeper towards the bottom. In the absence of photosynthesis, most of the organisms in this layer survive by feeding on dead plants and animals and their waste products, which sink down from the surface waters and also by making negatively geotropic migration to surface layer for the purpose of procuring food materials. Thus the transference of energy in the aphotic depth of the ocean is not a smooth process, and due to cut down of the initial input energy, the NPP and species diversity in this zone is extremely low.

A comparative picture of the estimates of primary production and fish production in three different zones of the oceans clearly reveals the low productivity and low efficiency of energy transfer in the open ocean in comparison to the coastal zone.

In 1977, a unique aquatic ecosystem was discovered deep in the abyssal zone around vents in the ocean floor of the Pacific Ocean. John Corliss and his co-workers explored the Galapagos Rift, a volcanic active boundary between the two Earth's crustal plates and observed that in that zone, the near-freezing seawater seeps into the fissures, becomes heated and is spewed out through vents at temperature exceeding 700 °F. This hydrothermal effluent deposits zinc, iron and copper sulphides as well as calcium and magnesium sulphates, all leached from rocks as pressure forces the heated waters upward. In contrast to most of the deep oceanic compartment, which is often called biological desert, these nutrient-rich warm water oases support diverse marine communities whose foundation lies on the unique type of energy source. In this zone, chemosynthetic bacteria and other microbes use the inorganic deposits as energy sources. They are the primary producers in the trophic level whose higher tiers are occupied by tube worms, clams, crabs and fishes.

4.4 Take Home Messages

[A] The concept of ecosystem is essential nowadays for two reasons. Firstly, it helps to understand the modification of environment caused by various types of organisms (including human beings) which has profound influence on other types of organisms. Secondly, it helps to understand the global biogeochemical cycles that support the foundation of life on Earth and the various human activities that are strongly influencing these cycles. Today the concept of ecosystem finds wide application in different spheres of life like town planning, resource conservation, pollution control, fisheries and even in feasibility study for constructing dams, hotels, bridges, power plants, tourism units, industries, etc.

[B] Ecosystems are dynamic, not static, and are composed of abiotic and biotic components. The biotic components encompass producers, consumers and decomposers. Producers are organisms that can prepare their own food and include all the green plants of the Earth. They can manufacture carbohydrates like glucose ($C_6H_{12}O_6$) through the process of photosynthesis from carbon dioxide (CO_2) and water (H_2O) utilizing sunlight as the source of energy. Consumers are organisms, mostly animals, that cannot manufacture or prepare their own food and depend directly or indirectly on producers for their nourishment. Decomposers constitute another important group of biotic component, which obtains nutrient breaking down complex molecules in the wastes and dead bodies of other organisms into simpler chemicals, some of which are returned to the soil and water for reuse by producers.

[C] In any ecosystem the food chain represents the sequence of transfer of energy in the form of food from organisms of one trophic level to organisms of another trophic level when one organism eats or decomposes another. In an ecosystem the producer–consumer relationship highlights one kind of structure called the trophic structure. Trophic levels in this structure are nothing but the steps or tiers through which food energy passes into an ecosystem. Trophic structure and also trophic function may be represented by means of ecological pyramids in which the first trophic level or producer forms the base and successive levels and the tiers make up the apex.

[D] The mechanism of energy flow through the marine ecosystem is a bit complicated as life in the oceans is unevenly distributed. The neritic zone, representing 10% of the total ocean area, contains 90% of all ocean plant and animal life and is the site of the commercial marine fisheries. The primary productivity due to solar energy in the marine ecosystem is limited in the photic zone (approximately up to 200 m deep from the surface) as the Sun rays fail to penetrate beyond this zone. This is one of the main reasons why a huge open sea has a relatively low average net primary productivity of plant life.

4.5 Brain Churners

This chapter is an introductory approach to understand ecosystem and its various types. The energy flow through the ecosystem is explained in details. The readers on acquiring knowledge on the flow of energy through various components and categories of ecosystems will gain confidence to face various competitive examinations as Ecology and Environment has been given thrust in recent times. The readers can consider this section as a sort of *self-appraisal approach* as correct answers to all the questions are filled in black in the Answer Box at the end of this section.

1. Primary consumers in the ecosystem feed on

   (a) Plants
   (b) Animals
   (c) Fishes
   (d) None of the above

2. Decomposers in the ecosystem

   (a) Combine simple molecules into complex compound
   (b) Form hydrocarbons
   (c) Degrade complex molecules into simpler forms
   (d) None of the above

3. Food chains are interlocked to form
   (a) Food web
   (b) Pyramid of numbers
   (c) Both a and b
   (d) None of the above
4. Trophic structure and functions are represented by
   (a) Pyramid of numbers
   (b) Pyramid of biomass
   (c) Pyramid of energy
   (d) All the above
5. Tropical rain forest comes under
   (a) Terrestrial water ecosystem
   (b) Marine ecosystem
   (c) Freshwater ecosystem
   (d) Desert ecosystem
6. The taiga is confined to
   (a) Southern Hemisphere
   (b) Equatorial region
   (c) Northern Hemisphere
   (d) None of the above
7. The lotic system encompasses
   (a) Running water bodies
   (b) Standing water bodies
   (c) Water bodies with low dissolved oxygen
   (d) None of the above
8. The lentic systems includes
   (a) Rivers
   (b) Ocean
   (c) Estuaries
   (d) Ponds
9. The zone of open water where light can penetrate is called
   (a) Profoundal zone
   (b) Limnetic zone
   (c) Thermocline zone
   (d) Hadal zone
10. Oligotrophic lakes are
    (a) Rich in nutrients and productivity

(b) Extremely cold in nature
(c) Medium in nutrients and productivity
(d) Poor in nutrients and productivity

11. The estuarine water is

(a) Fresh
(b) Highly saline
(c) Brackish
(d) None of the above

12. Mangroves are found in the

(a) Estuarine region
(b) Freshwater region
(c) Desert region
(d) None of the above

13. The continental margin in the marine ecosystem includes

(a) Bathyal and hadal zone
(b) Continental shelf and slope
(c) Continental shelf and bathyal zone
(d) Abyssal zone

14. Bathyal zone is

(a) Hot and illuminated
(b) Cold and illuminated
(c) Cold and dark
(d) None of the above

15. In the aquatic ecosystem the producer community includes

(a) Bacteria
(b) Fungi
(c) Zooplankton
(d) Phytoplankton

16. Zooplankton in the aquatic ecosystem are

(a) Consumers
(b) Decomposers
(c) Producers
(d) None of the above

17. The energy source of chemosynthetic bacteria is

(a) Sun
(b) Inorganic deposits
(c) Tides
(d) None of the above

18. The productivity is high in
    (a) Deep ocean
    (b) Bathyal zone
    (c) Coastal zone
    (d) Hadal zone

19. *Tilapia mozambica* is included under
    (a) Consumer in the freshwater ecosystem
    (b) Producer in the marine ecosystem
    (c) Decomposer in the estuarine ecosystem
    (d) Consumer in the marine ecosystem

20. The process of upwelling brings nutrients from
    (a) Adjacent landmasses to the sea
    (b) Bottom of the sea to the surface
    (c) Surface of the sea to the bottom
    (d) None of the above

Answer Sheet

S. No.	a	b	c	d
1.	■		■	
2.				
3.	■			
4.				■
5.	■			
6.			■	
7.	■			
8.				■
9.		■		
10.				■
11.			■	
12.	■			
13.		■		
14.			■	
15.				■
16.	■			
17.			■	
18.			■	
19.	■			
20.		■		

Annexure 4 A: Damage to Fish Juvenile Community Due to Wild Collection of Tiger Prawn Seeds in the Indian Sundarban Estuaries

Introduction

The ecosystem services of estuaries are wide and intricately related to the livelihood of the surrounding population. These services include production of fishes, molluscs and crabs for commercial/economic gain of the local people, navigational channel for transportation of local people, fishing vessels and trawlers, passenger vessels, cargo ship, etc. and regulatory services like flood control, bioremediation, tidal action, etc.

The flow of water along the estuaries is hindered by thousands of nets that are laid along the stretch for trapping seeds of tiger prawn (*Penaeus monodon*). This profession is extremely detrimental to the ecosystem services (Roy Chowdhury and Mitra 2017). The Government of West Bengal has already banned the wild fry collection. Lack of alternative livelihoods for prawn seed collectors is one of the principle constraints on implementing any policies, laws and/or regulations geared towards the development of a sustainable resource use strategy as well as conservation effort for the wise use of the marine, coastal and estuarine resources (Ahmed et al. 2012). For this, the Government placed the enforcement of the fry collection ban into abeyance (in many developing countries) pending further review of how the resource and biodiversity could be conserved at the same time as protecting the livelihoods of 300,000 fry collectors (Williams 2002).

On this background, the present paper attempts to capture an in-depth view of the effects of prawn seed collection on the stakeholders and ecosystem services of Indian Sundarban estuaries.

Material and Methods

Research Site

Sundarban delta were formed by the deposited sediments from three major rivers, i.e. the Ganga, Brahmaputra and Meghna, and their dense network of smaller rivers, channels. The maximum elevation is about 10 metres above the mean sea level. The Hooghly River demarcates the western limit of Indian Sundarbans, and the eastern boundary is the Raimangal River.

We undertook this study during March to July, 2016, in and around the Patharpratima block of Indian Sundarbans region, which is situated in the southeastern part of Kakdwip subdivision. It covers 484.50 sq.km.

Data Capture

A questionnaire was developed to collect data on socio-economic condition, livelihood status and alternative ways of livelihood during the season of fry collection from the study site and its adjacent rivers. Primary data on fry collector's activities were collected both by physical observation and interviewing with fry collectors. Secondary data and information related to the study was collected from several publications on Indian Sundarbans (Mitra et al. 1992; Mitra 1998, 2013; Mitra and Zaman 2014, 2015, 2016; Roy Chowdhury and Mitra 2017).

Result

(a) **Impact on socio-economic status of prawn seed collectors**

1. **Population**

This study was conducted at Patharpratima block, and 210 peoples were interviewed. Total family members were 1423, which means their average family size was about 7 persons per family. Out of 1423 people, 721 were male and 702 were female (Fig. 4.4A.1). This means that ratio of male and female was 1.03:1.

2. **Age structure**

About 23% of the prawn seed collectors were children whose age ranged between 10 and 14 years. About 20.0% were in the age between 15 and 20 years. 53% were in the age group of 21–50 years and 4.0% were aging above 50 years (Fig. 4.4A.2).

3. **Sex ratio**

Our team made an in-depth survey in the study area and documented that 71% of the prawn seed collectors were male and 29% were female. However, in many

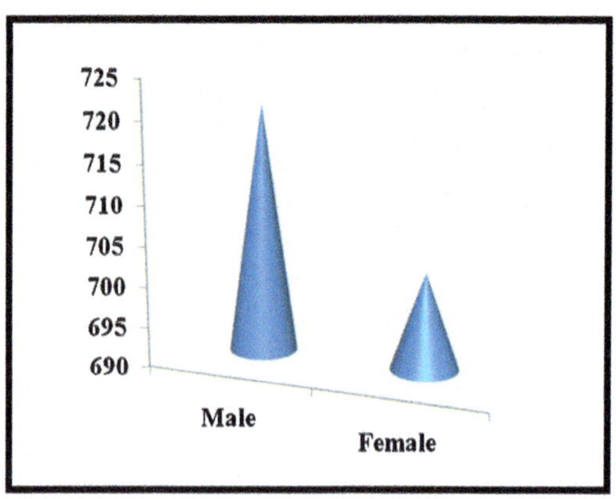

Fig. 4.4A.1 Male/female ratio in the family of prawn seed collectors at Patharpratima block

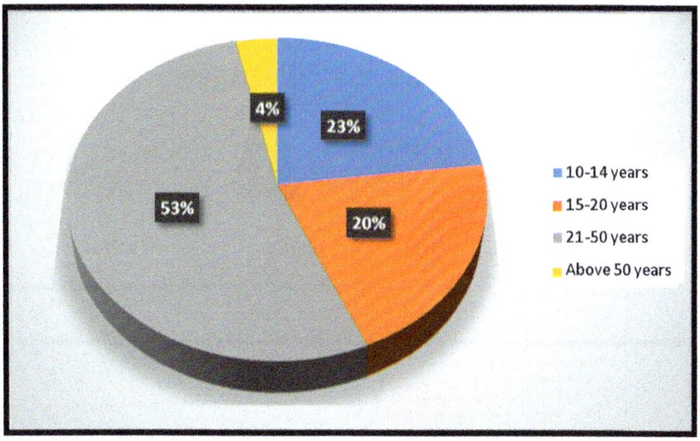

Fig. 4.4A.2 Age structure of prawn seed collectors at Patharpratima block

Fig. 4.4A.3 Male/female ratio of prawn seed collectors at Patharpratima block

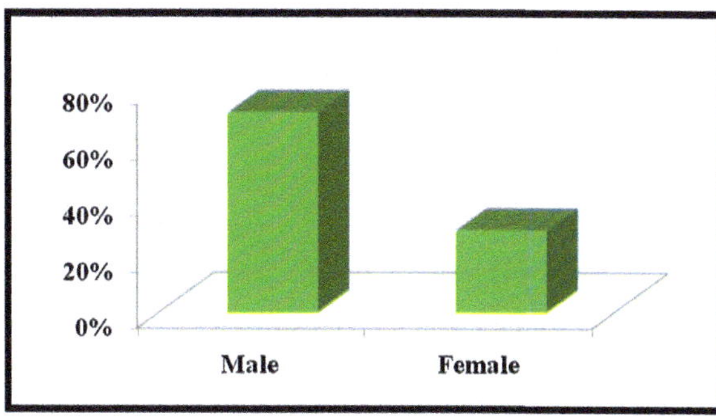

places like Kakdwip, Lot 8, Sagar South, the profession in dominated by females (Fig. 4.4A.3).

4. **Religion**

Both Muslim and Hindu are engaged in this profession without any conflicts. Rarely tribal communities (Bhumij, Garo, Kharria, Kol, Santhal, Oraon, Lodha–Savar, Munda, Kora, Mahilis, Malpaharias) participate in this profession.

Out of 1423 people, 1053 were Hindu, and 370 were Muslim which means 74% and 26% of the respondents in the study site were Hindu and Muslim, respectively (Fig. 4.4A.4). It is interesting to note that people of both these religions worship Bonobibi, which may be one of the reasons for religious convergence without any conflicts.

5. **Education**

The data captured by our team members revealed that 83% of the children (within the age group 10–14 years) dropped out their schools in order to get ready cash.

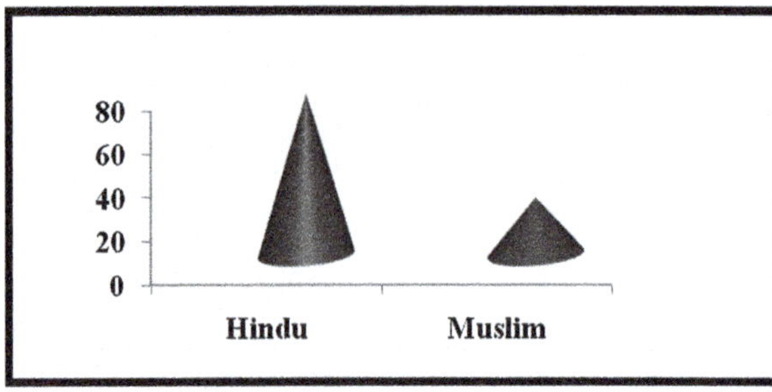

Fig. 4.4A.4 Hindu/Muslim population ratio of prawn seed collectors at Patharpratima block

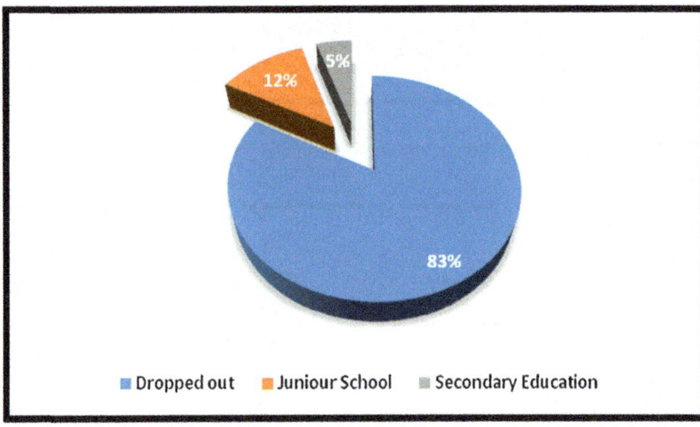

Fig. 4.4A.5 Educational profile of prawn seed collectors at Patharpratima block

Among the remaining 17%, 12.2% passed class VIII (Junior school), and the remaining 4.8% passed class X (Secondary education) Fig. 4.4A.5.

6. **Health and diseases**

The prawn seed collectors usually suffer from gastric and dermatological problems due to irregular feeding habit and continuous exposure to saline water. Many fry collectors also suffer from eye problem for screening the extremely thin tiger prawn seeds from the container containing variety of fish juveniles under the scorchy sunlight.

(b) **Impact on biodiversity spectrum**

1. **Fish juvenile loss**

Due to screening of estuarine water with nets of fine mesh size, many commercially important high valued fish juveniles are also caught in the nets, but they are simply thrown away and destroyed. We observed that an average of 313 juveniles of 37 species of finfish were destroyed only for 17 tiger prawn seeds.

2. **Uprooting of mangrove seedlings**

The dragging of nets by the fry catchers along the intertidal mudflats often uproots the mangrove seedlings and saltmarsh grass (*Porteresia coarctata*), which otherwise may generate several ecosystem services like control of erosion, beach stabilization, bioremediation, carbon sequestration, etc.

4.5.1 Discussion

Tiger prawn seed collection is a common profession, which is detrimental to the socio-economic and biodiversity profiles of Indian Sundarbans. Apart from destruction of finfish and shell fish juveniles (Roy Chowdhury and Mitra 2017), the practice of dragging the nets along the shore uproots the mangrove seedlings. These seedlings not only control soil erosion but also act as a major sink of carbon dioxide (Agarwal et al. 2016, Mitra et al. 2016a, b, 2017; Pal et al. 2016; Zaman et al. 2016). This practice is therefore a serious threat to ecosystem services of Sundarban estuaries (Mitra and Zaman 2014, 2015, 2016). Because of recession in traditional livelihood like agriculture (due to expansion of shrimp culture activities) and lack of alternative livelihood schemes at commercial scale, people of Sundarbans are accelerating their dependency on prawn seed collection. Considering the negative impact of this profession, it is of utmost importance to procure prawn seeds from hatcheries (available in plenty from the adjacent states like Odisha and Andhra Pradesh), develop seed bank and provide alternative livelihoods through ICZM (Integrated Coastal Zone Management) programme as a separate section for this mangrove dominated World Heritage Site.

4.5.2 Conclusion

Because of salinity problem, development of tiger prawn seed hatchery could not see the light of success in and around Indian Sundarbans due to which the profession of wild collection of prawn seed has sprouted. Considering the adverse impact of this occupation on economic profile, health of the fry collectors and fish biodiversity, it is important to provide an alternative livelihood for the fry catchers. In this context Techno India University, West Bengal, has planned to develop seed bank in and around Jharkhali region to reduce the pressure on this fragile ecosystem as a whole.

References

Agarwal S, Zaman S, Biswas S, Pal N, Pramanick P, Mitra A (2016) Spatial variation of mangrove seedling carbon with respect to salinity: a case study with *Bruguiera gymnorrhiza* seedling. Int J Adv Res Biol Sci 3(8):7–12

Ahmed F, Hossain MY, Fulanda B, Ahmed ZF, Ohtomi J (2012) Indiscriminate exploitation of wild prawn post larvae in the coastal region of Bangladesh, a threat to the fisheries resources, community livelihoods and biodiversity. Ocean Coast Manag 66:56–62

Churchman CW (1968) The systems approach. Dell Publishing Co., Onc, New York

Evans FC (1956) Ecosystems as the basic unit on ecology. Science 12:1127–1128

Mitra A (1998) Status of coastal pollution in West Bengal with special reference to heavy metals. J Ind Ocean Stud 5(2):135–138

Mitra A (2013) Sensitivity of mangrove ecosystem to changing climate. Springer, New Delhi, p 323. https://doi.org/10.1007/978-81-322-1509-7

Mitra A, Zaman S (2014) Carbon Sequestration by Coastal Floral Community; published by The Energy and Resources Institute (TERI) TERI Press, India, ISBN:978–81–7993-551-4

Mitra A, Zaman S (2015) Blue carbon reservoir of the blue planet, published by Springer, https://doi.org/10.1007/978-81-322-2107-4. ISBN:978–81–322-2106-7

Mitra A, Zaman S (2016) Basics of marine and Estuarine ecology, Springer, New Delhi. ISBN:978-81-322-2705-2

Mitra A, Choudhury A, Zamaddar YA (1992) Effects of heavy metals on benthic molluscan communities in Hooghly estuary. Proc Zool Soc 45:481–496

Mitra A, Saha A, Pal N, Fazli P, Zaman S (2016a) Allometry in *S. apetala* seedlings of Indian Sundarbans. Int J Rec Adv Mult Res 3:1903–1909

Mitra A, Pal N, Chakraborty A, Mitra A, Saha A, Trivedi S, Zaman S (2016b) Estimation of stored carbon in *Sonneratia apetala* seedlings collected from Indian Sundarbans. Ind J Geo Mar Sci 45(11):1598–1602

Mitra A, Sundaresan J, Banerjee K, Agarwal SK (2017) In: Environmental coastguards. Published by CSIR-National Institute of Science Communication and Information Resources (CSIR-NISCAIR), ISBN:978–81–7236-352-9

Odum EP (1971) Fundamentals of ecology., Published by Philadelphia, Saunders, 3rd edn

Pal N, Gahul A, Zaman S, Biswas P, Mitra A (2016) Spatial variation of stored carbon in *Avicennia alba* seedlings of Indian Sundarbans. Int J Trend Res Dev 3(4):100–103

Roy Chowdhury G, Mitra A (2017) Traditional halophytic medicine: a new era in the health care canvas. Org Med Chem Int J 2(2):555584. https://doi.org/10.19080/OMCIJ.2017.02.555584

Tansley AG (1935) The use and abuse of Vegetational concepts and terms. Ecol 16(3):284–307

Whittaker RH (1960) Vegetation of the Great Smoky Mountains. Ecol Monogr 26:1–80

Williams D (2002) Management options for the shrimp fry fishery, Regional stakeholder workshop in Khulna, Bangladesh, 8

Zaman S, Biswas S, Pal N, Datta U, Biswas P, Mitra A (2016) Ecosystem service of *Heritiera fomes* seedlings in terms of carbon storage. Int J Trend Res Dev 3(4):183–185

Chapter 5
Major Biogeochemical Cycles

Abstract Life on Earth would have been impossible without the cycling of inorganic elements through several physico-chemical and biological compartments. The chapter deals with the cycling of major elements in general, but the alteration of biogeochemical cycles through human intervention is a key issue of this chapter. The ground zero observation on the change of land-use pattern in the lower Gangetic delta has been addressed that has significant impact on the near-surface air temperature leading to alteration of biotic community and physico-chemical properties of the aquatic phase.

Keywords Hydrological cycle · Elemental cycle · Green House Gases (GHGs) · Indian Sundarbans

5.1 Hydrological Cycle

Life on Earth would be totally impossible without the support of inorganic substances. All these inorganic substances tend to move in cycles from the physical environment, to the biotic compartment and then back to the environment. These circular movements of the essential inorganic substances are referred to as biogeochemical cycles. Living organisms require six major elements in relatively large quantities: carbon, hydrogen, oxygen, nitrogen, phosphorous and sulphur. Elements needed in smaller amounts include sodium, potassium, manganese, calcium, iron, magnesium, chlorine, iodine, cobalt and boron. These elements are present in rocks, usually as salts. They are released, by erosion and weathering, into soil, rivers, lakes and the oceans. The movements of nutrient elements through the biosphere, or through any particular ecosystem, by the physical processes of erosion, deposition, evaporation and precipitation and by the biological processes of uptake and release are the important events of the biogeochemical cycles. They are called cycles because nutrient elements, unlike solar energy, may be used over and over again by living systems.

A. Mitra, S. Zaman, *Environmental Science - A Ground Zero Observation on the Indian Subcontinent*, https://doi.org/10.1007/978-3-030-49131-4_5

Sometimes the nutrients are recycled rapidly through living systems, as in grasslands, where the above-ground vegetation dies away every year and its nutrients are made available again. In some cases, nutrients are locked in natural reservoirs for many years apart from the activities of the biological world, e.g. remains of marine organisms may sink to the bottom and get incorporated into sedimentary rocks that are uplifted and exposed to erosion and nutrient released only after millions of years. Every nutrient element has a somewhat different fate, depending on its physical and chemical properties and on its necessity in living organisms. In the present chapter, the concept of cycling of inorganic elements is discussed below with some few simplified examples. The hydrological cycle is also discussed here as it regulates the pattern of biota on the Earth.

Solar energy and gravity continuous move water among several compartments like the ocean, air, land and biota through evaporation, condensation, precipitation and runoff back to the sea to begin the cycle again.

When incoming solar energy warms on or near the surfaces of oceans, rivers, lakes, soil and plant leaves, the water evaporates and enters the atmosphere as water vapour, leaving behind dissolved impurities. Water and various dissolved compounds in the soil are drawn up through the roots of plants into leaves and other surfaces. When warmed by the sun, water in the exposed parts of the plants passes through lenticels and stomata by the process of transpiration and then evaporates into the atmosphere. This transfer of water from the plant tissue and soil to water vapour in the atmosphere through a combination of transpiration and evaporation is called evapotranspiration.

Fresh water removed from the oceans and other bodies of water is returned from the atmosphere to the land and other water bodies as precipitation in the forms of rain, sleet, hail and snow. Some of this fresh water becomes locked in glaciers, and some sinks downwards or percolates through the soil, where it may remain for hundreds to thousands of years as groundwater in slow flowing, slowly renewed and underground reservoirs. When the rate of rainfall is faster than the rate of infiltration downwards into the soil, it collects in puddles and ditches and eventually flows into the nearby streams, rivers, lakes, from where it is carried back to the oceans thus completing the cycle (Fig. 5.1).

5.2 Major Elemental Cycles

5.2.1 Carbon Cycle

Chains of carbon atoms make up most of the organic molecules that are the backbones of living organisms. The source of carbon for the synthesis of these molecules is carbon dioxide from the atmosphere, ocean water and fresh water. Carbon moves through rapid short-term cycles (food pyramids and respiration) and through slower geological cycles.

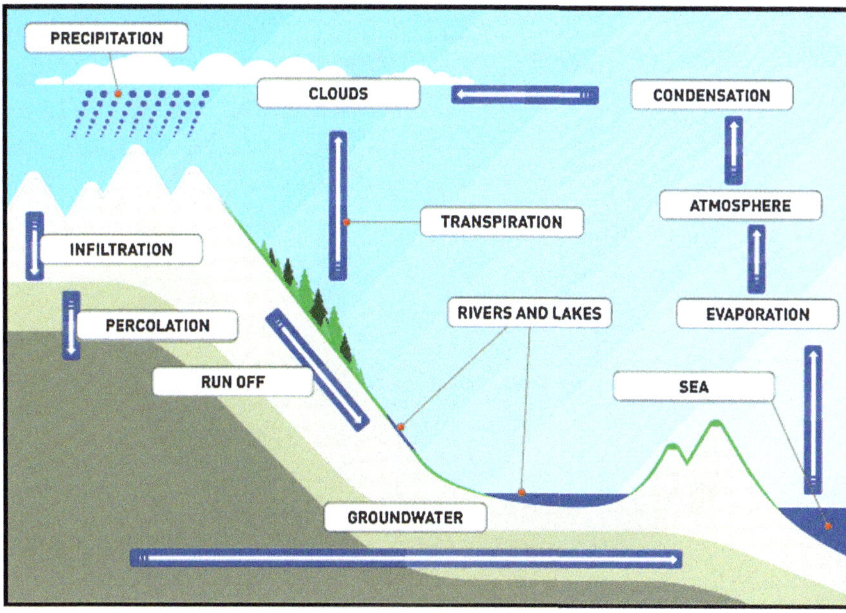

Fig. 5.1 Diagrammatic representation of hydrological cycle

During photosynthesis, chlorophyll bearing plants synthesize glucose from carbon dioxide and water utilizing sunlight as the source of energy. These carbohydrates may be converted to other forms of organic molecules by plant cells. The animal world is dependent on the green tier (producers) to obtain the organic nutrients needed for their growth and reproduction.

Carbon dioxide is constantly recycled in ecosystems by the respiratory processes of living organisms. Plant and animal respiration degrades such compounds as simple sugars into water and carbon dioxide. This carbon dioxide is returned to the atmosphere and to the water. When plants and animals die, decomposers like bacteria and fungi use the dead bodies for their own nutritional needs. They also respire and release carbon dioxide. The entire carbon cycle is represented in Fig. 5.2.

In aquatic ecosystems, the corresponding exchange is between living organisms and carbon dioxide or bicarbonates (HCO_3^-) dissolved in the surrounding water. In some ecosystems, however, carbon may accumulate either as undecomposed organic matter, as in peaty layers of bogs and moorlands, or as insoluble carbonates, which accumulate in bottom sediments in aquatic systems. Such carbon has entered a cycle of longer duration and may take a long period to be released. Deposits of coal, oil and natural gas represent part of the net production of a bygone *era*. These fossil fuels are organic compounds that were buried before they could be decomposed and were subsequently transformed by time and geological processes. As they are burned, their carbon content is released back into the atmosphere as carbon dioxide (CO_2). Sedimentary carbon eventually turns into sedimentary rocks such as limestone and dolomite. After millions of years, these rocks may be lifted above sea

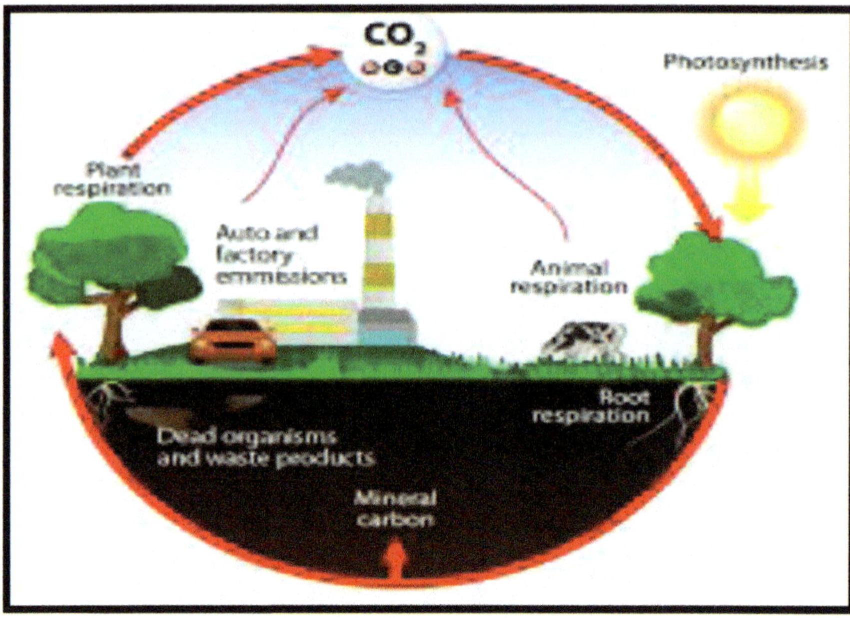

Fig. 5.2 Diagrammatic representation of carbon cycle

level and exposed to erosion, releasing their carbon as carbonates ($CO_3^{=}$) and bicarbonates into streams and rivers.

Many other limestone and chalk deposits represent extinct coral reefs. During the life of a coral reef, millions of organisms secrete calcium carbonate shells and skeletons that remain after the death of animals and plants that produced them.

The carbon balance of the biosphere as a whole is moderated by exchange of carbon dioxide between the atmosphere and the oceans. The carbon dioxide content of the atmosphere is about 0.03%, but is steadily increasing due to excessive burning of fossil fuels and clearing down of vegetations, both of which are very keenly related to industrialization and urbanization. A study conducted by researchers of Techno India University (Kolkata, India) shows that the carbon dioxide level in the city of Kolkata and Bhubaneshwar has crossed the permissible level (Figs. 5.3 and 5.4) in certain sampling stations.

The rate of carbon dioxide increase in the atmosphere, however, is less than half of what would be expected from these human activities. The remainder is almost certainly absorbed by the oceans, which thus act as a global buffer for carbon dioxide (CO_2), moderating its concentration in the atmosphere and accounting for the basic stability of the global carbon budget.

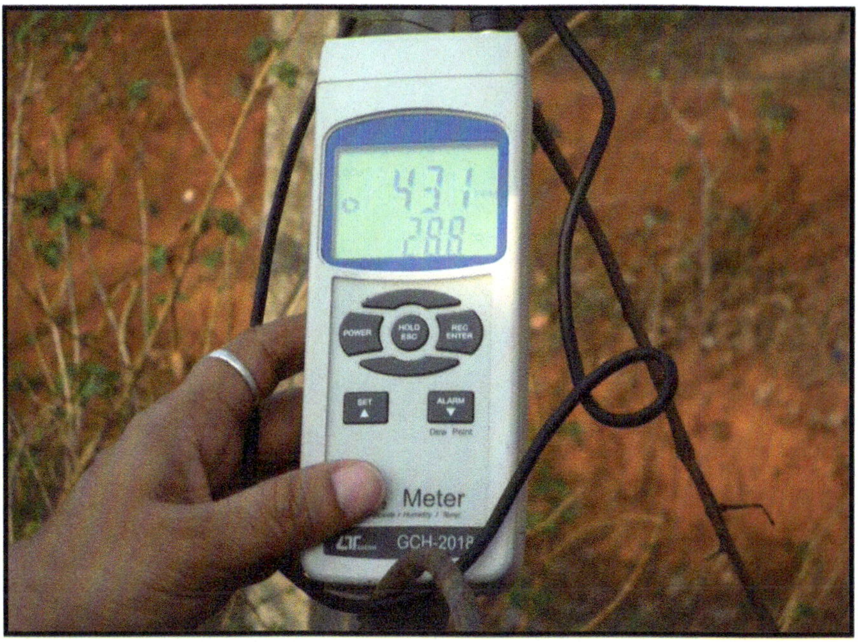

Fig. 5.3 CO_2 level in the city of Kolkata

5.2.2 *Oxygen Cycle*

Most living organisms require oxygen for performing the process of respiration. The source of this oxygen is the molecular oxygen from air or water. During cellular respiration, oxygen combines with hydrogen to produce water. The primary means of renewing and maintaining oxygen levels is through photosynthesis. During this process green plants give out oxygen from the water molecules. Some of this oxygen is used by the plant cells themselves, but much of it released into the atmosphere and water. Yet another role is played by oxygen. Ozone (O_3), another molecular form of oxygen, forms a screen in the upper atmosphere that absorbs large amount of the ultraviolet radiation produced by the sun. If it were not for this protective shield, life would be restricted to areas protected from destructive radiation. The diagrammatic representation of oxygen cycle is represented in Fig. 5.5.

Recently it has been reported that fluorocarbons, the coolant in refrigerators and air conditioners and the propellant gas in many aerosol spray cans, react with ozone and may reduce the critical ozone layer. This is an alarming situation in relation to life on Earth.

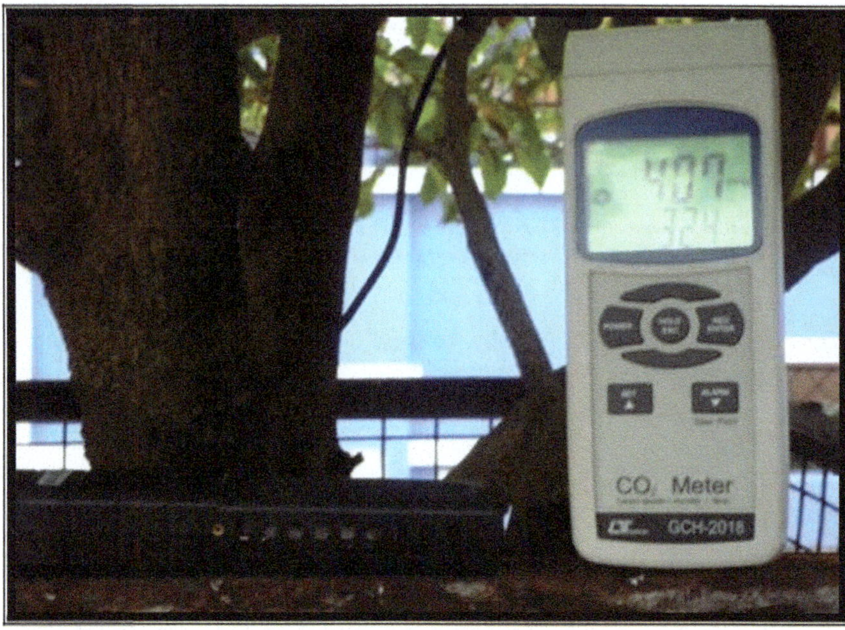

Fig. 5.4 CO_2 analyser-monitoring atmospheric carbon dioxide in the heart of the Bhubaneswar city

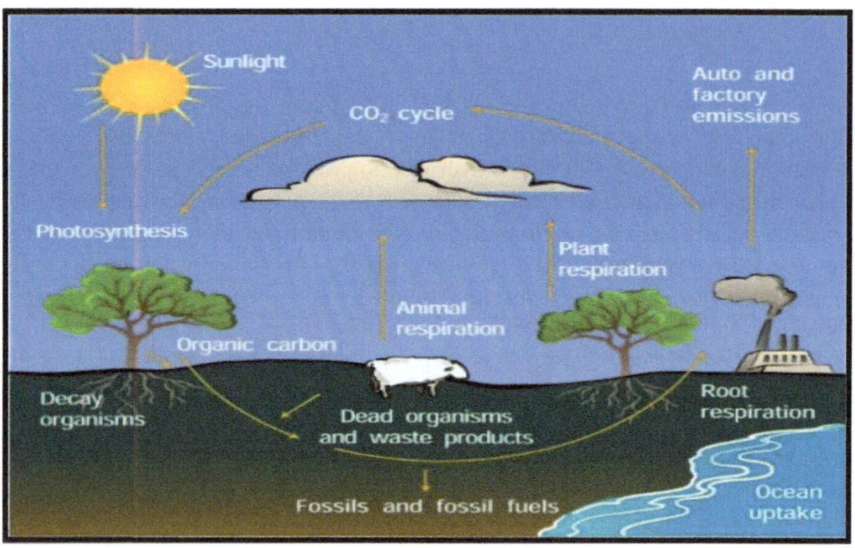

Fig. 5.5 Diagrammatic representation of oxygen cycle

5.2.3 Nitrogen Cycle

The cycling of nitrogen through an ecosystem differs from that of carbon in that most of the enormous pool of molecular nitrogen gas is available only to those bacteria and cyanobacteria that can fix gaseous nitrogen by reducing it to ammonia. The entire nitrogen cycle is depicted in Fig. 5.6. Plants acquire this vital nutrient either directly from nitrogen-fixing microorganisms that are restricted to the root nodules of leguminous plants or indirectly from the limited pool of nitrogen found as ammonium or nitrate ions in soil water, rivers, lakes and oceans. Some common nitrogen-fixing microbes are *Azotobacter* sp., *Clostridium* sp., *Oscillatoria* sp., etc. Other inputs to this pool include nitrate formed from atmospheric nitrogen by lightning, ammonium or nitrate from the decay of organic matter, and nitrate from the erosion of nitrate-rich rocks.

Nitrifying bacteria, also in the soil, oxidize some of the ammonia to nitrite. Finally, there are other bacteria that oxidize nitrite to nitrate, another form of nitrogen that can be absorbed by plants.

Nitrogen fixation requires a great deal of energy, which nitrogen-fixing microorganisms must obtain from organic matter. Some decompose organic matter in the soil, some live as symbionts in the roots of higher plants and some are photosynthetic. Thus nitrogen fixation occurs at a slow rate relative to the rate of uptake of fixed nitrogen by plants. Consequently, despite the fact that nitrogen makes up 78%

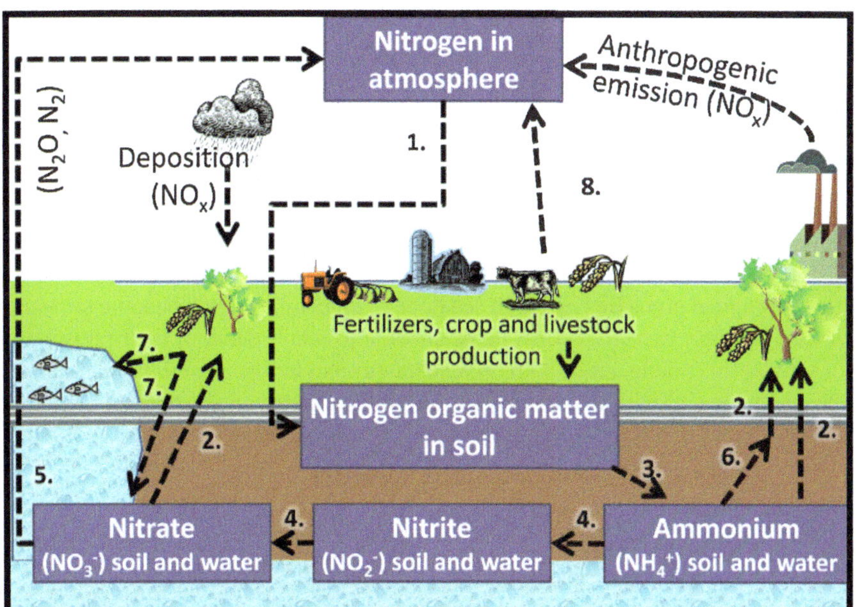

Fig. 5.6 Diagrammatic representation of nitrogen cycle

of the atmosphere, it is often the limiting nutrient in plant growth. Due to this reason nitrogen is most commonly applied in crop fertilizers.

Although many ecosystems, such as tropical rain forests, are remarkably efficient at retaining their nitrogen, almost all terrestrial ecosystems continually lose some nitrogen as nitrate and organic matter dissolved in groundwater and streams. In the final phase of the cycle, nitrogen is returned back to the atmospheric 'reservoir' by denitrifying bacteria living in the anaerobic mud of fertile lakes, bogs, estuaries and selective zones of the ocean basin.

5.2.4 Phosphorus Cycle

Phosphorus is a major constituent of ATP, DNA, RNA and biological membranes and many animals need large quantities of this element for shells, bones, and teeth formations. Since phosphorus never occurs as a gas, its cycle, unlike the atmospheric cycles of nitrogen and carbon, is basically a sedimentary cycle. When rocks are eroded by weather, minute amounts of phosphorus dissolve, usually as phosphate and become available to plants. Unlike nitrogen, which must be released from organic compounds before plants can take it up again, much of the phosphorus excreted by animals is in the form of phosphate, which is immediately reusable by plants. Thus cycling of phosphorus in terrestrial ecosystems is usually very efficient, although small amounts are continually lost down to aquatic phase like oceans. The ocean compartment also receives phosphorus through 'guano', which are phosphate rich droppings of marine birds (Fig. 5.7). Since some phosphate salts are not very soluble in water, the oceans continually deposit phosphate minerals in their sediments. The main reason for the infertility of the open oceans is their low phosphorus content.

At the level of the biosphere, the phosphorus cycle is, in the short run, a one-way flow from rocks to land ecosystem to the oceans from where it is finally transferred to the sediment compartment. The entire phosphorous cycle is represented in Fig. 5.8. Apart from trivial amounts deposited on land in the droppings of oceanic birds (Guano), the only natural way for phosphorus to return to land is via immensely slow geological processes in which sea floor sediment may again become terrestrial rocks. However, terrestrial ecosystems are able to retain much of their phosphorus for short-term cycling, since both organic humus and inorganic soil particles absorb phosphate ions, providing a local reservoir of this element in the soil and helping to promote a steady supply of it into and out of the living organisms of the ecosystem.

5.2.5 Sulphur Cycle

Sulphur is one of the components that make up proteins and vitamins. Proteins consist of amino acids that contain sulphur atoms. Sulphur is important for the functioning of proteins and enzymes in plants, and in animals that depend upon plants

Fig. 5.7 Phosphate-rich guano from marine birds

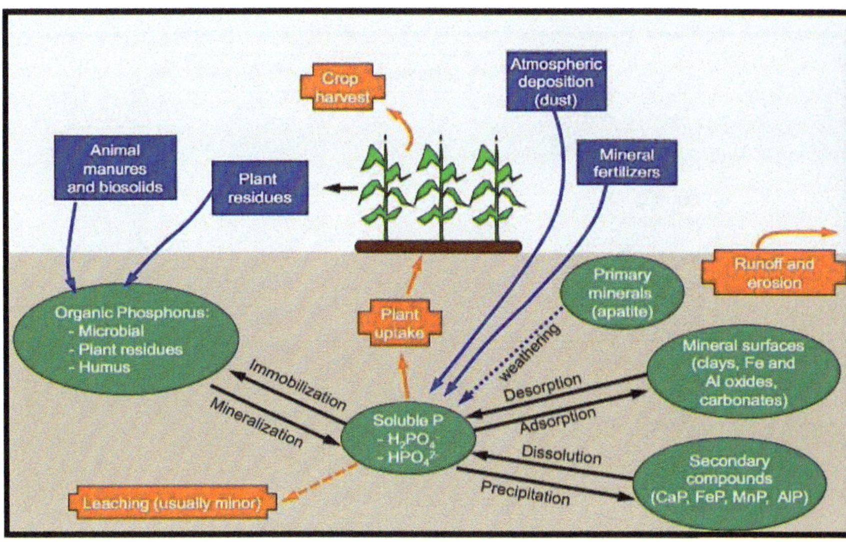

Fig. 5.8 Diagrammatic representation of phosphorous cycle

(producer community) for acquiring sulphur. Plants absorb sulphur when it is dissolved in water. Animals consume these plants, so that they take up enough sulphur to maintain their health.

Most of the Earth's sulphur is tied up in rocks and salts or buried deep in the ocean in oceanic sediments. Sulphur can also be found in the atmosphere. It enters the atmosphere through both natural and human sources. Natural resources can be for instance volcanic eruptions, bacterial processes, evaporation from water or decaying organisms. When sulphur enters the atmosphere through human activity, this is mainly a consequence of industrial processes where sulphur dioxide (SO_2) and hydrogen sulphide (H_2S) gases are emitted on a wide scale.

When sulphur dioxide enters the atmosphere it will react with oxygen to produce sulphur trioxide gas (SO_3), or with other chemicals in the atmosphere, to produce sulphur salts. Sulphur dioxide may also react with water to produce sulphuric acid (H_2SO_4). Sulphuric acid may also be produced from dimethylsulphide, which is emitted to the atmosphere by phytoplankton that are tiny free-floating floral particles found in the aquatic ecosystem.

All these particles will settle back onto earth, or react with rain and fall back onto earth as acid deposition. The particles will then be absorbed by plants again and are released back into the atmosphere, so that the sulphur cycle will start over again. The cycling of sulphur is represented in Fig. 5.9.

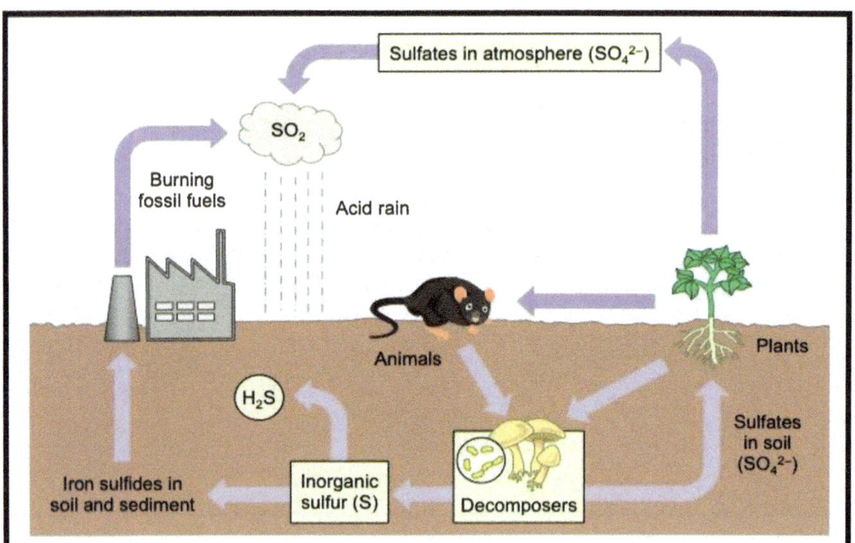

Fig. 5.9 Diagrammatic representation of sulphur cycle

5.3 Alteration of Biogeochemical Cycles by Human Beings

Biogeochemical cycles involve the fluxes of chemical elements among different parts of the Earth: from living to nonliving, from atmosphere to land to sea and from soils to plants. They are called 'cycles' because matter is always conserved and because elements move to and from major pools via a variety of two-way fluxes, although some elements are stored in locations or in forms that are differentially accessible to living things. Human activities have mobilized Earth elements and accelerated their cycles, for example, more than doubling the amount of reactive nitrogen that has been added to the biosphere since pre-industrial times. Reactive nitrogen is any nitrogen compound that is biologically, chemically or radioactively active, like nitrous oxide and ammonia, but not nitrogen gas (N_2). Global-scale alterations of biogeochemical cycles are occurring, from human activities with impacts and implications now and into the future. Global carbon dioxide emissions are the most significant driver of human-caused climate change. The recent level of atmospheric carbon dioxide has exceeded 400 ppm, which is primarily due to intense industrialization, rapid urbanization, deforestation and change of land-use pattern. A study done by the present authors in the Sagar Island of Indian Sundarbans during 2018 exhibited a value of 401.22 ppm of atmospheric carbon dioxide, in postmonsoon season (Fig. 5.10).

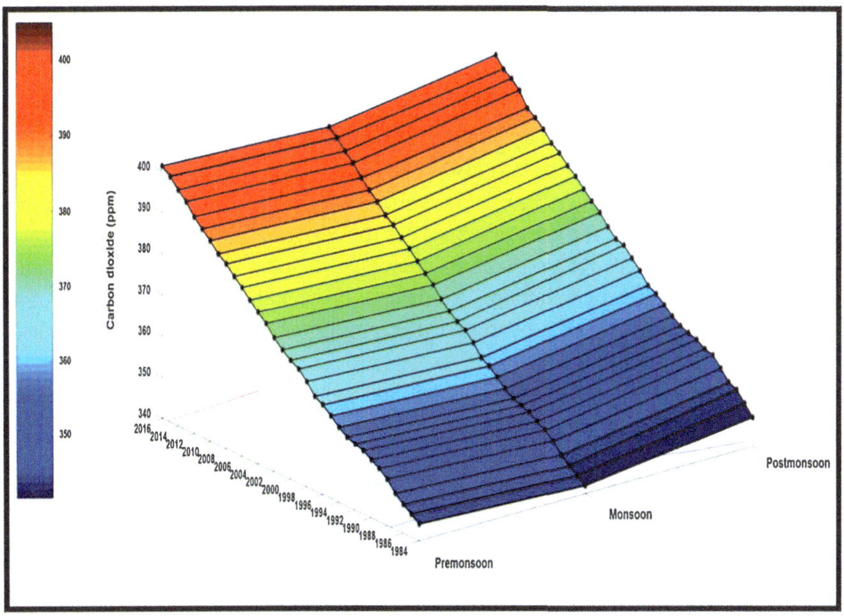

Fig. 5.10 Temporal variation of carbon dioxide in Sagar Island

Fig. 5.11 Aquaculture in Sagar Island at the cost of mangrove forest: An example of carbon sinks destruction

Expansion of shrimp farms and tourism units coupled with modernization of ports and harbours in this region have reduced the area occupied by blue carbon, which otherwise could be a major sink of carbon dioxide (Fig. 5.11).

The destruction of blue carbon also triggers the rising of air temperature at the local level as evidenced by the authors while carrying out a study on Sagar Island. Interested readers may consult Annexure 5A for detailed report.

Human-accelerated cycles of other elements, especially nitrogen, phosphorus and sulphur, also influence climate. These elements can affect climate directly or act as indirect factors that alter the carbon cycle, amplifying or reducing the impacts of climate change.

Climate change is having, and will continue to have, impacts on biogeochemical cycles, which will alter future impacts on climate and affect our capacity to cope with coupled changes in climate, biogeochemistry and other factors.

The human mobilization of carbon, nitrogen and phosphorus from the Earth's crust and atmosphere into the environment has increased 36, 9 and 13 times, respectively, compared to geological sources over pre-industrial times. Fossil fuel burning, land-cover change, cement production and the extraction and production of fertilizer to support agriculture are major causes of these increases. Carbon dioxide is the most abundant of the heat-trapping greenhouse gases that are increasing due to human activities and its production dominates atmospheric forcing of global climate change. However, methane and nitrous oxide have higher greenhouse-warming potential per molecule than carbon dioxide, and both are also increasing in the atmosphere. In the United States and Europe, sulphur emissions have declined over the past three decades, especially since the mid-1990s, because of efforts to reduce air pollution. During some festivals the burning of crackers causes a hike in sulphur dioxide level. This is because sulphur along with various other toxic compounds and heavy metals are used to make crackers which on burning emits the oxides of these elements. Laser show seems to be a better alternative to burning of crackers during the festive months to maintain the stability of biogeochemical cycles at local level.

5.4 Take Home Messages

(A) Life on Earth would be totally impossible without the support of inorganic substances. All these inorganic substances tend to move in cycles from the physical environment to the biotic compartment and then back to the environment. These circular movements of the essential inorganic substances are referred to as biogeochemical cycles.

(B) The carbon balance of the biosphere as a whole is moderated by exchange of carbon dioxide between the atmosphere and the oceans. The carbon dioxide content of the atmosphere is about 0.03%, but is steadily increasing due to excessive burning of fossil fuels and clearing down of vegetations, both of which are very keenly related to industrialization and urbanization.

(C) Fluorocarbons, the coolant in refrigerators and air conditioners and the propellant gas in many aerosol spray cans, react with ozone and may reduce the critical ozone layer. This is an alarming situation in relation to life on Earth as ozone forms a screen in the upper atmosphere that absorbs large amount of the ultraviolet radiation produced by the Sun.

(D) The human mobilization of carbon, nitrogen and phosphorus from the Earth's crust and atmosphere into the environment has increased 36, 9 and 13 times, respectively, compared to geological sources over pre-industrial times. Fossil fuel burning, land-cover change, cement production and the extraction and production of fertilizer to support agriculture are major causes of these increases. Carbon dioxide is the most abundant of the heat-trapping greenhouse gases that are increasing due to human activities and its production dominates atmospheric forcing of global climate change. However, methane and nitrous oxide have higher greenhouse-warming potential per molecule than carbon dioxide, and both are also increasing in the atmosphere.

5.5 Brain Churners

This chapter highlights the biogeochemical cycles operating in the planet Earth and the processes by which they are often altered by human activities. It is expected that the readers will find interest in answering the questions asked on the cycling of most common elements of the Earth. The readers can consider this section as a sort of '*self-appraisal approach*' as correct answers to all the questions are marked as black in the box at the end of this section.

1. The hydrological cycle encompasses
   (a) Evaporation, condensation and sublimation
   (b) Evaporation, condensation, precipitation and runoff to the aquatic ecosystem
   (c) Evaporation and precipitation
   (d) Evaporation and condensation
2. The transfer of water from the plant tissue and soil to water vapour in the atmosphere is called
   (a) Condensation
   (b) Transpiration
   (c) Evaporation
   (d) Evapotranspiration
3. Precipitation includes
   (a) Rain, hail and snow
   (b) Only snowfall
   (c) Runoff from adjacent landmasses
   (d) None of the above
4. The bottom sediment in the aquatic ecosystem contains
   (a) Volatile substances
   (b) Insoluble carbonates
   (c) Coal
   (d) All the above
5. Ozone absorbs
   (a) Carbon dioxide
   (b) Nitrogen
   (c) GHG's
   (d) UV radiation

6. The coolants in the refrigerators react with ozone due to which
   (a) The critical ozone layer increases
   (b) The ozone layer remains constant
   (c) The ozone layer reduces
   (d) None of the above
7. Limestone and dolomite are
   (a) Igneous rock
   (b) Sedimentary rock
   (c) Metamorphic rock
   (d) None of the above
8. The nutrients recycle very rapidly in
   (a) Desert
   (b) Ocean
   (c) Lakes
   (d) Grasslands
9. The constituents of fossil fuel are
   (a) Mostly organic in nature
   (b) Inorganic in nature
   (c) Elements
   (d) None of the above
10. The excessive burning of fossil fuel
    (a) Increases GHGs in the atmosphere
    (b) Decreases GHGs in the atmosphere
    (c) Makes the GHG level constant
    (d) None of the above
11. Six major elements required by living organisms are
    (a) C, H, O, N, P and Hg
    (b) C, H, O, N, Co and P
    (c) C, H, Na, O, Co and N
    (d) C, H, O, N, P and S
12. Boron is required by living organisms in
    (a) Small amount
    (b) Large amount
    (c) Medium amount
    (d) None of the above

13. The coolant in the refrigerator is
    (a) Hydrogen
    (b) Nitrogen
    (c) Fluorocarbon
    (d) Chlorine
14. Chemosynthesis occurs in
    (a) Trees
    (b) Microbes
    (c) Herbs
    (d) Phytoplankton
15. Limestone and chalk deposits represent
    (a) Live coral reefs
    (b) Volcano
    (c) Extinct coral reefs
    (d) None of the above
16. The pH of seawater is around
    (a) 8.2
    (b) 7.6
    (c) 7.0
    (d) 6.5
17. Cobalt is required by living organisms in
    (a) Large amount
    (b) Medium amount
    (c) Small amount
    (d) None of the above
18. Guano is rich in
    (a) Sulphur
    (b) Zinc
    (c) Magnesium
    (d) Phosphorous
19. The major constituent of DNA, RNA and ATP is
    (a) Phosphorous
    (b) Sulphur
    (c) Sodium
    (d) Chromium
20. The fertility of the soil is increased by bacteria like
    (a) *E. coli*

(b) *Lactobacillus*
(c) *Azotobacter*
(d) None of the above

Answer Sheet

S. No.	A	b	c	D
1.		■		
2.				■
3.	■			
4.		■		
5.				■
6.			■	
7.		■		
8.				■
9.	■			
10.	■			
11.				■
12.	■			
13.			■	
14.		■		
15.			■	
16.	■			
17.			■	
18.				■
19.	■			
20.			■	

Annexure 5A: Impact of Anthropogenic Activities on Near-Surface Air Temperature of Sagar Island in Indian Sundarbans

Introduction

Air temperature is basically a measurement to detect the hotness or coldness of the air. It describes the kinetic energy of the gas molecules that constitute the air. Air temperature regulates several weather parameters like the rate of evaporation, relative humidity, wind speed/direction and precipitation. It is a function of the amount and strength of the sunlight heating the earth and atmospheric conditions such as cloud cover and humidity that trap the heat energy. It has also been documented that greenhouse gases trap heat near the Earth surface. People are adding several types of greenhouse gases to the atmosphere, and each gas's effect on climate change depends on three main factors, namely, the amount/quantum of gas added, the duration of gas addition and the potential of the added gas in terms of warming (https://archive.epa.gov/climatechange/kids/basics/today/greenhouse-gases). Carbon dioxide is one of the major greenhouse gases that are emitted by vehicles, industries, power plants, etc., which traps heat in the air (Mitra 2013).

The planet Earth is getting warmer due to climate change caused by anthropogenic activities or natural variability. However, the rate has steadily increased since the beginning of the industrial revolution. A study conducted by the researchers of NASA's Goddard Institute of Space Studies (GISS) reveals that the average global temperature on Earth has increased by about 0.8 °C. Two-thirds of the warming has occurred since 1975 at a rate roughly between 0.15% and 0.20% per decade. Generally warming is greater over land than over the ocean because water is slower to absorb and release heat (Thermal inertia) (https://earthobservatory.nasa.gov).

The Indian Sundarbans at the apex of Bay of Bengal is a deltaic complex dominated by mangrove vegetation. Temporal variation of air temperature in Sundarbans has been recorded from 11.96 to 37.0 °C (Pitchaikani et al. 2017). This deltaic complex experiences maximum air temperature value during premonsoon followed by monsoon and postmonsoon (Mitra 2013; Mitra and Zaman 2014, 2015, 2016). The present paper is a first-order analysis of the temporal variation of near-surface air temperature in Sagar Island over a period of more than three decades (1984–2016).

Materials and Methods

Data Sources and Quality

In this paper, a data set of more than three decades on near-surface air temperature in Indian Sundarbans region has been considered for a first-order analysis. This time frame is in accordance with the minimum standard norms of climate-related researches. The World Metrological Organization (WMO) and the Intergovernmental Panel on Climate Change (IPCC) define 'climate' as the average state of the weather over time with the period generally being 30 years, (although for some marine climate parameters such as storminess, longer averages are required (Zhang et al. 2000). More than three decades of data (1984–2016) were compiled from the archives of the Department of Marine Science, University of Calcutta and Techno India University, West Bengal, for this study. A number of studies on different aspects of the Sundarban complex have been published over the years, which include description of the data (and methods) at different times over the past three decades (Mitra et al. 1987, 1992, 2009; Sengupta et al. 2013; Chakraborty and Choudhury 1985; Mitra and Choudhury 1994; Saha et al. 1999; Banerjee et al. 2002a, b, 2013; Mondal et al. 2006; Pitchaikani et al. 2017).

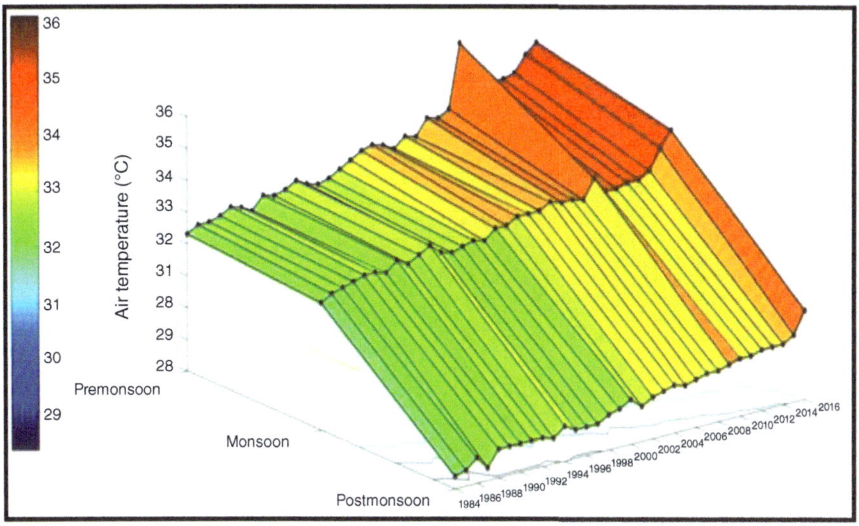

Fig. 5.A.1 Temporal variation of near-surface air temperature in Sagar Island

Results

The present pattern of near-surface air temperature clearly reflects an increasing trend through seasons in the study site (Fig. 5.A.1).

In premonsoon, the increment is 0.09 °C/year, and, while in both monsoon and postmonsoon, seasons the increments are 0.07 °C/year.

The near-surface air temperature exhibited an increasing trend with the passage of time. This may be attributed to change of land-use pattern, deforestation (preferably mangrove deforestation for the sake of shrimp culture), unplanned urbanization, intense industrialization and expansion of tourism units. In addition, the rise of temperature is also attributed to gradually increase of atmospheric carbon dioxide in this deltaic complex (Agarwal et al. 2018).

Discussion

Sagar Island is the largest island of Indian Sundarbans with an area of 224.3 sq.km. (marked with 'S' in Fig. 5.A.2).The island was dominated by mangroves, and about 25 species of mangroves were documented in the southernmost tip of Sagar Island near the lighthouse area (Mitra 2013). However, due to expansion of shrimp farms in an unplanned manner, particularly in the Chemaguri region (88°09′46.64″E; 21°38′54.86″N), there has been a massive deforestation of mangroves. The intense rates of urbanization, mushrooming of tourism activities and construction of a proposed harbour have reduced the mangrove patches in Sagar Island to a great extent (Fig. 5.A.2).

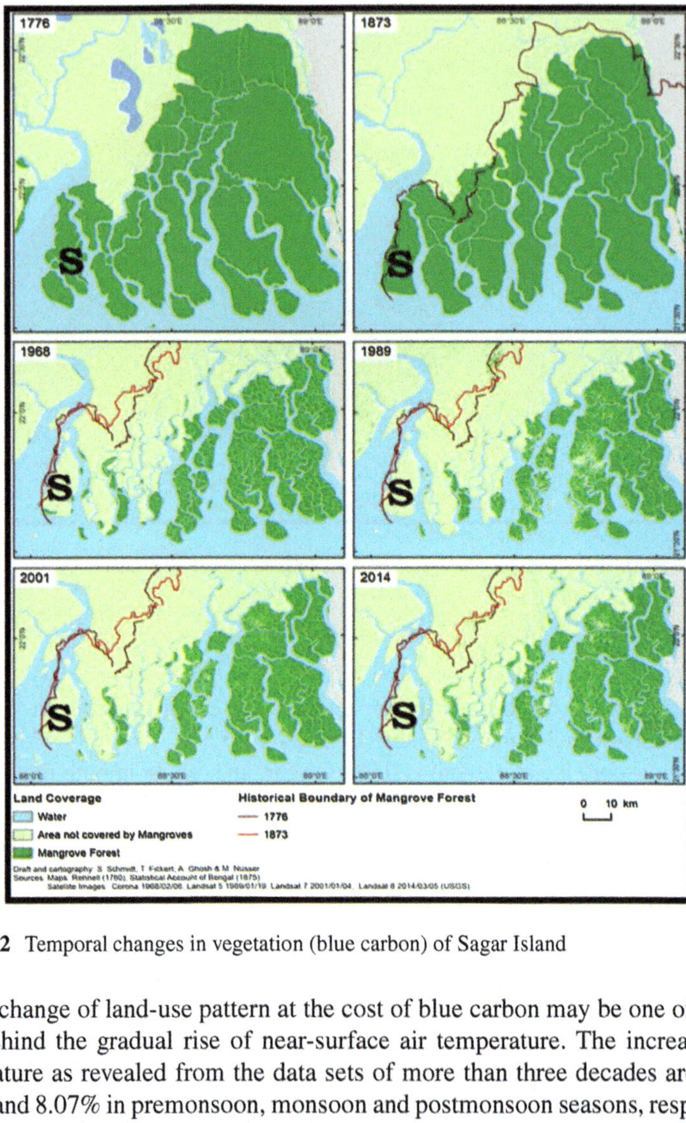

Fig. 5.A.2 Temporal changes in vegetation (blue carbon) of Sagar Island

The change of land-use pattern at the cost of blue carbon may be one of the reasons behind the gradual rise of near-surface air temperature. The increase of air temperature as revealed from the data sets of more than three decades are 9.23%, 7.12% and 8.07% in premonsoon, monsoon and postmonsoon seasons, respectively. Another important feature in the graph of premonsoon air temperature is an abrupt spike in 2009, which may be attributed to super cyclone Aila that passed across lower Gangetic delta during May 25 and 26, 2009 (Mitra et al. 2011).

Global atmospheric temperatures and CO_2 concentrations have risen throughout the last 50 years (Trenberth 1997; Quay 2002), which has triggered the rise of air temperature. Simultaneously, the world's oceans have experienced a net warming (Levitus et al. 2000; Sheppard 2001; Fukasawa et al. 2004). Regional increases in temperature have been documented in the southwest Pacific Ocean and North Atlantic Ocean (Bindoff and Church 1992; Parrilla et al. 1994). For the last

20–30 years, the western Mediterranean Sea temperatures have been rising (Bethoux et al. 1990), which is reflected in the presence and abundance of ectothermic marine life (Francour et al. 1994). For example, off the coast of France, two thermophilic algal species, several thermophilic echinoderm species, and some thermophilic fishes have increased in abundance, while other thermophilic species are being observed for the first time (Francour et al. 1994).

Estimates of future surface air temperature increases range from 1 to 7 °C, depending on the hypothesized atmospheric carbon dioxide contents (Daniels et al. 1993; Kwon and Schnoor 1994; Manabe et al. 1994; Woodwell et al. 1998). Air temperatures are expected to increase ocean warming, most significantly in the upper 500–800 m (Bernal 1993). However, even slight warming of deeper oceanic layers will have a huge impact on the Earth's energy budget due to the mass of water they contain (Bernal 1993; Levitus et al. 2000; Stevenson et al. 2002). Ocean circulations are predicted to shift, possibly interacting with land masses, creating a north-south thermal asymmetry (Bernal 1993). For example, the northern boundary of the Gulf Stream has shifted slightly northwards in recent decades (Taylor and Stephens 1998).

Climate change will affect individuals, populations and communities through the individual's physiological and behavioural responses to environmental changes (Boesch and Turner 1984). Extremes in environmental factors, such as elevated water temperature, low-dissolved oxygen or salinity and pH, can have deleterious effects on fishes (Moyle and Cech Jr 2004). Sub-optimal environmental conditions can decrease foraging, growth and fecundity; alter metamorphosis; and affect endocrine homeostasis and migratory behaviour (Barton and Barton 1987; Donaldson 1990; Pörtner et al. 2001). These organisms' changes directly influence population and community structure by their associated effects on performance, patterns of resource use and survival (Ruiz et al. 1993; Wainwright 1994).

Estuarine and coastal regions are extremely productive because they receive inputs from several primary production sources and detritus food webs. Yet, these systems present the biota with a harsh environment, forcing organisms to evolve physiological or behavioural adaptations to cope with wide-ranging physical and chemical variables (Horn et al. 1999). Due to water circulation and oceanic volume changes, estuarine and coastal systems are predicted to experience a loss of marsh and intertidal habitat, a greater marine intrusion or freshwater plumes and increased eutrophication, hypoxia and anoxia (Officer et al. 1984; Kennedy 1990; Ray et al. 1992; Schwartz 1998). The close relationship between laboratory-based measurements of fish's responses to temperature and the same species thermal distributional limits in their native habitats have been demonstrated (Cech Jr et al. 1990). Because many native organisms currently live near their tolerance limits, estuarine and coastal ecosystems will likely exhibit responses earlier to regional changes, including native species loss and exotic species increases (Kennedy 1990; Carlton 1996).

Marine pelagic systems are susceptible to climate change through extreme events and the contraction or expansion of oceanic zones. For example, sea temperature changes driven by variations in the North Atlantic Oscillation (NAO) have been linked to fluctuations in cod (*Gadus morhua*) recruitment and habitat shifts off

Fig. 5.A.3 Point sources of GHGs in and around Indian Sundarbans

Labrador and Newfoundland (Rose et al. 2000). On the west coast of Canada and Alaska, the Gulf of Alaska is exhibiting increased temperature and decreased salinity levels. The result is seen in shallower mixed layers, which lead to a reduced nutrient supply (Whitney and Freeland 1999), impacting primary production levels and altering food webs (Bjørnstad and Grenfell 2001; Zabel et al. 2003; Richardson and Schoeman 2004).

The rising trend of near-surface air temperature in the mangrove-dominated Indian Sundarbans region (as revealed from the data bank generated from Sagar Island) has the possibility of altering the biotic community in this designated World Heritage Site of Indian sub-continent. Hence a proper management action plan is of great important to minimize and control the emission from brick kilns, industries, fishing vessels and trawlers, which are the main point sources of GHG in this deltaic complex (Fig. 5.A.3). Also there is a need to develop nursery of blue carbon on large-scale basis, which can act as GHG and temperature regulators.

References

Agarwal SK, Fazli P, Zaman S, Mitra A (2018) Near surface air temperature and carbon dioxide in Indian Sundarbans: a time series analysis. Parana J Sci Edu 4(1):10–15

Banerjee K, Mitra A, Bhattacharyya DP, Choudhury A (2002a) Role of nutrients on phytoplankton diversity in the north–east coast of the Bay of Bengal. In: Kumar A (ed) Ecology and ethology of aquatic biota. Daya Publishing House, New Delhi, pp 102–109

Banerjee K, Mitra A, Bhattacharyya DP (2002b) Phytopigment level of the aquatic subsystem of Indian Sundarbans at the apex of Bay of Bengal. Sea Exp 6:39–46

Banerjee K, Sengupta K, Raha AK, Mitra A (2013) Salinity based allometric equations for biomass estimation of Sundarban mangroves. Biomass Bioenergy 56:382–391

Barton M, Barton AC (1987) Effects of salinity on oxygen consumption of *Cyprinodon variegatus*. Copeia 1987:230–232

Bernal PA (1993) Global climate change in the oceans: a review. In: Mooney HA, Fuentes ER, Kronberg BI (eds) Earth system responses to global change: contrast between North and South America. Academic Press, San Diego, pp 1–15

Bethoux JP, Gentili B, Raunet J, Tailliez D (1990) Warming trend in the western Mediterranean deep water. Nature 347:660–662
Bindoff NL, Church JA (1992) Warming of the water column in the southwest Pacific Ocean. Nature 357:59–62
Bjørnstad ON, Grenfell BT (2001) Noisy clockwork: time series analysis of population fluctuations in animals. Science 293:638–643
Boesch DF, Turner RE (1984) Dependence of fishery species on salt marshes: the role of food and refuge. Estuaries 7:460–468
Carlton JT (1996) Pattern, process, and prediction in marine invasion ecology. Biol Conserv 78:97–106
Cech JJ Jr, Mitchell SJ, Castleberry DT, McEnroe M (1990) Distribution of California stream fishes: influence of environmental temperature and hypoxia. Environ Biol Fish 29:95–105
Chakraborty SK, Choudhury A (1985) Distribution of fiddler crabs in Sundarbans mangrove estuarine complex, India. In: Proceedings of National Symposium on Biology, Utilization and Conservation of Mangroves, pp 467–472
Daniels RC, White TW, Chapman KK (1993) Sea-level rise: destruction of threatened and endangered species habitat in South Carolina. Environ Manag 17:373–385
Donaldson EM (1990) Reproductive indices as measures of the effects of environmental stressors. Am Fish Soc Symp 8:145–166
Francour P, Boudouresque CF, Harmelin JG, Harmelin-Vivien ML, Quignard JP (1994) Are the Mediterranean waters becoming warmer? information from biological indicators. Mar Pollut Bull 28:523–526
Fukasawa M, Freeland H, Perkin R, Watanabe T, Uchida H, Nishina A (2004) Bottom water warming in the North Pacific Ocean. Nature 427:825–827
Horn MH, Martina KLM, Chotkowski MA (1999) Intertidal fishes: life in two worlds. Academic Press, San Diego, p 399
Kennedy VS (1990) Anticipated effects of climate changes on estuarine and coastal fishes. Fisheries 15:16–24
Kwon OY, Schnoor JL (1994) Simple global carbon model: the atmosphere-terrestrial biosphere-ocean interaction. Global Biochem Cycles 8:295–305
Levitus S, Antonov JI, Boyer TP, Stephens C (2000) Warming of the world ocean. Science 287:2225–2229
Manabe S, Stouffer RJ, Spelman MJ (1994) Response of a coupled ocean atmosphere model to increasing atmosphere carbon dioxide. Ambio 23(1):44–49
Mitra A (2013). In: *Sensitivity of mangrove ecosystem to changing climate* by Dr. Abhijit Mitra. Publisher *Springer* New Delhi Heidelberg New York Dordrecht London, 2013 edition (August 31, 2013); ISBN-10: 8132215087; ISBN-13: 978-8132215080., copyright Springer India 2013; ISBN 978-81-322-1509-7 (eBook)
Mitra A, Choudhury A (1994) Heavy metal concentrations in oyster *Crassostrea cucullata* of Henry's Island, India. J Eco-Biol 6(2):157–159
Mitra A, Zaman S (2014) Carbon sequestration by coastal floral community. The Energy and Resources Institute (TERI) TERI Press, New Delhi
Mitra A, Zaman S (2015). Blue carbon reservoir of the blue planet published by Springer, ISBN 978-81-322-2106-7 (Springer DOI https://doi.org/10.1007/978-81-322-2107-4)
Mitra A, Zaman S (2016) Basics of marine and estuarine ecology, Springer, ISBN 978-81-322-2705-2
Mitra A, Ghosh PB, Choudhury A (1987) A marine bivalve *Crassostrea cucullata* can be used as an indicator species of marine pollution. In: Proc Nat Seminar on Estua Manag, pp 177–180
Mitra A, Choudhury A, Zamaddar YA (1992) Effects of heavy metals on benthic molluscan communities in Hooghly estuary. Proc Zool Soc 45:481–496
Mitra A, Banerjee K, Sengupta K, Gangopadhyay A (2009) Pulse of climate change in Indian Sundarbans: a myth or reality. Nat Acad Sci Lett 32:1–7

Mitra A, Halder P, Banerjee K (2011) Changes of selected hydrological parameters in Hooghly estuary in response to a severe tropical cyclone (Aila). Ind J Geo Mar Sci 40(1):32–36
Mondal SK, Mukhopadhyay N, Biswas H, De TK, Jana TK (2006) Fluxes of nutrients from the tropical River Hooghly at the land–ocean boundary of Sundarbans, NE Coast of Bay of Bengal, India. J Mar Syst 62:9–21
Moyle PB, Cech JJ Jr (2004) Fishes: an introduction to ichthyology, 5th edn. Prentice Hall, Upper Saddle River, p 726
Officer CB, Biggs RB, Taft JL, Cronin LE, Tyler MA, Boynton WR (1984) Chesapeake Bay anoxia: origin, development, and significance. Science 223:22–27
Parrilla G, Lavin A, Bryden H, Garcia M, Millard R (1994) Rising temperatures in the subtropical North Atlantic Ocean over the past 35 years. Nature 369:48–51
Pitchaikani JS, Sen Sarma K, Bhattacharyya S (2017) First time report on the weather patterns over the Sundarban mangrove forest, East Coast of India. Ind J Geo Mar Sci 46(04):766–770
Pörtner HO, Berdal B, Blust R, Brix O, Colosimo A, De Wachter B, Giuliani A, Johansen T, Fischer T, Knust R, Lannig G, Naevdal G, Nedenes A, Nyhammer G, Sartoris FG, Serendero I, Sirabella P, Thorkildsen S, Zakhartsev M (2001) Climate induced temperature effects on growth performance, fecundity and recruitment in marine fish: developing a hypothesis for cause and effect relationships in Atlantic cod (*Gadus morhua*) and common eelpout (*Zoarces viviparus*). Cont. Shelf Res 21:1975–1997
Quay P (2002) Ups and downs of CO_2 uptake. Science 298:2344
Ray GC, Hayden BP, Bulger AJ Jr, McCormick-Ray MG (1992) Effects of global warming on the biodiversity of coastal-marine zones. In: Peters RL, Lovejoy TE (eds) Global warming and biological diversity. Yale University Press, New Haven, pp 91–104
Richardson AJ, Schoeman DS (2004) Climate impact on plankton ecosystems in the north-east Atlantic. Science 305:1609–1612
Rose GA, deYoung B, Kulka DW, Goddard SV, Fletcher GL (2000) Distribution shifts and overfishing the northern cod (*Gadus morhua*): a view from the ocean. Can J Fish Aquat Sci 57:644–663
Ruiz GM, Hines AH, Posey MH (1993) Shallow water as a refuge habitat for fishes and crustaceans in non-vegetated estuaries: an example from Chesapeake Bay. Mar Ecol Prog Ser 99:1–16
Saha SB, Mitra A, Bhattacharyya SB, Choudhury A (1999) Heavy metal pollution in Jagannath canal, an important tidal water body of the north Sundarbans aquatic ecosystem of West Bengal. Ind J Environ Prot 19:801–804
Schwartz FJ (1998) Fishes affected by freshwater and/or marine intrusions in North Carolina. J Elisha Mitchell Sci Soc 114:173–189
Sengupta K, Roy Chowdhury M, Bhattacharya SB, Raha AK, Zaman S, Mitra A (2013) Spatial variation of stored carbon in *Avicennia alba* of Indian Sundarbans. Disco Nature 3(8):19–24
Sheppard C (2001) The main issues affecting coasts of the Indian and Western Pacific Oceans: a meta-analysis from sea at the millennium. Mar Pollut Bull 42:1199–1207
Stevenson JC, Kearny MS, Kock EW (2002) Impacts of sea level rise on tidal wetlands and shallow water habitats: a case study from Chesapeake Bay. In: McGinn NA (ed) Fisheries in a changing climate, American Fisheries Society Symposium, vol 32. American Fisheries Society, Bethesda, pp 23–36
Taylor AH, Stephens JA (1998) The North Atlantic oscillation and the latitude of the Gulf Stream. Tellus 50A:134–142
Trenberth KE (1997) The use and abuse of climate models. Nature 386:131–133
Wainwright PC (1994) Functional morphology as a tool in ecological research. In: Wainwright PC, Reilly SM (eds) Functional morphology: integrative organism biology. Chicago University Press, Chicago, pp 42–59
Whitney FA, Freeland HJ (1999) Variability in upper-ocean water properties in the NE Pacific. Deep-Sea Res 46:2351–2370

Woodwell GM, Mackenzie FT, Houghton RA, Apps MA, Gorham E, Davidson E (1998) Biotic feedbacks in the warming of the earth. Clim Chang 40:495–518
Zabel RW, Harvey CJ, Katz SL, Good TP, Levin PS (2003) Ecologically sustainable yield. Am Sci 91:150–157
Zhang X, Vincent LA, Hogg WD, Niitsoo A (2000) Temperature and precipitation trends in Canada during the 20th century. Atmosphere-Ocean 38:395–429

Internet Reference

https://archive.epa.gov/climatechange/kids/basics/today/greenhouse-gases.html
https://earthobservatory.nasa.gov

Chapter 6
Biodiversity and Its Conservation

Abstract Biodiversity acts as a fuel to run the engine of biosphere where a wide spectrum of flora and fauna is bound since long evolutionary period of time. The matrix of biodiversity is, however, affected through several man-made and natural factors which have been highlighted in the chapter with relevant case studies. The survival of human beings is keenly related to conservation of floral and faunal diversity as all the basic raw materials starting from food items to energy are sourced from the biotic components. The ecosystem services offered by common medicinal plants are salient feature of the chapter.

Keywords Genetic diversity · Species diversity · Community and ecosystem diversity · Threats · Medicinal plants

6.1 Introduction to Biodiversity

The term 'biodiversity' refers to the totality of genus, species habitat and ecosystem of a region. The total number of races, varieties or species, i.e. the sum total of various types of plants, animals and microbes present in a system is known as biological diversity or biodiversity (Table 6.1). Whether or not a species can occur on a site depends on the environmental conditions of the site and range of tolerance of the species. Thus, biodiversity varies from place to place. Taking into consideration, the total habitats of plants and animals, one can arrive at the conclusion that the living world is full of biodiversity. In a nutshell, the term denotes whole varieties of life on Earth.

Levels of Biodiversity

Biological diversity can be studied at three different hierarchical levels. These levels of diversity are interrelated but should be studied separately to understand the interconnections that support life on Earth.

A. Mitra, S. Zaman, *Environmental Science - A Ground Zero Observation on the Indian Subcontinent*, https://doi.org/10.1007/978-3-030-49131-4_6

Table 6.1 Approximate numbers of species which have been described and identified all over the world

Group	Numbers of species
Higher plants	2,70,000
Algae	40,000
Fungi	72,000
Bacteria (including cyanobacteria)	4000
Viruses	1550
Mammals	4650
Birds	9700
Reptiles	7150
Fishes	26,959
Amphibians	4780
Insects	10,25,000
Crustaceans	43,000
Molluscs	70,000
Nematodes and worms	25,000
Protozoa	40,000
Others	1,10,000

1. Genetic Diversity

All the species present on Earth like blue-green algae and bacteria to higher plants and most developed animals carry lots of genetic information. Species can be differentiated into many varieties, races or strains. These varieties, races or strains differ from each other in one or more traits (characters), e.g. size, shape, resistance to insect and pest, total yield, etc. These differences are because of differences in genetic information. Thus, the diversity in the genetic organization of a species is termed as genetic diversity. This difference of genes within species could be because of difference in alleles (different variants of same genes), in entire genes (the traits determining particular characteristics) or in chromosomal structure.

2. Species Diversity

Numbers of biotic components present in an ecosystem may vary from a few species to a large number of species of microorganisms, plants and animals. They react with each other and with the abiotic components of the environment. Thus, species play a vital role in an ecosystem, and loss of species affects the ecosystem as a whole. The richness of a species in an ecosystem is known as species diversity. The number of species increases with the area of the site; the greater the species richness, the greater is the species diversity. But, the number of individuals among the species may differ resulting into differences in evenness. In nature, the number and kind of species and number of individuals per species differ, leading to more and more diversity.

3. Community and Ecosystem Diversity

An ecosystem develops its own characteristic community of living organisms (biotic components) which depends largely upon the quantum of abiotic resources and conditions of the environment. Ecosystem diversity has three aspects:

(a) **Alpha diversity (within community diversity):** It includes the diversity of organisms sharing the same habitat or community. It is a mixture of species richness and evenness, which is used to represent diversity within a community or habitat.
(b) **Beta diversity (between community diversity):** It is the turnover or replacement of species while moving from one habitat to another within a given geographical area. Communities differ in species composition along environmental gradients of altitudinal gradient, moisture gradient, etc. The greater the dissimilarity between communities, the higher is the beta diversity.
(c) **Gamma diversity:** It refers to the diversity of the habitats over the total landscape or geographical area. In other words, the term denotes the rate of turnover or replacement of species between similar habitats in different geographical areas.

6.2 Threats to Biodiversity

Biodiversity is affected by habitat loss, climate change, overexploitation, invasive alien species and pollution. Each of these causes is diverse that puts an adverse impact on the biodiversity, and in many cases they occur simultaneously in the same ecosystem or environment.

6.2.1 Habitat Loss

The most common types of habitat loss include cutting down forest for timber and opening of land for agriculture, destroying wetlands to make way for new development projects or construction of dams for agriculture and providing water to the urban units. Habitat loss leads to disruption of complex interactions among species. If a big forest area is fragmented into cropland, orchards and plantation of urban areas, species found in deep forests disappear first because when a habitat is fragmented into small patches, its capacity to support populations of different species reduces considerably, and as a result, the excess number of species is removed.

6.2.2 Climate Change

The rapid pace of industrialization and urbanization without any proper planning has resulted in the release of greenhouse gases like carbon dioxide, methane, nitrous oxide, ozone, chlorofluorocarbons and hydrofluorocarbons. This has resulted in the rise of temperature, extreme weather events, sea level rise, acidification of water, etc. All these effects of climate change have resulted in the loss of biodiversity and

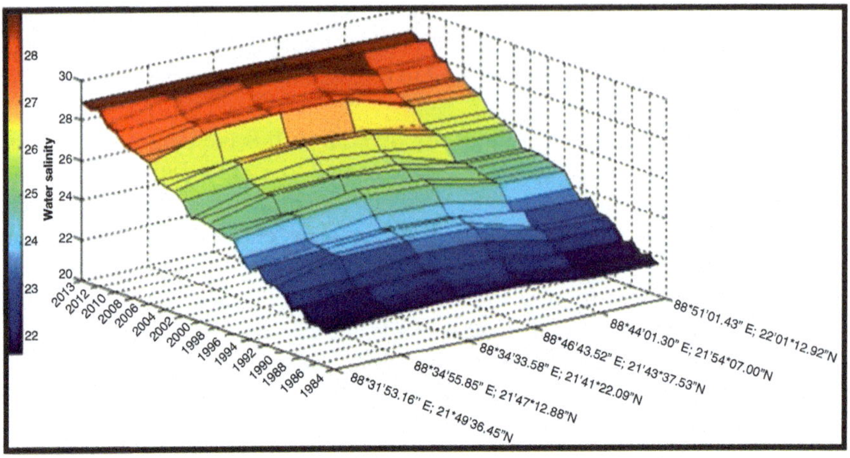

Fig. 6.1 Spatio-temporal variation of salinity in central Indian Sundarbans

in many cases extinction of species. A study conducted by the present authors in central Indian Sundarbans shows that the freshwater loving mangrove flora *Heritiera fomes* (locally called Sundari) has become almost extinct in the central part of Indian Sundarbans due to rise of salinity (Fig. 6.1).

In the lower Gangetic region, acidification of water has resulted in the thinning of the oyster shell (Fig. 6.2).

Thus climate change is not only eroding the species from the ecosystem but at the same time posing a negative impact on their physiology, metabolism and life cycle pattern.

6.2.3 Overexploitation

Overexploitation happens when biodiversity is removed faster than it can be replenished and, over the long term, can result in the extinction of the species. Rare plants of *Nepenthes khasiana*, *Drosera* sp., *Gnetum* sp., *Psilotum* sp. and *Isoetes* sp. are carelessly used for collection, teaching and laboratory purposes. Some other rare medicinal plants like *Podophyllum* sp., *Coptis* sp., *Aconitum* sp., *Rauwolfia* sp. and *Saussurea lappa* are disappearing very fast because of over-collection. The population of world's famous orchid plants is also dwindling because of overexploitation and trade in these beautiful flowers. It is to be mentioned in this context that some rare occupations also result in the loss of biodiversity, e.g. in Sundarbans during the collection of prawn seed, lots of fish juveniles are caught in the net, but they are thrown away and wasted as they are not the target species for the shrimp culturists. The prawn seed are collected from wild in the Sundarban estuaries as there is no hatchery in this region due to low salinity of the water. Such destructive collection

Fig. 6.2 Oyster on the substratum in the estuaries of Indian Sundarbans

of prawn seed causes a loss of different species of fish juveniles. It has been reported that for collecting one prawn seed, about 65 species of finfish juveniles are wasted.

6.2.4 Invasive Alien Species

An exotic species is any species, which is not a natural inhabitant of the locality but is accidentally or deliberately introduced into a geographical region. These species, when introduced into a system, may cause disappearance of native species through changed biotic interactions in their struggle for existence; they become adaptive to stressful environment and hence become tough, tenacious and resistant. When such plants are introduced in mild conditions, they thrive well, and in this process of establishment, they suppress the natives or local population. Some examples are:

(i) **Nile perch** an exotic predatory fish introduced into Lake Victoria (South Africa) threatened the lake ecosystem by consuming and destroying many native species of small cichlid fish species endemic to this freshwater aquatic ecosystem.

(ii) Water hyacinth was introduced as an exotic species, but it started clogging rivers and lakes and threatened the survival of many species in several tropical countries, including India.

(iii) *Lantana camara* has invaded many areas, including forestlands, in different parts of India, and competes with the native species.
(iv) *Parthenium hysterophorus* (the congress grass) and *Agrostemma githago* (the corn cockle) came to India along with contaminated wheat from the United States and invaded Indian fields. They delete not only nutrients from the soil but huge amount of pollens produced by *Parthenium* showing allelopathic effects on cultivated plants like tomato, chillies, brinjal, etc.
(v) In addition to this, alien species are also introduced through ballast waters.

6.2.5 Pollution

All forms of pollution pose a negative impact on biodiversity. The nutrients like nitrate and phosphate released from the municipal and agricultural waste cause algal bloom, which is referred to as eutrophication. Under this condition, the dissolved oxygen of the aquatic system drops down to a minimum level in the night. This results in the mortality of aquatic organisms. The heavy metals get bioaccumulated within the body tissues of the organisms and lead to metabolic disorders (Interested readers may have a look at Annexure 6A for a case study on bioaccumulation of heavy metals). Oil pollution in the coastal region and marine waters results in the death of phytoplankton as oil layers on the surface of the water cut off the solar radiation and inhibit the process of photosynthesis in the phytoplankton. Finally this results in the disruption of aquatic food web.

6.3 *In situ*, *Ex situ* and Policy-Level Approaches to Conserve Biodiversity

Conservation may be defined as '…*the management of human use of the biosphere (i.e., all living things) so that it may yield the greatest sustainable benefit to present generations while maintaining its potential to meet the need and aspiration of future generations……*'……

Biodiversity conservation aims at conservation of species diversity, genetic diversity and ecosystem diversity present in different parts. But species diversity should get the preference. The most effective and efficient mechanism for conserving biodiversity is to prevent further destruction or degradation of habitats by us. There are two basic strategies of biodiversity conservation:

(a) In situ (on site)
(b) Ex situ (off site)

(a) In situ Conservation

The term '*In situ* Conservation' denotes conservation of species in its natural habitat, i.e. the place where the species is normally found. It emphasizes protection of total ecosystems through a network of protected areas. Here, we simply eliminate factors which are detrimental to the existence of species concerned, and rest is left to the nature which takes care of its own.

The in situ strategies stress on protection of total ecosystems through a network of protected areas like National Parks, Sanctuaries and Biosphere Reserves.

(i) **National Parks**: A National Park is an area which is strictly reserved for the betterment of the wildlife and where activities like forestry, grazing and cultivation are not permitted.

(ii) **Sanctuary**: Sanctuaries can be defined as a protected area which is reserved for the conservation of only animal and human activities like harvesting of timber and collection of minor forest products. Here, private rights are allowed as long as they do not interfere with the well-being of animals.

(iii) **Biosphere Reserves**: A Biosphere Reserve is a specified area in which multiple use of the land is permitted by dividing it into certain zones being specified for a particular activity. They are large protected areas of usually more than 5000 km^2. Their boundaries are delimited by legislation.

(b) Ex situ Conservation

Ex situ conservation refers to conservation outside their habitats by perpetuating sample populations in genetic resource centre, culture, collection, zoos, botanical gardens, etc., or as gamete or gene pool storage and germplasm banks for pollens, seeds, semen, ova, cells, etc. Seed gene banks are the easiest way to store germplasm of cultivated and wild plants at low temperature in cold room.

The methods of ex situ conservation are given here:

(i) Use of Seed Banks, Gene Banks or Germplasm Reserves

Seeds show variable periods of dormancy. Most of seed plants, therefore, can be preserved in the form of their seeds in small packets for long durations. Places where seeds are stored are called seed banks or gene banks or sometimes germplasm banks. The term germplasm of a plant refers to any of its living organ or a part of it from which new plants can be generated. They utilize the technique of cryopreservation in liquid nitrogen at a temperature of -196 °C.

(ii) Long-Term Captive Breeding

The method involves capture, maintenance and breeding in captivity on long-term basis of individuals of the endangered species which have lost their habitats permanently or certain highly unfavourable conditions are present in their habitat. Our zoos and botanical gardens house a number of such animals and plants. Examples are Siberian tiger, Pere David's deer, lion-tailed macaque, European bison, Przewalski's horse, etc.

(iii) Short-Term Propagation and Release

Short-term maintenance and captive breeding followed by release of the affected animals in their natural habitat are adopted when the population of a species declines due to some temporary setback in their living conditions. There is no permanent change in the habitat, and the animal concerned can survive in its natural home after the factors causing the setback are removed.

(iv) Animal Translocations

Translocations encompass release in a new locality of animals which come from anywhere else, i.e. other than the locality in which they are being released. The capture, transfer and release of animals from one place to another also include maintenance of the animal in captivity for some time.

(v) Animal Reintroduction

It implies release of animals from wild and grown in captivity into an area from which they have disappeared or declined due to some reason (e.g. epidemic or hunting). Examples are American Bison (*Bison bison*) which were bred in captivity and were reintroduced in North America.

(vi) Cryopreservation of Gametes and Embryos

This method aims at preservation of large amount of biodiversity in viable state in a small space and for infinite time, e.g. preservation of semen and preservation of ova and embryos.

Botanical gardens and zoological gardens are the ex situ conservation sites.

Policy-Level Approaches to Conserve Biodiversity

The human population in the present world is approximately 7.7 billion. To sustain this huge population, establishments of industries and factories are needed to provide job opportunities. This opens the windows of overexploitation, deforestation, biodiversity loss and change of land-use pattern as industries are set up in regions which may be rich in floral and faunal diversity. To protect the natural resources from deterioration and safeguard the biodiversity, the government has introduced the concept of Environmental Impact Assessment (EIA) and Environmental Management Plan (EMP). A brief account of these concepts is discussed here.

EIA is a system developed by the Government, which deals with the systematic of the likely impacts of a particular project on the environment prior to the beginning of any kind of activity related to the project.

The term EIA is derived from Section 102 (2) of the National Environmental Policy Act (NEPA) of 1969 which marked a turning point in environmental legislation in the United States. People proposing to set up a project have to show that their project is not going to significantly affect the environment and residents of that area. The impacts of the project on the area, including wildlife in the area, water and air quality, etc. are identified in the decision-making process only. Presently the disaster management is also considered in the domain of EIA.

Thus, EIA aims to achieve sustained development causing very little harm to the environment and prevent any adverse effects on the environment in the future.

Aims and Objectives of EIA

1. Development of the project without causing harm to the environment.
2. Maintaining air, water and soil quality.
3. Usefulness and advantage of particular project to the people.
4. Preventing environmental degradation.
5. Environment-friendly decisions regarding the project to be taken in the planning stage only.
6. The project must ensure sustainable development so that needs of present and future generations are satisfied.

To serve as a proper tool, Environmental Impact Assessment has to be done at the planning stage only, so that the site selection, use of proper technology, proper layout, use of proper raw materials, etc. can be done in such a way so as to minimize environmental degradation.

Before starting a new project, many countries including India have made it mandatory that the respective companies prepare an Environmental Impact Statement (EIS).

It is mandatory for development projects like dams, iron and steel mines, mining and oil exploration plants, paper and wood industry, etc., to submit an EIS before commencement of the project.

EIS aims to show the nature and scale of the project, its effect on the environment and its likely impacts and measures taken to reduce or monitor the impacts.

Thus, EIS looks at all the positive and negative impacts of a project on the environment.

The Environmental Impact Statement (EIS) contains the following data:

1. Effect of the project on the land, i.e. the land where it is going to be located
2. Effect on the surface water and groundwater
3. Possible effect on wildlife (flora and fauna) endangered species, commercially important fish species
4. Possibility of earthquakes and other natural disasters like cyclones, tsunamis, etc.
5. Effect of the project on level of noise (noise pollution)
6. Effect on health of local people
7. Waste disposal system-recycling and reduction of waste
8. Efficient use of inputs, i.e. raw materials
9. Risk factors if any and disaster management
10. Effect of the project on soil and air of the region

EIA in India

The Ministry of Environment and Forests (MOEF) is directly responsible for Environment Impact Assessment reports that are sent by the various companies involved in setting up developmental projects. Developmental projects such as major irrigation projects, river valley projects, power projects, ports, harbours, etc. have to secure clearance from the Ministry of Environment and Forests. Environment appraisal is carried out by the various committees of the Ministry.

On January 27, 1994, the Ministry of Environment and Forests made it mandatory to submit EIA reports, under rule 5 of the Environmental (Protection) Rules of 1986 for 29 designated projects. The notification made it obligatory to prepare and submit an EIA, an Environment Management Plan (EMP) and a project report to an Impact Assessment Agency. Public hearings were mandatory, and any member of the public could have access to the summary of the project report and detailed EMPs. Thus, a comprehensive scheme regarding EIA came into existence.

The Department of Environment came into being in 1980 within the Ministry of Science and Technology under the charge of the then Prime Minister. In 1989 the subjects of Wildlife and Forestry were added to the list, and the new Ministry of Environment and Forests was created with the prime Minister in charge. Since then the Ministry has issued various guidelines on EIA for many different projects (Table 6.2).

Table 6.2 Procedures and methodology of EIA (The Ministry of Environment and Forest, Government of India, has laid down the following phases for the process of EIA in India)

Sl. No	Phases	Process
1	Screening	Initial screening and scrutiny of the project is done to see what will be the effects on the environment by the concerned project. If the project is not likely to affect the environment, then it is cleared immediately
2.	Scoping and identification of alternatives	If the initial screening shows that the project needs to be studied and could have environmental implications, then it is passed on to an expert committee. This committee is referred to as the Impact Assessment Agency. Alternative techniques are suggested by the agency, and its impact is evaluated
3.	Comprehensive impact assessment	When the screening and scoping of a project is done and alternatives are suggested, then a detailed, complete analysis of the project is done. This study critically analyses each and every aspect of the project and its impact on the environment
4.	Impact prediction and data collection	Data is collected regarding the site of the project, suitability of land area, availability of water, waste disposal, labour, etc. On the basis of the data collected, 'mapping' of the environmental consequences of the significant aspects of the project and its alternatives is predicted
5.	Decisions and monitoring of the project	Once the decision is made, a report is drawn up about the project, and guidelines are given for its implementation by the expert committee
6.	Public hearing and participation	After the completion of the EIA report, it is necessary to have a public hearing. The law requires that the public must be informed and consulted on various aspects of the project. This is done after the completion of the EIA report
7.	Regular monitoring	Monitoring is done regularly during both the construction and operation phase of the project. This is done to ensure that the commitments made are carried out as well as to check whether the predictions made in the EIA reports were correct or not

Table 6.3 Advantages and disadvantages of EIA

Advantages of EIA	Disadvantages of the EIA
It provides a better understating of the impact of the project on the environment	The assessment is very time-consuming
It helps to identify the impact and find alternative solutions	Illiteracy among local people causes delays
It brings a sense of cooperation between the company putting up the project and the local people of the area	Time and cost involved in the assessment is high
It help to resolve issues and dispute between the company and the local people	Experts may not always be available to offer alternative solutions

Thus, EIA is an important tool of the Government that makes sure that the best use is made of the natural resources and leads to sustainable development without damaging the environment (Table 6.3).

In this context it is worthy to mention that during setting up of any industries, many species of flora and fauna are destroyed and damaged. In this context it is mandatory to plant similar species in botanical garden or any other space and grow them with care so that the diversity is maintained and there is no genetic erosion. A list of common medicinal plants available in the tropical savannah climatic zone in Indian sub-continent along with their uses is highlighted as Annexure 6B.

6.4 Take Home Messages

[A] The term 'biodiversity' refers to the totality of genus, species, habitats and ecosystem of a region. The total number of races, varieties or species, i.e. the sum total of various types of plants, animals and microbes present in a system may be referred to as biological diversity or biodiversity. Biodiversity is affected by habitat loss, climate change, overexploitation, invasive alien species and pollution. Each of these causes is diverse that puts an adverse impact on the biodiversity, and in many cases they occur simultaneously in the same ecosystem or environment.

[B] Biological diversity can be studied at three different hierarchical levels. These levels of diversity are interrelated but should be studied separately to understand the interconnections that support life on Earth. These three levels are (i) genetic diversity, (ii) species diversity and (iii) community and ecosystem diversity.

[C] Biodiversity conservation aims at conservation of species diversity, genetic diversity and ecosystem diversity present in different parts. Species diversity, however, should get the preference. The most effective and efficient mechanism for conserving biodiversity is to prevent further destruction or degradation of habitats. There are two basic strategies of biodiversity conservation, namely, (i) in situ (on site) conservation mechanism and (ii) ex situ (off site) conservation mechanism. The in situ strategies stress on protection of total ecosystems through a network of protected areas like National Parks, Sanctuaries and Biosphere Reserves. Ex situ conservation refers to conservation outside their habitats by perpetuating sample populations in genetic resource centre, culture, collection, zoos, botanical gardens, etc., or as gamete or gene pool storage and germplasm banks for pollens, seeds, semen, ova, cells, etc.

[D] To protect the natural resources from deterioration and safeguard the biodiversity, Government of India has introduced the concept of Environmental Impact Assessment (EIA) and Environmental Management Plan (EMP).

[E] During setting up of any industries, several species of flora and fauna are destroyed and damaged. In this context it is mandatory to plant similar species in botanical garden and grow them with care so that the diversity is maintained and there is no genetic erosion. On this background, many countries including India have made it compulsory that the respective companies prepare an Environmental Impact Statement (EIS) before starting a new project. It is mandatory for development projects like dams, iron and steel mines, mining and oil exploration plants, paper and wood industry, etc., to submit an EIS before commencement of the project.

6.5 Brain Churners

This chapter provides the readers a basic idea on the floral and faunal diversity. The chapter has immense importance on the backdrop of natural resource conservation, which is gradually depleting due to anthropogenic and natural factors. The section will also help the readers to face various competitive examinations as biodiversity has been given thrust in recent times. The readers can consider this section as a sort of '*self-appraisal approach*' as correct answers to all the questions are marked in black in the Answer Box at the end of this section.

1. Biodiversity is effected by
   (a) Climate change
   (b) Habitat loss
   (c) Overexploitation
   (d) All of the above
2. Overexploitation happens when
   (a) Biodiversity is removed faster than it can be replenished
   (b) Biodiversity is removed slower than it can be replenished
   (c) When both biodiversity removal and replenishment are equal
   (d) None of the above

3. Exotic species are those species which are

   (a) Not natural inhabitant of the locality
   (b) Natural inhabitant of the locality
   (c) Exported from one habitat to another
   (d) None of the above

4. Gamma diversity includes

   (a) Diversity of organisms sharing different habitats
   (b) Diversity of organisms sharing the same habitat
   (c) Diversity of organisms at genetic level
   (d) None of the above

5. Acidification results in

   (a) Thickening of the oyster shell
   (b) Thinning of the oyster shell
   (c) Results in no effect on the oyster shell
   (d) All of the above

6. Nile Perch is an

   (a) Endemic species
   (b) Exotic fish with symbiotic nature
   (c) Exotic predatory fish introduced into Lake Victoria
   (d) None of the above

7. *Alpha* diversity includes

   (a) Diversity of organism sharing different habitats
   (b) Diversity of organism sharing the same habitat
   (c) Diversity of organism at genetic level
   (d) None of the above

8. Water hyacinth is

   (a) An endemic species
   (b) An exotic species
   (c) Both a and b
   (d) None of the above

9. The common name of *Parthenium hysterophorus* is

   (a) Salt marsh grass
   (b) Sea grass
   (c) Congress grass
   (d) None of the above

10. In situ conservation means

   (a) On site conservation
   (b) Off site conservation

(c) Both on site and off site conservations
(d) None of the above

11. Genetic diversity refers to the

(a) Diversity of ecosystem
(b) Diversity of population
(c) Diversity in the genetic organism
(d) All of the above

12. The *beta* diversity becomes high when

(a) Greater is the similarity between communities
(b) Greater is the dissimilarity between communities
(c) When both the communities are equal in structure
(d) None of the above

13. *Rauwolfia* sp. is a

(a) Common medicinal plant
(b) Medicinal plant
(c) Coastal plant
(d) Desert plant

14. Prawn seed collection is

(a) Good ecological practice
(b) Destructive ecological practice
(c) Costly fishing activity
(d) None of the above

15. Alien species are introduced through

(a) Drinking water
(b) Rain water
(c) Ballast water
(d) None of the above

16. Heavy metals are

(a) Biodegradable in nature
(b) Degradable in nature
(c) Organic in nature
(d) None of the above

17. National Parks come under

(a) In situ conservation
(b) Ex situ conservation
(c) Sensitive zone
(d) None of the above

18. Indian Sundarbans is a

    (a) Marine habitat
    (b) Freshwater habitat
    (c) Biosphere Reserve
    (d) None of the above

19. Cryopreservation technology uses

    (a) Dry ice
    (b) Liquid nitrogen
    (c) Pesticides
    (d) None of the above

20. Lemon grass is used for

    (a) Erosion control in the coastal area
    (b) Curing asthma
    (c) Curing heart blockage
    (d) Erosion control in mining zone

Answer Sheet

S. No.	A	B	C	d
1.				■
2.	■			
3.	■			
4.	■			
5.		■		
6.			■	
7.		■		
8.		■		
9.			■	
10.	■			
11.			■	
12.		■		
13.		■		
14.		■		
15.			■	
16.		■		
17.	■			
18.			■	
19.		■		
20.				■

Annexure 6A: Bioaccumulation of Heavy Metals in Commercially Important Finfishes from the Lower Gangetic Delta

Introduction

The Gangetic delta, at the apex of Bay of Bengal, is recognized as one of the most diversified and productive ecosystems of the tropics. The deltaic lobe is unique for its wilderness, mangrove gene pool and tiger habitat. However due to intense industrial activities in the upstream zone, and several anthropogenic factors, the western part of the deltaic complex is exposed to pollution from domestic sewage and industrial effluents leading to serious impacts on biota (Mitra and Choudhury 1992). The presence of Haldia port-cum-industrial complex in the downstream region of the River Ganga (also known as the Hooghly River) has accelerated the pollution problem with a much greater dimension (Mitra 1998). The organic and inorganic wastes released from these industries and urban units contain substantial concentrations of heavy metals. The central part of the delta (encompassing the surroundings of Matla River) is relatively less stressful in terms of industrial discharge. Due to siltation of the Bidyadhari channel, the area does not receive any water supply from the Hooghly River in the western sector and is therefore tide fed in nature receiving the tidal flux from the Bay of Bengal (average salinity = ~32 psu). 85 percent of the people in this area consume fishes that are caught from the Gangetic delta region. The present paper aims to highlight the concentration of selective heavy metals (Zn, Cu, Pb and Cd) in the muscle tissue of five common finfish species, namely, *Polynemus paradiseus*, *Tenualosa ilisha*, *Liza parsia*, *Liza tade* and *Stolephorus commersonii* from four stations distributed in two sectors (western and central Indian Sundarbans) of the lower Gangetic region.

Methods

Description of the Study Site

Four sampling sites were selected two each in the western and central sectors of Indian Sundarbans, a Gangetic delta at the apex of the Bay of Bengal. The western sector of the deltaic lobe receives the snowmelt water of mighty Himalayan glaciers after being regulated through several barrages on the way. The central sector, on the other hand, is fully deprived from such supply due to heavy siltation and clogging of the Bidyadhari channel since the late fifteenth century (Chaudhuri and Choudhury 1994). The western sector also receives wastes and effluents of complex nature from multifarious industries concentrated mainly in the upstream zone. On this background four sampling stations (two each in western and central sectors) were selected (Table 6.6A.1) to analyse the concentrations of heavy metals in the common edible finfish and shellfish species inhabiting the zone.

Sampling of Specimen

Five commonly edible finfish species (*Polynemus paradiseus*, *Tenualosa ilisha*, *Liza parsia*, *Liza tade* and *Stolephorus commersonii*) were collected during low-tide condition from the selected stations (Table 6.6A.1) during a rapid EIA study from October 20 to November 30, 2019. The collected samples were stored in a container, preserved in crushed ice and brought to the laboratory for further analysis. Similar-sized specimens of each species were sorted out for analysing the metal level in the muscle of finfish, shrimps, crabs and soft body parts of oyster.

Analysis

Inductively coupled plasma–mass spectrometry (ICP-MS) is nowadays accepted as a fast, reliable means of multielemental analysis for a wide variety of sample types (Date and Gray 1988). A Perkin-Elmer Sciex ELAN 5000 ICP mass spectrometer was used for the present analysis. A standard torch for this instrument was used with an outer argon gas flow rate of 15 L/min and an intermediate gas flow of 0.9 L/min. The applied power was 1.0 kW. The ion settings were standard settings recommended, when a conventional nebulizer/spray is used with a liquid sample uptake rate of 1.0 mL/min. A Moulinex Super Crousty microwave oven of 2450 MHz frequency magnetron and 1100 Watt maximum power polytetrafluoroethylene (PTFE) reactor of 115 ml volume, 1 cm wall thickness with hermetic screw caps, were used for the digestion of the collected biological samples. All reagents used were of high purity available and of analytical reagent grade. High purity water was obtained with a Barnstead Nanopure II water purification system. All glasswares were soaked in 10% (v/v) nitric acid for 24 h and washed with deionised water prior to use.

The analyses were carried out on composite samples of ten specimens of each species having uniform size. This is a measure to reduce possible variations in metal concentrations due to size and age. 20 mg composite samples from each species of

Table 6.6A.1 Sampling stations with coordinates and salient features

Station	Coordinates	Salient features
Nayachar Island (Stn.1)	88° 15′ 24″ E 21° 45′ 24″ N	It is located in the Hooghly estuary in the western sector of the lower Gangetic delta and faces the Haldia port-cum-industrial complex that houses a variety of industrial units
Sagar South (Stn.2)	88° 01′ 47″ E 21° 39′ 04″ N	Situated at the confluence of the River Hooghly and the Bay of Bengal in the western sector of Indian Sundarbans, the station is an important navigational channel for the major ports of the area
Gosaba (Stn. 3)	88° 39′ 46″ E 22° 15′ 45″ N	Located in the Matla Riverine stretch in the central sector of Indian Sundarbans
Annpur in Satjelia Island (Stn. 4)	88° 50′ 43″ E 22° 11′ 52″ N	Located in the central sector of Indian Sundarbans. Noted for its wilderness and mangrove diversity; selected as our control zone

finfish and shellfish were weighed and successively treated with 4 ml aqua regia, 1.5 mL HF and 3 ml H_2O_2 in a hermetically sealed PIFE reactor, inside a microwave oven, at power levels between 330 and 550 Watt, for 12 min to obtain a clear solution. The use of microwave-assisted digestion appears to be very relevant for sample dissolution, especially because it is very fast (Nadkarni 1984; Matusiewicz and Sturgeon 1989; De la Guardia 1990). After digestion, 4 ml H_2BO_3 was added and kept in a hot water bath for 10 min, diluted with distilled water to make up the volume to 50 ml. Taking distilled water in place of biological samples and following all the treatment steps described above, the blank process was prepared. The final volume was made up to 50 ml. Finally, the samples and process blank solutions were analysed by ICP-MS. All analyses were done in triplicate, and the results were expressed with standard deviation.

The accuracy and precision of our results were checked by analysing standard reference material (SRM, Dorm-2). The results indicated good agreement between the certified and the analytical values (Table 6.6A.2).

Statistical Analysis

A logarithmic transformation was done on the data to improve normality. Analysis of variance (ANOVA) was performed to assess whether heavy metal concentrations varied significantly between sites and species; possibilities less than 0.05 ($p < 0.05$) were considered statistically significant. All statistical calculations were performed with SPSS 21.0 for Windows. Superscripts were used to show the statistically significant difference between stations and species.

Results

The species-wise variation was not uniform for all the metals. Zn accumulated as per the order *Liza parsia* > *Liza tade* > *Tenualosa ilisha* > *Polynemus paradiseus* > *Stolephorus commersonii* (Table 6.6A.2). Cu accumulated as per the order *Liza parsia* > *Tenualosa ilisha* > *Polynemus paradiseus* > *Liza tade* > *Stolephorus commersonii* (Table 6.6A.2). Pb accumulated as per the order *Liza parsia* > *Liza*

Table 6.6A.2 Zn concentrations (in ppm dry wt.) in finfish muscles

Species	Stn. 1	Stn. 2	Stn. 3	Stn. 4
Polynemus paradiseus	73.45 ± 1.40	63.61 ± 1.09	37.23 ± 1.34	11.90 ± 0.88
Tenualosa ilisha	89.22 ± 1.44	74.89 ± 1.67	42.14 ± 1.87	14.00 ± 0.98
Liza parsia	119.66 ± 1.53	90.55 ± 1.54	56.22 ± 1.51	23.67 ± 1.22
Liza tade	99.45 ± 1.38	87.77 ± 1.88	49.89 ± 1.49	21.54 ± 1.21
Stolephorus commersonii	41.10 ± 1.09	21.33 ± 1.09	15.67 ± 1.29	12.00 ± 0.66
WHO (1989) level for Zn in food	100 ppm			
FAO (1992) level for Zn in fish	30–100 ppm			

Table 6.6A.3 Cu concentrations (in ppm dry wt.) in finfish muscles

Species	Stn.1	Stn.2	Stn.3	Stn.4
Polynemus paradiseus	60.21 ± 0.74	41.00 ± 1.45	19.18 ± 0.39	12.43 ± 0.41
Tenualosa ilisha	63.09 ± 0.66	44.65 ± 1.58	23.75 ± 0.30	15.89 ± 0.30
Liza parsia	73.22 ± 0.83	52.44 ± 1.44	26.11 ± 0.55	19.10 ± 0.30
Liza tade	41.22 ± 0.50	47.38 ± 1.29	25.99 ± 0.57	15.56 ± 0.44
Stolephorus commersonii	13.17 ± 0.19	12.09 ± 0.29	11.19 ± 0.38	10.89 ± 0.19
WHO (1989) level for Cu in food	30 ppm			
FAO (1992) level for Cu in fish	10–100 ppm			

Table 6.6A.4 Pb concentrations (in ppm dry wt.) in finfish muscles

Species	Stn.1	Stn.2	Stn.3	Stn. 4
Polynemus paradiseus	9.99 ± 0.45	8.17 ± 0.55	7.04 ± 0.40	6.02 ± 0.45
Tenualosa ilisha	12.70 ± 0.51	11.22 ± 0.78	10.03 ± 0.50	8.93 ± 0.56
Liza parsia	18.88 ± 0.45	14.61 ± 0.50	13.58 ± 0.66	11.29 ± 0.47
Liza tade	14.77 ± 0.54	12.50 ± 0.44	14.77 ± 0.99	10.56 ± 0.41
Stolephorus commersonii	3.66 ± 0.40	2.99 ± 0.99	3.99 ± 0.41	2.33 ± 0.09
WHO (1989) level for Pb in food	2 ppm			
FAO (1992) level for Pb in fish	0.5–6.0 ppm			

BDL means below detectable level

Table 6.6A.5 Cd concentrations (in ppm dry wt.) in finfish muscles

Species	Stn.1	Stn.2	Stn.3	Stn. 4
Polynemus paradiseus	BDL	BDL	BDL	BDL
Tenualosa ilisha	1.21 ± 0.28	BDL	BDL	BDL
Liza parsia	3.12 ± 0.10	BDL	BDL	BDL
Liza tade	1.76 ± 0.31	BDL	BDL	BDL
Stolephorus commersonii	BDL	BDL	BDL	BDL
WHO (1989) level for Cd in food	1 ppm			
FAO (1992) level for Cd in fish	0.05–5.5 ppm			

BDL means below detectable level

tade > *Tenualosa ilisha* > *Polynemus paradiseus* > *Stolephorus commersonii* (Table 6.6A.3). Cd was BDL in all the stations except Station 1, where the order was *Liza parsia* > *Liza tade* > *Tenualosa ilisha* (Table 6.6A.3). For Zn and Cu, accumulated metal concentration in Stn. 4 was significantly higher than accumulated metal concentrations in Stn 1 and Stn 2. For Pb, significantly difference between stations was not found. Between all studied fish species, lowest metal accumulation values was found for *S. commersonii* ($p < 0.05$) (Tables 6.6A.4 and 6.6A.5).

Discussion

Heavy metals are stable and persistent environmental contaminants of aquatic environments. They occur in the environment both as a result of natural processes and as pollutants from human activities (Garcia-Montelongo et al. 1994; Jordao et al. 2002). Some metals like Zn and Cu, which are required for metabolic activity in organisms, lie in the narrow 'window' between their essentiality and toxicity. Other heavy metals like Cd and Pb may exhibit extreme toxicity even at low levels under certain conditions, thus necessitating regular monitoring of sensitive aquatic environments (Cohen et al. 2001; Fergusson 1990; Peerzada et al. 1990). From an environmental point of view, coastal zones can be considered as the geographic space of interaction between terrestrial and marine ecosystems that is of great importance for the survival of a large variety of plants, animals and marine species (Castro et al. 1999). The coastal zone receives a large amount of metal pollution from agricultural and industrial activity (Usero et al. 2005). Adverse anthropogenic effects on the coastal environment include eutrophication, heavy metals, organic and microbial pollution and oil spills (Boudouresque and Verlaque 2002). The discharge of these wastes without adequate treatment often contaminates the estuarine and coastal water with conservative pollutants (like heavy metals), many of which accumulate in the tissues of resident organisms like fishes, oysters, crabs, shrimps, seaweeds, etc. In many parts of the world, especially in coastal areas and on smaller islands, fish is a major part of food, which supplies all essential elements required for life processes in a balanced manner (Iyengar 1991). Hence, it is important to investigate the levels of heavy metals in these organisms to assess whether the concentration is within the permissible level and will not pose any hazard to the consumers (Krishnamurti and Nair 1999).

Heavy metal contamination of the environment has been occurring for centuries, but its extent has increased markedly in the last 50 years due to technological developments and increased consumer use of materials containing these metals. Pollution by heavy metals is a serious problem due to their toxicity and ability to accumulate in the biota (Islam and Tanaka 2004). There is still a general concern about the impact of metals in the aquatic environment (Grosell and Brix 2005). Heavy metals have contaminated the aquatic environment in the present century due to intense industrialization and urbanization. The Gangetic delta is no exception to this usual trend. The rapid industrialization and urbanization of the city of Kolkata (formerly known as Calcutta), Howrah and the newly emerging Haldia complex in the maritime state of West Bengal has caused considerable ecological imbalance in the adjacent coastal zone (Mitra and Choudhury 1992; Mitra 1998). The Hooghly estuary, situated on the western sector of the Gangetic delta, receives drainage from these adjacent cities, which have sewage outlets into the estuarine system. The chain of factories and industries situated on the western bank of the Hooghly estuary is a major cause behind the gradual transformation of this beautiful ecotone into stinking cesspools of the megapolis (Mitra and Choudhury 1992). The lower part of the estuary has multifarious industries such as paper, textiles, chemicals, pharmaceuticals, plastic, shellac, food, leather, jute, tyres and cycle rims (UNEP 1982). These units are point sources of heavy metals in the estuarine water. Due to toxic nature of certain heavy metals, these chemical constituents interfere with the ecology of a particular environment, and on entering into the food chain, they cause potential health hazards, mainly to human beings. It was reported by several workers that the discharge of heavy metals into the sea through rivers and streams results in the accumulation of pollutants in the marine environment especially within fishes (Yusof et al. 2004). Thus fishes can be used for monitoring potential risk to humans because these are directly consumed by a large population (Subramanian and Sukumar 1988).

Romeo et al. (1999) pointed out that the affiance of metal uptake from contaminated water and food may differ in relation to ecological needs, metabolism and the contamination gradients of water, food and sediment, as well as other factors, such as salinity, temperature and interacting agents. The selected finfish species in the present study have different food preference and different behavioural pattern, e.g. *Liza parsia*, *Liza tade*, *Polynemus paradiseus* and *Stolephorus commersonii* are resident fish species in the study area, while *Tenualosa ilisha* exhibit migration from coastal region (~ salinity = 20 psu) to freshwater system in the upstream zone of the River Ganga for breeding. These factors may be attributed to species-wise variation of heavy metals in the study zone. The spatial variation of bioaccumulation followed the order Station 1 > Station 2 > Station 3 > Station 4, which may be related to different degree of contamination in different locations. All the heavy metals (except Zn in *Liza parsia* in station 1) were found to be lower than the recommended maximum level allowed in food as prescribed by the World Health Organization (WHO 1989). Furthermore the selected heavy metals in finfish muscle (except Zn in *Liza parsia* in station 1) were also within the permissible limits for human consumption as indicated by the Food and Agricultural Organization (FAO 1992).

Of the four metals studied in the present work, Zn and Cu are essential elements, while Pb and Cd are non-essential elements for most of the living organisms. The concentrations of Zn and Cu in all the finfish and shellfish species were relatively higher, compared to the concentration of other metals in same samples. It can be explained because these metals (Cu and Zn) are essential elements required by animals for metabolic process. Zn and Cu appear to diffuse passively (probably as a soluble complex) the gradients created by adsorption of membrane surfaces and are found in blood proteins metallothioneins. Carbonell and Tarazona (1994) concluded that different tissues of aquatic animals provide and/or synthesize non-exchangeable binding sites resulting in different accumulation levels. The primary sources of Zn in the present geographical locale are the galvanization units, paint manufacturing units and pharmaceutical processes, which are mainly concentrated in the Haldia industrial sector (opposite to station 1). Reports of high concentrations of Zn were also highlighted in the same environment by earlier workers (Mitra and Choudhury 1992, 1993; Mitra 1998).

The main sources of Cu in the coastal waters are antifouling paints (Goldberg 1975), particular type of algaecides used in different aquaculture farms, paint manufacturing units, pipe line corrosion and oil sludges (32–120 ppm). Ship bottom paint has been found to produce very high concentration of Cu in seawater and sediment in harbours of Great Britain and Southern California (Bellinger and Benham 1978; Young et al. 1979). In the present study area, the major source of Cu is the antifouling paints used for conditioning fishing vessels and trawlers apart from industrial discharges (i.e. predominant around station 1). This is the reason why Cu was detected in the fish samples of Stations 3 and 4, even if there is no existence of industries. The complete siltation of the Bidyadhari River also does not permit the industrial effluents released in the Hooghly River to mix with the rivers in the central sector of the deltaic complex (location zone of Stations 3 and 4).

Pb is a toxic heavy metal, which finds its way in coastal waters through the discharge of industrial waste waters, such as from painting, dyeing, battery manufacturing units and oil refineries (Mitra 1998). Antifouling paints used to prevent growth of marine organisms at the bottom of the boats and trawlers also contain Pb as an important component. These paints are designed to constantly leach toxic metals into the water to kill organisms that may attach to the bottom of the boats, which ultimately is transported to the sediment and aquatic compartments. Lead also enters the oceans and coastal waters both from terrestrial sources and atmosphere, and the atmospheric input of lead aerosols can be substantial. Station 1 is exposed to all these activities being proximal to the highly urbanized city of Kolkata, Howrah and the newly emerging Haldia port-cum-industrial complex, which may be attributed to high Pb concentrations in the finfish species.

The main sources of Cd in the present geographical locale are electroplating, manufacturing of Cd alloys, production of Ni-Cd batteries and welding (Mitra 1998). No trace of Cd was recorded in the fish muscle from Stations 3 and 4, which are located almost in industry-free zone surrounded by mangrove vegetation.

Apart from industrial discharges, significant difference in metal concentrations in biological samples between stations ($p < 0.01$) may also be related to contrasting physico-chemical characteristics between the western and central part of the Gangetic delta. The western part of the Gangetic delta is connected to Himalayan glacier through Bhagirathi River. Researchers pointed out that the glaciers in the Himalayan range are melting at the rate of 23 m/year (Hasnain 1999, 2000, 2002). This along with Farakka discharge has resulted in gradual freshening of the system (Mitra et al. 2009), which has role in elevation of dissolved metal level in the aquatic phase by way of lowering the salinity, pH and enhancing the process of dissolution of metallic species (Mitra 1998). The central sector on contrary is deprived from freshwater supply of Ganga-Bhagirathi system, and the Matla River is now tide fed with an increasing trend of salinity. This results in the precipitation of dissolved metals on the sediment bed (an event of compartmentation due to variation in metallic species) making the availability of metal low to the tissue of fish species thriving in the system (Mitra and Choudhury 1992, 1993; Mitra 1998).

The international official regulatory agencies have set limits for heavy metal concentrations above which the fish is considered unsuitable for human consumption. However in the Indian sub-continent, there is no safety level of heavy metals in fish tissue (recommended by the pollution control authority or any other regulatory body) although the Indian population is the major fish consumer in the tropics with a weekly average consumption of 1050 g and an annual rate of 55 kg per person. The presence of high concentrations of Zn, Cu, Pb and Cd in fish muscle of Station 1, in combination with the fact that fish is consumed by 85% of the people in the study area, calls for effective preventive measures to safeguard public health.

Annexure 6B: Common Types of Medicinal Plants with Salient Features and Ecosystem Services

Scientific name: ***Cymbopogon citratus***
Common name: Lemon grass/Gandhabena

Salient Features

- Aromatic, evergreen, clump-forming, perennial grass.
- Numerous stiff stems arising from a short rhizomatous rootstock and growing around 1.5 m tall.
- Green-coloured leaves turning red in the fall and emit a strong lemon fragrance when damaged/squeezed/scratched/crushed.

Ecosystem Services

- The grass is useful for soil improvement and erosion control.
- The heart of young shoots may be cooked and consumed as a vegetable.
- Leaves may also be used to make lemon grass tea.
- Oil extract of leaf is used in insect repellents, perfumes and soaps.

Scientific name: ***Cyperus rotundus***
Common name: Purple nutsedge

Salient Features

- Perennial plant that spread by extensive, horizontal, slender network of rhizomes which are white and fleshy and covered with scale leaves when young, becoming brown, fibrous or 'wiry' when old and grows up to 40 cm in height.
- Leaves are dark green above, with reddish-brown sheaths, clustered at the base of stem.
- Spikelet flowers are red brown to almost black in colour, and flower stems have a triangular cross-section.
- The nut fruit is yellow-coloured, ovate having three angles and becomes black on ripe.

Ecosystem Services

- According to the Ayurveda, root is pungent, acrid, cooling, astringent and appetizer and is used in the treatment of leprosy, thirst, fever, blood diseases, biliousness, dysentery, vomiting, epilepsy, ophthalmia, erysipelas, etc.
- The leaves are used in basketry and for weaving hats, matting, etc.

Scientific name: ***Stevia rebaudiana***
Common name: Sugar leaf

Salient Features

- Perennial herb.
- 30.5–80 cm (1–2.5 ft) in height.
- Leaves are oblong and aromatic.
- Flowers have five white petals.

Ecosystem Services

- Health benefiting plant; the derived phytochemical compounds help in controlling blood sugar, cholesterol and blood pressure in addition to its worth as a natural sweetener.
- Leaves are used as colour and flavour enhancer.
- Leaves have a pleasantly sweet and refreshing taste.

Scientific name: *Datura stramonium*
Common name: Jimson weed/datura

Salient Features

- Herbaceous green.
- Leaves are simple and alternate and have teeth at the edge of leaf blade.
- The leaves have an unpleasant scent when crushed or bruised.
- The fragrant flowers are trumpet-shaped, white to creamy or violet.
- The fruit is an ovoid, erect capsule.

Ecosystem Services

- Used as herbal medicine, especially in case of Ayurveda for asthma and bone setting; datura seeds and leaves are used as antiasthmatic, antispasmodic, hypnotic and narcotic.
- The leaves of datura are used to treat heart problems like palpitations and hypertension.

Scientific name: ***Pluchea lanceolata***
Common name: Sugandha/rasna

Salient Features

- Medicinal stout herb with 0.33–2 m in height.
- Stem is grooved, rough and very hairy.
- Leaves are elliptical, large and sessile with a tiny point at the tip.
- Flowers are yellow and many in heads that bloom during March–August.
- Fresh root (resemble in aroma of camphor) is brown and becomes greyish on drying.

Ecosystem Services

- The plant is used for the inflammations and bronchitis, psoriasis, cough and piles.
- It is also used as antipyretic, analgesic, laxative and nerve tonic.
- The leaves are aperients and used as a laxative, analgesic and antipyretic.

Scientific name: ***Elettaria cardamomum***
Common name: Choti elaichi

Salient Features

- Pungent, aromatic, herbaceous plant, growing to 2–4 m in height.
- Leaves are alternate in two ranks and dark green in colour.
- Pale green-coloured flowers are borne on flowering stalk.
- Fruits are pale green to yellow and elongate oval-shaped that has three chambers filled with small aromatic seeds.

Ecosystem Services

- In Ayurvedic medicine, it is used to treat disorders of the stomach and urinary system, asthma, bronchitis and heart problems.
- It is used by mixing neem and camphor as a nasal preparation in the treatment of cold and cough.
- It is added to tea to make a tonic to relieve the symptoms of stress.
- Its oil has cooling properties, and therefore it is used in cosmetics.
- It is highly valued in Kashmir as an essential ingredient of the fragrant drink *kahwa* and sweet Kashmiri black tea.

Scientific name: ***Zingiber officinale***
Common name: Ginger/adrak/ada

Salient Features

- Aromatic herb with an underground rhizome (edible) and an erect stem, up to 75 cm in height.
- The rhizome is brown with a corky outer layer and pale yellow-scented centre.
- Leaves are simple, alternate, linear-lanceolate, sheathing at the base, sessile, acuminate at apex.
- Flowering heads, borne on separate shorter stems, are cone-shaped spikes and composed of a series of greenish to yellowish leaf-like bracts.

Ecosystem Services

- Ginger is stimulant, carminative and diaphoretic.
- It is popularly used as a spice in cooking and to flavour beverages.
- It is grilled and used to flavour fish and meats or for making ginger tea.

Scientific name: ***Asparagus racemosus***
Common name: Satamuli/buttermilk root/shatavari

Salient Features

- Perennial climbing plant with straight and hooked spines, growing to 1–2 m in height.
- Leaves are thin and pine needles.
- White-coloured flowers having fragrance with small spikes.
- Small berries like fruits with purple to black in colour.

Ecosystem Services

- In Ayurvedic medicine, the root of shatavari is used in the form of juice, paste, decoction and powder to treat intrinsic haemorrhage, diarrhoea, piles, hoarseness of voice, cough, arthritis, poisoning, diseases of female genital tract, erysipelas and fever. It is an important medicinal plant that increases the white blood cell and reduces the risk of cardiovascular diseases.
- Saponin is the main component of shatavari that help in the oestrogen regulation and therefore help to control menstrual cycles.

Scientific name: ***Swertia chirayita***
Common name: Chirata

Salient Features

- An erect, annual herb whose stem is about 2–3 ft long, and the middle portion is round, while the upper is four-angled.
- The stems are orange brown or purplish in colour.
- Leaves are alternate, subsessile broadly lanceolate, acute at both ends.
- Flowering in Chirata is in the form of numerous small, axillary, opposite, lax cymes arranged as short branches.
- Flowers are small, stalked, green yellow, tinged with purple colour.

Ecosystem Services

- Chirata is a valuable bitter tonic.
- It has a great medicinal value as it possesses antidiarhoetic properties, and it is laxative and an appetizer.
- It is especially beneficial in the treatment of malarial fevers.
- It is also used in treating piles, skin diseases, ulcers and diabetes.
- Chirata has a huge role in the human system to create blood and thus can prevent anaemia.

Scientific name: ***Gymnema sylvestre***
Common name: Gudumari/gurmar

Salient Features

- Perennial, large climber and woody medicinal herb.
- Leaves are elliptic, acuminate, base acute to acuminate, glabrous above sparsely or densely tomentose beneath.
- Yellow-coloured flowers are small, bell-shaped and lateral umbel-like cymes that bloom during October–January.
- Fruits are slender, follicles and glabrous.

Ecosystem Services

- Leaves and roots are important in folk, Ayurvedic and homeopathic systems of medicine.
- Asthma, eye complaints and inflammations are cured by these leaves and also used for curing cough and cold.
- Roots are used as antidote to snakebite.
- Noted for high content of stored carbon.

Scientific name: *Mentha piperita*
Common name: Peppermint/pudina

Salient Features

- Strongly aromatic perennial herb.
- Square stem that grows to a height of between 40 cm and 80 cm and branches profusely at the top.
- Dark green-coloured leaves are lanceolate to elliptical, opposed and toothed margin.
- Flowers are purple in colour in the midsummer to late summer season.

Ecosystem Services

- Peppermint is used in medical science as nerve tonic and also to treat the symptoms such as nausea, vomiting, abdominal pain, indigestion, irritable bowel and bloating.
- Peppermint (*Mentha piperita*) leaves can be used as dried or fresh in teas for refreshing.
- Peppermint flowers are large nectar producers, and honeybees as well as other nectar harvesting organisms forage them heavily.
- Peppermint oil is widely used as remedy for skin irritation and itchiness.

Scientific name: *Grewia asiatica*
Common name: Pharosakoli/phalsa

Salient Features

- Shrub or small tree which can grow up to 12 ft high.
- Bark is greyish white or greyish brown.
- Leaves with serrated margins vary from broadly heart-shaped to obliquely ovate.
- Yellow-coloured flowers are observed.
- Fruit is edible, rounded, small dark blue to almost black when ripe.

Ecosystem Services

- The fresh leaves are important as fodder and cure pustular eruptions.
- Fruit is supposed to possess astringent, cooling and stomachic properties.
- Fruits are freshly consumed as dessert which is made into syrup and extensively employed in the manufacture of soft drinks.
- Yellow white-coloured wood is fine-grained, strong and flexible which is used for archers' bows, spear handles, shingles and poles for carrying loads on the shoulders.

Scientific name: ***Bougainvillea glabra***
Common name: Baganbilas/paper flower

Salient Features

- Evergreen, climbing shrub with thorny stems.
- Usually grows up to 3–4 m (10–12 feet) tall.
- Leaves are oval, sparse and coloured, but they have decorative papery bracts surrounding small cream white-coloured flowers.

Ecosystem Services

- Leaves are used in medical science for the treatment of anti-inflammation, diarrhoea, stomach and diabetes.
- Flowers are involved in reducing the blood pressure.
- It is used as indoor house plant.

Scientific name: ***Alangium salviifolium***
Common name: Ankol

Salient Features

- Deciduous small medicinal shrub grows up to 3–10 m in height.
- Bark is pale brown colour with rough surface and trunk with numerous holes.
- Leaves are long, alternate, simple, without stipules, narrowly oblong or ovate-lanceolate and glabrous.
- Flowers are bisexual, regular and white in colour.
- Fruits are small, nearly globular, purplish red.

Ecosystem Services

- Leaves, roots and fruits are used for the treatment of rheumatism, leprosy and haemorrhoid. Also it is used for the treatment of bites by rats and dogs.
- Fruits are sweet, cooling and purgative which used as a poultice for treating burning sensation and haemorrhage.
- Stem is used in curing vomiting and diarrhoea.

Scientific name: *Holarrhena antidysenterica*
Common name: Indrajao/kurchi/dhulkari

Salient Features

- Large shrub that is found throughout the deciduous forest area at low elevation.
- Bark is rough, pale brownish or greyish in colour and releases milk.
- Leaves are opposite oval which are arranged in reverse directions.
- White-coloured flowers are borne in corymbose cymes with mild fragrance during April–July.
- Fruits are paired and cylindrical shape, dehiscing follicles hang out from the stalk during August–October.

Ecosystem Services

- It is medicinal plant and therefore used in the treatment of anaemia, colic pain, diarrhoea, haematuria and obstetric conditions.
- Powdered seeds mixed with honey given in chronic chest affection, asthma and colic pain.
- Leaves are used in curing chronic bronchitis.

Scientific name: ***Acacia torta***
Common name: Twisted acacia

Salient Features

- Large climbing shrub that grows up to 5–6 m.
- Stem is prickly and pubescent at young stage and prominently fluted and angled when old.
- Leaves are double-pinnate, and leaflets are close enough to touch each other.
- White-creamish spherical flower heads like bract.
- Pod is velvety brown when young and glabrous; at later stage it becomes flat and reddish brown.

Ecosystem Services

- Medicinal shrub that is used in cough, bronchitis, measles and tubercular fistula and also in the treatment of menstrual disorders.
- Vegetative parts like leaves, shoot and woods are used as the raw material of the medicines for skin diseases, haemorrhoids, bleeding and wounds.

Scientific name: ***Clianthus puniceus***
Common name: Kaka beak

Salient Features

- Soft woody shrub with 2 m height.
- Leaves are imparipinnate with 15–20 pairs of subsessile leaflets which are dull green to grey green.
- Flowers are scarlet, pink or entirely white with standard ovate-acuminate.

Ecosystem Services

- Attracts insects and butterflies
- Unique storehouse of carbon and nitrogen and has considerable importance in the domain of climate change mitigation

Scientific name: ***Solanum ptycanthum***
Common name: Eastern black nightshade

Salient Features

- Annual plant growing to 0.6 m in height.
- The stems are glabrous to slightly short-hairy, but not prickly.
- Leaves are alternate, a dull green to dark green, long ovate, with pointed tips and long slender grooved stalks and glabrous to slightly short-hairy.
- Flowers are 1/4 to 1/3 inch wide and each on a 1 inch stalk. The corolla is white to purple tinged, wheel or star shaped with lobes that reflex backward at maturity but not to the extent of climbing nightshade.

Ecosystem Services

- The whole plant acts as purgative and sedative.
- Leaves, stems and roots are used externally as a poultice, wash, etc. in the treatment of cancerous sores, boils, leucoderma and wounds.
- The plant has been used in the manufacture of locally analgesic ointments.

Scientific name: ***Commiphora caudata***
Common name: Guggul

Salient Features

- Small- to medium-sized deciduous medicinal plant.
- Smooth and succulent green bark which partly flakes off with age.
- Greenish to cream yellow flowers are pedestal with pink to red petals.
- Fruit is a globose fleshy drupe with two to six valves and contains single seed that is black and has four wings.

Ecosystem Services

- Rasinous gum from the tree is astringent which is used in remedy of wound, ulcer and skin disease.
- Fruits are rich in carotenoid.

Scientific name: ***Plectranthus barbatus***
Common name: Patharchur

Salient Features

- Perennial plant that grows to about 45–60 cm in height.
- Leaves are ovate to oblong, blunt, rounded-toothed, short-stalked and arranged in opposite pairs perpendicular to each other.
- Pale blue flowers arranged in whorls, forming long leafless interrupted spikes.
- The root is typically golden brown, thick, fibrous and radially spreading.

Ecosystem Services

- It is a remedy for malaria, and to cure liver and stomach diseases as the they have antibacterial, antiviral and antifungal activity, and antiprotozoic activity against protozoa, such as Giardia.
- The leaves are used for curing conjunctivitis and earache.
- Plant stores substantial amount of carbon and nitrogen.

Scientific name: ***Ocimum kilimandscharicum***
Common name: Karpura tulsi

Salient Features

- This species has a strong but less pleasant flavour. It is an aromatic under shrub.
- Perennial, aromatic evergreen under shrub.
- Pale green leaves are glandular, ovate or oblong in shape, oppositely arranged, deeply serrated and pubescent on both surfaces.
- Flowers are creamy white in colour.

Ecosystem Services

- Economically important perennial herb in different sectors like medicinal, agriculture, industrial and ornamental sectors.
- Leaves are good remedy for cold, cough, malaria, abdominal pains and diarrhoea.
- Local applications on sprains and in various dental and oral preparations of drugs are cured by the oil that extracted from leaves.
- Essential oils possess biologically active constituents that act as insect repellents, particularly against mosquitoes and storage pests.

Scientific name: ***Cascabela thevetia***
Common name: Yellow oleander/kolkaphul

Salient Features

- Large shrub or small tree that grows up to 10 to 20 ft. in height.
- Linear and oleander-like leaves are closely spaced and spirally arranged along the stem.
- Large and yellow-coloured flowers are funnel-shaped and sweetly scented which open in the morning and close in the afternoon.
- Triangular fruit is initially green but turns black, soft and juicy like a plum at maturity and contains a stone with two kernels.

Ecosystem Services

- Used as ornamental plant.
- Planted as flowering shrub (aesthetic value).
- Seed oil is used to make a paint having antifungal, antibacterial and anti-termite properties.

Scientific name: ***Ziziphus mauritiana***
Common name: Ber

Salient Features

- Small- or medium-sized subtropical tree.
- Leaves are ovate, petiolate with very fine serration.
- Dark green and shining from above part of leaf and white tomentose from beneath part.
- Flowers are yellowish.
- Fruits are fleshy drupes, ovoid or oblong and dark reddish brown when ripe.

Ecosystem Services

- Used as a folk remedy for curing anxiety, asthma, diarrhoea, fatigue, fever and blood pressure.
- The pulp of dried fruit is recommended in Ayurveda for treating burning sensation, blood impurities, excessive thirst and anorexia.

Scientific name: ***Cinnamomum zeylanicum***
Common name: Daruchini

Salient Features

- Evergreen tree up to 30 ft. tall.
- Strong branches and thick scabrous yellow-coloured bark.
- Leaves are dark green on top and lighter green underneath.
- It has small yellowish-white flowers.
- Fruit is ovate or oblong.

Ecosystem Services

- Bark is used in cookery as a condiment and flavouring material.
- The paste made of bark with lemon juice is used in the treatment of pimples as herbal medicine.
- The chewing of bark is very helpful to prevent foul smell from the mouth.

Scientific name: ***Pongamia pinnata***
Common name: Karanj

Salient Features

- Medium-sized evergreen glabrous shrub or tree up to 25 m height.
- Bark is greyish brown and smooth.
- Leaves are alternate, imparipinnate with long slender leafstalk.
- Flowers are 2–4 together, short-stalked and pea-shaped.

Ecosystem Services

- Used as an ornamental plant.
- It is capable of fixing atmospheric nitrogen; thus it can be used as green manure.
- Flowers are good source of pollen for honeybees in India, and they yield adequate nectar.
- Root juice is a good remedy for cleaning the teeth.
- The wood is used in cabinet making, agricultural implements, tool handles, paper making, etc.

Scientific name: ***Annona reticulata***
Common name: Custard apple

Salient Features

- Small deciduous or semievergreen tree with 10 m in height.
- Leaves are deciduous, alternate and oblong with conspicuous veins.
- Light-green flowers (in drooping clusters) are fragrant, slender, with 3 outer fleshy, narrow petals and never fully open.
- Fruits are compound, symmetrically heart-shaped, lopsided, or irregular or nearly round, or oblate, with a deep or shallow depression at the base.

Ecosystem Services

- The leaf decoction is given as a vermifuge.
- Crushed leaves are poulticed on boils, abscesses and ulcers.
- Tannin-enriched unripe fruit is employed against diarrhoea and dysentery.
- The bark is very astringent, and the decoction is taken as a tonic.
- Fragments of the root bark are packed around the gums to relieve toothache.

Scientific name: ***Putranjiva roxburghii***
Common name: Putranjiva/jioysuta

Salient Features

- Evergreen tree that grows up to 12 m in height.
- Bark is smooth, corky, grey-coloured and white-dotted.
- Leaves are simple, alternately arranged, shiny dark green and distantly serrated.
- Male flowers are found with short-rounded axillary clusters, and female flowers are in 1–3 in leaf axil.
- Fruit is a yellowish olive-shaped or rounded drupe.

Ecosystem Services

- Decoction of leaves and fruits is used in the treatment of liver complaints, colds, fevers and rheumatism.
- The wood is used for construction, turnery and tool handles.
- Necklaces made from hard and white seed of the fruit are traditionally used for the protection from liver diseases.

Scientific name: *Celtis laevigata*
Common name: Sugar berry

Salient Features

- Very large, broad, fast-growing deciduous tree.
- Bark with grey brown to silver colour has some warty projections or corky ridges.
- Leaves are simple, alternate and serrate green-coloured.
- Fruits are red- and black-coloured with round shape.

Ecosystem Services

- This tree may be used as a lawn tree or street tree.
- The tiny, berry-like, sweet fruits attract many birds and are used in any natural landscape setting for aesthetic beauty.
- Potential sink of carbon and nitrogen.

Scientific name: ***Abrus precatorius***
Common name: Gunchi

Salient Features

- Slender, perennial climber that twines around trees.
- Leaves are glabrous with long internodes.
- Flowers are small and pale violet in colour with a short stalk, arranged in clusters.
- Fruit is flat, oblong and truncate-shaped with a sharp-deflexed beak.

Ecosystem Services

- The purified *Abrus* is a nerve stimulant and is used to cure disorders like joint pains and paralysis.
- The paste of seeds is applied locally in alopecia and skin diseases.
- Digestive disorder is treated by its root.
- The paste of the leaves is used in medicinal sector in the treatment of leukoderma, swellings, boils, rheumatism, etc.

Scientific name: ***Acacia catechu***
Common name: Cutch tree/khayer/black cutch

Salient Features
- Deciduous tree with a light feathery crown
- Dark brown, glabrous, slender, thorny, shining branchlets

Ecosystem Services
- It is a fodder tree for goat and cattle.
- It is used as fuel and is a source of charcoal of good quality.
- It is used in food and beverages as a flavouring agent.
- Catechu is most commonly used by mouth for stomach problems such as diarrhoea, swelling of the colon (colitis) and indigestion.
- It is chiefly used as a source of katha and kutch and also serves as a useful timber.
- Source of a very good quality gum.
- Storehouse of carbon.

Scientific name: ***Aegle marmelos***
Common name: Bel

Salient Features

- Slow-growing deciduous, spinous and aromatic tree.
- Pale green leaves are alternate, trifoliate and leaflet at terminal.
- Sweet-aromated flowers are greenish white.
- Fruits are large, woody, greenish yellow and globular in shape.
- Seeds are compressed, oblong and numerous found in aromatic pulp.

Ecosystem Services

- Bel leaves is widely used in worship of Lord Shiva.
- It is a medicinal plant that used in curing diabetes, dyspepsia, chronic diarrhoea, heart diseases and dysentery.
- The fruit is eaten fresh as juice (to make drink and squashes) or dried and used commercially in jam making.
- It is used in the preparation of candy, toffee, pulp powder and other edible products.
- It cleans the atmosphere as it helps to purify the atmosphere by absorbing the harmful gases.

Scientific name: *Terminalia elliptica*
Common name: Asan/sahaju

Salient Features

- Deciduous tree, growing up to 32 m.
- Bark is deep longitudinal fissures and transverse cracks, grey black.
- Leaves are elliptic to ovate with one or two glands at lamina and petiole junction.
- Flowers in panicles, terminal and axillary, small and dull yellow.
- Fruits a drupe with five broad wings.

Ecosystem Services

- Plants have medicinal properties like antifungal, anti-hyperglycaemic, antidiarrhoeal and antileucorrhoeal.
- Wood is used for furniture, cabinetwork, joinery, paneling, specialty items, boat building, railroad crossties and decorative veneers.
- Its bark is an important resource in leather industry.
- Leaves are used in sericulture, for feeding the silkworms.
- Potential storehouse of carbon.

Scientific name: ***Madhuca longifolia***
Common name: Mahua

Salient Features

- Medium-sized deciduous tree, ~15 m height with short, stout trunk.
- The bark is grey, vertically cracked and wrinkled.
- Leaves are alternate and clustered at the end of branchlets.
- Flowers are borne on green or pink, furry bunches, each bunch consisting of 12 fragrant cream-coloured flowers.
- The flowers live only one night and then fall to the ground.
- Pollinated flowers develop into a fleshy, greenish ovoid fruit containing 1–4 shiny, oily brown seeds.

Ecosystem Services

- It is a medicinal tree whose bark is used to cure leprosy and to heal wounds. The flowers are prepared to relieve coughs, biliousness and heart trouble, while the fruit is given in cases of consumption and blood diseases.
- Its flowers produce a nectar that is very valuable to honeybees in periods of scarcity.
- The sweet, fleshy flowers are used as sweetener or fermented to produce alcohol.

Scientific name: ***Cinnamomum tamala***
Common name: Tejpata

Salient Features

- Moderate-sized evergreen tree attaining a height of ~8 m.
- Bark is thick, smooth and reddish brown and produces mucilage.
- Leaves are opposite, ovate or ovate-lanceolate, hard and coriaceous and glabrous with strong odour when it crushed.
- Flowers are found in cluster and have a strong odour during January.
- Purple-coloured fruits are round in shape with a single seed inside, seen during May–August.

Ecosystem Services

- Dried leaves are used as a common ingredient of Indian cooking.
- Inner bark of the tree constitutes drug cinnamon which is used in curing diarrhoea, nausea and vomiting.
- Leaves and bark have aromatic, astringent, stimulant and carminative qualities and therefore used in curing rheumatism, colic, diarrhoea, nausea and vomiting.
- Bark is used as herbal medicine in the treatment of pimples by its paste with lemon juice.
- Seed oil is used as flavouring agent in food, toothpaste and perfume.

Scientific name: *Bauhinia tomentosa*
Common name: Yellow orchid tree

Salient Features

- Medium-sized tree grows up to 4 m in height.
- Bark is grey or brown.
- Leaves are divided into two lobes with a leathery texture.
- Large bell-shaped, bright yellow flowers that bloom during December–March.
- Light-green fruits are pea-like pods, slender and velvety which turn into pale brown with age.

Ecosystem Services

- Flowers are used in curing dysentery and diarrhoea.
- Seed is also used by pasting with vinegar as a local application to the wounds produced by venomous animals.
- The wood is used to make rafters for huts.
- The swelling of liver can be treated by consuming an extract of *Bauhinia tomentosa.*
- Unique storehouse of carbon and nitrogen and has considerable importance in the domain of climate change mitigation.

Scientific name: ***Dracontomelon dao***
Common name: Dahu

Salient Features

- Deciduous and large evergreen tree up to 43–55 m high.
- Bark is greyish brown.
- Leaves are compound-pinnate, spirally arranged around branchlet.
- Flowers are white, small and fragrant.
- Fruits are slightly fleshy drupes that produced in clusters.

Ecosystem Services

- Tree is used as ornamental tree which could be used effectively for avenues.
- The bark has the great medicinal value and used in the treatment of dysentery.
- Flowers and leaves are cooked and eaten as a vegetable or used as a food flavouring.
- The timber is soft and not very durable. It is used for veneer, paneling, furniture, flooring, matches and boxes.

Scientific name: ***Terminalia chebula***
Common name: Harida/chebulic myrobalan/haritaki

Salient Features

- Medium- to large-sized deciduous tree with umbrella-shaped crown and crowded branches and height reaches up to 25–30 m.
- Bark is dark brown with longitudinal cracks.
- Leaves are oval or elliptic and sub-opposite.
- Flowers are yellowish white with strong and unpleasant odour.
- Fruit (drupe) is yellowish green, obovoid or ellipsoid and hard.
- Seed is globose and pale yellow in colour.

Ecosystem Services

- It is a medicinal herb that uses in skin disorders, anaemia, marcosis, piles, heart diseases, diarrhoea, cough and excessive secretion of mucus.
- Used in constipation, abdominal pain and distension.
- Fruit stimulates the digestion and increases intelligence and also purifies the digestive tract.
- Fruit paste is used externally to reduce inflammation, cleansing and quick healing of wounds.

Scientific name: ***Isoberlinia doka***
Common name: Doka

Salient Features

- Woodland tree, grows up to 20 m in height.
- The leaflets are arranged in three or four pairs.
- The flowers are small and white that form large open inflorescences which are held conspicuously above the leafy crown.
- The pods are oblong, flat and quite large.

Ecosystem Services

- It is potentially useful for land reclamation and reforestation schemes on account of its vigour and ability to colonize abandoned land.
- Economically this is valuable tree as its timber is used in carpentry and making furniture.
- In traditional medicine it has an essential role in the treatment of muscular-skeletal system disorders, infectious diseases, etc.

Scientific name: ***Alstonia scholaris***
Common name: Chatim/devil's tree

Salient Features

- Glabrous tree, grows up to 40 m in height.
- The upper side of the leaves is glossy, while the underside is greyish.
- The leaves are leathery and narrowly obovate to very narrowly spathulate, base cuneate, apex usually rounded.

Ecosystem Services

- Used in medical science as a tonic and drug for digestive system.
- Bark is bitter, astringent that lowers fevers, relaxes spasms, stimulates lactation and expels intestinal worms.
- Bark paste can be applied on wounds, whereas its powder is applied to relieve pain in the abdomen and lumps.
- The twigs of the plant have the antibacterial properties and therefore used as a toothbrush.
- The fruits of the plant relieve psychosis as well as epilepsy.
- The powder of the flower is applied in case of headaches.

Scientific name: ***Adina cordifolia***
Common name: Karam/haldu/karma

Salient Features
- Large deciduous tree up to 40 m tall.
- Leaves opposite and heart-shaped.
- Flowers are yellow-coloured with round heads on long stalks.
- Fruits are numerous with round heads.

Ecosystem Services
- The tree is an important source of timber.
- Trees serve as source of fuel.
- Bark possesses medicinal properties and acts as an antiseptic.
- Noted for carbon storage potential.

Scientific name: ***Bauhinia purpurea***
Common name: Butterfly tree/purple orchid tree

Salient Features

- Small- to medium-sized deciduous tree with bushy crown.
- The bark is pale grey brown with fairly smooth to slightly fissured and scaly.
- Leaves are simple, alternate, long and broad, rounded and bilobed at the base and apex.
- Flowers are conspicuous, pink and fragrant, with five petals.
- The fruit is a long pod in which 12–16 seeds are found.

Ecosystem Services

- The bark is the remedy for diarrhoea (acts as an astringent). A decoction of this astringent bark is also recommended to wash ulcers; a paste of the bark mixed with flowers and rice water is used to treat boils and abscesses.

Scientific name: ***Diospyros melanoxylon***
Common name: Bada kendu

Salient Features

- Medium-sized tree or shrub that grows up to 25 m.
- The bark is pelican in colour.
- Leaves are opposite, alternate and coriaceous.
- Male flowers are mauve in colour, tetramerous to sextamerous.
- Female flowers are mauve, mostly extra-axillary.
- Olive green-coloured fruits are ovoid or globose.
- Oblong-shaped seeds are compressed and shiny.

Ecosystem Services

- The leaves are raw material for the making of bidis.
- The dried fruit contents high percentage of tannin which helps in the treatment of urinary, skin and blood diseases.
- Leaves are used in the treatment of scabies and old wounds and act as laxative and carminative medicine.

Scientific name: ***Semecarpus anacardium***
Common name: Anchu

Salient Features

- Medium- to large-sized medicinal tree with grey bark exfoliating in small irregular flakes and grows up to 15–25 m.
- Leaves are simple, alternate and oblong, rounded at the apex.
- Flowers are greenish white, in panicles.
- Fruit nut is a drupe, ovoid and smooth lustrous black.

Ecosystem Services

- The fruit is useful in the treatment of leucoderma, scaly skin, allergic, dermatitis, poisonous bites, leprosy, cough and asthma.
- It is extremely beneficial in the diseases like piles, colitis, diarrhoea, dyspepsia, ascites, tumours and worms.
- The topical application of its oil on swollen joints and traumatic wounds effectively controls the pain.
- Seed (known as godambi) is eaten by Indians in winter and was commonly used as a method of birth control for women.
- Stores carbon and nitrogen in considerable levels in the vegetative parts.

Scientific name: *Ficus racemosa*
Common name: Dumur/cluster fig/gular

Salient Features

- Evergreen and large-sized spreading, lactiferous, deciduous tree.
- Its surface is reddish brown with smooth, coarsely flaky, fibrous, milky latex, white hairy young shoots and twigs.
- Leaves are simple, alternate and dark green-coloured with ovate oblong, or elliptic-lanceolate.
- Flowers are unisexual and inflorescence a syconia, on short leafless branches or warty tubercles of trunk or on larger branches.
- Fruits are receptacles shortly pedunculate, on short leafless branches. Ripe fruit is smooth or pubescent obovoid and reddish.

Ecosystem Services

- Medicinal plant whose bark is astringent and infusion of bark is employed as mouthwash in spongy gum condition, dysentery, menorrhea and diabetes.
- Paste of stem bark is used in the treatment of burns, swelling and leucorrhoea.
- Root helps in the treatment of diarrhoea and diabetes.

Scientific name: ***Anogeissus latifolia***
Common name: Axle wood tree/dhaoya/dhavada

Salient Features

- Medium-sized tree with a straight and cylindrical hole.
- Smooth bark with scales pale to dark grey in colour and branches are drooping.
- Leaves are opposite or sub-opposite, simple with greyish yellow or whitish hairs below.
- Small and pale yellow stalkless flowers are borne in dense, spherical heads on stalks in leaf axils or at the end of branches.
- The fruit is a two-winged pseudo-achene, packed into a dense head with a single seed.

Ecosystem Services

- Used as fodder for cattle, buffaloes and other ruminants.
- Its timber gives a heavy hardwood, good charcoal and firewood.
- Bark powder is applied as paste to treat wounds.
- It also involved in the medical sector to treat toothache, dysuria, diarrhoea, cold, cough, spleen enlargement, cholera, piles, urinary disorders and poisonous bites (scorpion sting and snakebite) as have the antimicrobial, antibacterial and antifungal properties.

Scientific name: ***Diospyros chloroxylon***
Common name: Kinnu/kala tendu

Salient Features

- Medium-sized deciduous tree.
- Dark brown-coloured bark, peeling-off in small scales, branchlets woolly.
- Leaves are simple, bright green-coloured above and paler beneath with leathery and velvet hairy on both sides.
- Flowers are small, white and borne in cymes in leaf axils.
- Fruit is globose, size of a cherry and purplish when ripe, is eaten and is very palatable.

Ecosystem Services

- The leaves are used as the wrapping around tobacco to make bidi, cigarettes, etc.
- The wood is hard, whitish pink, tough and fairly durable which is used for building, shoulder poles, mine props and shafts of carriages.
- The fruits are edible as yellow pulp and are soft and sweet.
- The dried flowers are useful in treating urinary, skin and blood diseases.

Scientific name: ***Tecoma stans***
Common name: Yellow bells/chandaprabha

Salient Features

- Small tree that grows up to 4 m in height.
- Bark on the main stem is pale brown to grey and roughens.
- Leaves are opposite and imparipinnate with five to seven leaflets which are lanceolate, pointed at both ends, and toothed on the margins.
- Pale to bright yellow-coloured flowers with faint orange stripes at the throat (in clusters at the ends of the branches) are trumpet-shaped with five-rounded lobes.
- Fruits are large, elongated, flattened capsules containing many winged seeds and turn from green to brown in colour as they mature.

Ecosystem Services

- It is an ornamental and herbal plant.
- Bark has smooth muscle relaxant, mild cardiotonic and chloretic activity.
- Roots are powerful diuretic and vermifuge.

Mixture of the grinded root and lemon juice is used as an external application and also taken internally in small quantities as a remedy for snakebites, rat bites and scorpion sting.

Scientific name: ***Corymbia calophylla***
Common name: Marri

Salient Features

- Trees may grow up to 40 m depending on the edaphic factors.
- The bark is rough (tessellated), brown to grey brown and often has exudations of reddish gum.
- Leaves are disjunct, glossy, green, thick and concolorous with a broad lanceolate to ovate shape that is basally tapered or basally rounded.

Ecosystem Services

- It is not suitable for most gardens or street plantings, but it is an excellent tree for paddock plantings providing shade for animals.
- Stores considerable levels of carbon and nitrogen in the vegetative parts.

References

Bellinger E, Benhem B (1978) The levels of metals in dockyard sediments with particular reference to the contributions from ship bottom paints. Environ Poll Assess 15(1):71–81

Boudouresque CF, Verlaque M (2002) Biological pollution in the Mediterranean Sea: invasive versus introduced macrophytes. Mar Poll Bull 44:32–38

Carbonell G, Tarazona JV (1994) Toxicokinetics of cu. Aqua Toxicol 29:213–221

Castro H, Aguilera PA, Martinez JL, Carrique EL (1999) Differentiation of clams from fishing areas an approximation to coastal quality assessment. Environ Monit Assess 54:229–237

Chaudhuri AB, Choudhury A (1994) Mangroves of the Sundarbans, India, vol-I, 1st edn. IUCN, Gland

Cohen T, Hee S, Ambrose R (2001) Trace metals in fish and invertebrates of three California coastal wetlands. Mar Poll Bull 42:232–242

Date AR, Gray AL (eds) (1988) Applications of inductively coupled plasma source mass spectrometry. Blackie, Glassgow

De la Guardia M (ed) (1990) Empleo de losHornos de Microondas en Quimica. University of Valencia, Valencia

FAO (1992) FAO/WHO, Food standard programme. 2nd ed. Codex Alimentarius Commission 1:114–190

Fergusson JE (1990) The heavy elements: chemistry, environmental impact and health effects. Pergamon Press, New York, p 614

Garcia-Montelongo F, Díaz C, Galindo L, Larrechi MS, Rius X (1994) Heavy metals in three fish species from the coastal waters of Santa Cruz de Tenerife (Canary Islands). Sci Mar 58:179–183

Goldberg ED (1975) The mussel watch – a first step in global marine monitoring. Mar Poll Bull 6:111

Grosell M, Brix KV (2005) Introduction to the special issue on mechanisms in metal toxicology. Aqua Toxicol 72:3–4

Hasnain SI (1999) Himalayan glaciers: hydrology and hydrochemistry, 1st edn. Allied Publication Limited, New Delhi

Hasnain SI (2000) Status of the glacier research in the HKH region. ICIMOD, Kathmandu

Hasnain SI (2002) Himalayan glaciers meltdown: impact on south Asian Rivers. IAHS 7:274

Islam MD, Tanaka M (2004) Impact of pollution on coastal and marine ecosystems including coastal and marine fisheries and approach for management: a review and synthesis. Mar Poll Bull 48:624–649

Iyengar GV (1991) Milestones in biological trace elements research. Sci Total Environ 1:100

Jordao CP, Pereira MG, Bellato CR, Pereira JL, Matos AT (2002) Assessment of water systems for contaminants from domestic and industrial sewages. Environ Monit Assess 79:75–100

Krishnamurti JA, Nair RV (1999) Concentration of metals in shrimps and crabs from thane-Bassein creek system. Maharastra Ind J Mar Sci 28:92–95

Matusiewicz H, Sturgeon RE (1989) Present status of microwave sample dissolution and decomposition for elemental analysis. Prog Anal Spectrom 12:21

Mitra A (1998) Status of coastal pollution in West Bengal with special reference to heavy metals. J Ind Ocean Stud 5(2):135–138

Mitra A, Choudhury A (1992) Trace metals in macrobenthic molluscs of the Hooghly estuary. India Mar Poll Bull UK 26(9):521–522

Mitra A, Choudhury A (1993) Seasonal variations in metal content in the gastropod *Nerita articulata* (*Gould*). Ind J Environ Health NEERI 35(1):31–35

Mitra A, Banerjee K, Sengupta K, Gangopadhyay A (2009) Pulse of climate change in Indian Sundarbans: a myth or reality? Natl Acad Sci Lett 32(2):1–2

Nadkarni RA (1984) Applications of microwave oven sample dissolution in analysis. Anal Chem 56:22–33

Peerzada N, McMorrow L, Skiliros S, Guinea M, Ryan P (1990) Distribution of heavy metals in Gove harbour, Northern Territory, Australia. Sci. Total Environ 92:1–12

Romeo M, Siaub Y, Sidoumou Z, Gnassia-Barelli M (1999) Heavy metal distribution in different fish species from the Mauritania coast. Sci. Total Environ 232:169–175

Subramanian R, Sukumar A (1988) Biological reference materials and analysis of toxic elements. Fresenius Z Anal Chem 332:623–626

UNEP (1982) Pollution and the marine environment in the Indian Ocean. UNEP Regional Seas Programme Activity Centre. Geneva, Switzerland

Usero J, Morillo J, Graccia I (2005) Heavy metal concentrations in molluscs from the Atlantic coast of southern Spain. Chemosphere 59:1175–1181

World Health Organization (1989) Heavy metals environmental aspects, Environmental Health Criteria. No. 85. Switzerland, Geneva

Young DR, Alexander GV, McDermott-Ehrlich D (1979) Vessel related contamination of southern California harbours by copper and other metals. Mar Pollut Bull 10:50–56

Yusof AM, Yanta NF, Wood AKH (2004) The use of bivalves as bio-indicators in the assessment of marine pollution along a coastal area. J Radioanal Nuclear Chem 259(1):119–127

Chapter 7
Air Pollution and Its Mitigation

Abstract Air is vital for sustaining life on Earth. It is composed of several gases with a definite proportion. The proportionality is disrupted due to intense industrial activities, change of land-use pattern, deforestation, unplanned urbanization, etc. that lead to air pollution. Polluted air has several negative impacts on the survival and health of human beings. All these issues have been discussed in the chapter along with the carbon storage potential of producer community that performs the ecosystem service perfectly by balancing the carbon dioxide budget in the atmosphere.

Keywords Pollutants · Suspended particulate matter (SPM) · Human health · Carbon sequestration

7.1 Air Pollution: An Overview

The vast expanse of air which envelopes the Earth is called atmosphere. The atmosphere is a mixture of many discrete gases, in which varying quantities of tiny solid particles are suspended. With increasing height above sea level, the atmospheric pressure gradually decreases. It consists of a mixture of gases having a relatively uniform composition in the lower layers close to the Earth. Pure dry air on an average constitutes mainly of nitrogen (78%) and oxygen (21%) making thus 99% of the air by volume, and the remaining 1% accounts for gases like argon (0.93%), carbon dioxide (0.03%), hydrogen, helium and ozone. The relative proportions of different gases are shown in Table 7.1.

Besides gases, water vapours, dust particles, smoke, salts and other impurities, microorganisms, pollen grains, etc. are also present in the air in varying quantities in the lower layers of atmosphere. As a result, the composition of air is never constant and varies from time to time and place to place. However, if these variable elements are removed from the atmosphere, its composition will remain fairly constant all over the Earth, at least in the lower atmosphere. Of the many constituents, carbon dioxide, dust particles, water vapour and ozone are of great importance for the Earth's climatic conditions. However the concentrations of different gases

A. Mitra, S. Zaman, *Environmental Science - A Ground Zero Observation on the Indian Subcontinent*, https://doi.org/10.1007/978-3-030-49131-4_7

Table 7.1 Relative proportion of various gases (by volume) in the atmosphere

Gases	%	Gases	%
Nitrogen	78.0841	Krypton	0.00011
Oxygen	20.9486	Xenon	0.00009
Argon	0.9340	Hydrogen	0.00006
Carbon dioxide	0.0318	Methane	0.00020
Neon	0.0018	Nitrous oxide	0.00005
Helium	0.0005	Ozone	0.000004

exhibit variations due to deterioration of the air quality. This leads to air pollution which makes the air unfit for the survival of the organisms.

According to WHO, air pollution is defined as '...Phenomenon in which substances put into air by the activity of mankind into concentration sufficient to cause harmful effect to his health, vegetables, property or interfere with the enjoyment of his property...'

The various sources of air pollution are highlighted here:

7.1.1 Industrial Pollutants

CO, CO_2, SO_2, H_2S and hydrocarbons are most common air pollutants which are discharged into air from industrial chimneys and power houses. These gases are produced due to burning of fossil fuels and by combustion of lignite at thermal power stations. Various fluorine compounds (HF, F_2, SiF_4, H_2SiF_6, etc.) are emitted from phosphate fertilizer manufacturing plants (since large amounts of fluorine are present in phosphate rock), aluminium extraction (where cryolite, i.e. Na_3AlF_6, is used to lower the melting point of the mixture of CaF_2, Al_2O_3 and Na_3AlF_6), ceramic firing, steel making and some chemical processing. Many metallurgical processes release dust and fumes which are loaded with Pb, Cr, Ni, etc. Many chemical industries release hydrochloric acid, chlorine, oxides of nitrogen, Zn, Pb, As, oxides of Cu, in addition to SO_2, CO and H_2S. It has been observed that industrial processors like metallurgical plants and smelters, chemical plants, petroleum refineries, pulp and paper mills, sugar mills and cotton mills and synthetic rubber manufacturing plants are responsible for about one-fifth of the air pollution.

Apart from these, cement manufacturing is one of the largest causes of human-induced carbon dioxide emissions. The gas is produced when calcium carbonate ($CaCO_3$) is heated to produce the calcium oxide (CaO, also called quicklime), which is the main ingredient for cement ($CaSiO_3$) production. The reaction steps in this manufacture are:

$$CaCO_3 \rightarrow CaO + CO_2$$

$$CaO + SiO_2 \xrightarrow{\Delta} CaSiO_3$$

CO_2 produced in the first step has considerable contribution to global warming. While fossil fuel combustion and deforestation each produce significantly more carbon dioxide in the Earth's atmosphere, cement production alone is responsible for approximately 2.5% of total worldwide emissions from industrial sources.

7.1.2 Automobiles

Automobiles (cars, scooters, motor cycle, etc.) have been regarded as the greatest sources of air pollution. They produce nearly two-thirds of the carbon monoxide and one-half of the hydrocarbons and nitrous oxides. The automobile exhaust has also leaded gas and particulate lead. The combustion of petroleum emits particulate lead compounds. The particulate lead compounds emitted from petroleum combustion have profound adverse influence on biotic community.

Benzene, toluene and xylene are the three volatile organic compounds (VOC) present in urban atmosphere mainly due to use of petrol and diesel in automobiles. Out of these three organic compounds, the concentrations of benzene in air showed strong correlations with the incidence of cancer. The limit value of benzene in air for health safety should not exceed 10 $\mu g/m^3$.

7.1.3 Burning of Fuels

Fossil fuels are the sources of energy for cooking, heating, lightening our houses, washing clothes through washing machine, for running TV, etc. Coal and a variety of hydrocarbons including methane and soot, ashes and SO_2 are also the products of coal burning. It has been recorded that electrical power plants, burning fossil fuels, particularly coal and sometimes petrol or diesel produce two-thirds of the SO_2.

In majority of the Indian cities, towns and villages, cooking is done by burning colas, which is an important cause behind the increment of benzene concentration in the atmosphere. While in most of the developed countries, the most toxic benzene is much less than toluene and xylene in the urban air, in the third world countries, the picture is totally reverse. In Kolkata, more than 70% of the families still use coal chullas which account for high benzene load in the city air.

7.1.4 Aircraft Emissions

The pollution from the aircraft is also an important component of the total air pollution problem in the world. Aircrafts are responsible for about 2.5% of the CO emissions and about 1% of the hydrocarbon emissions but only negligible amounts of the other major air pollutants. The smoke emitted from jet aircraft is composed largely of fine particles approximately 0.5 μm in diameter, which are completely burnt. As these particles scatter light quiet well, they often reduce visibility.

7.1.5 Agricultural Activities

Burning of forest areas, grasslands, etc. for pastures and croplands produces about 60 to 65% of CO_2. About 40% of methane is produced from paddy fields, guts of livestocks and also from burning of biomass.

Crop spraying and dusting for pest and weed control are responsible for emitting organic phosphates, chlorinated hydrocarbons, arsenic and lead into air.

7.1.6 Ionizing Radiation

These radiations are characterized by sufficiently great energy to ionize atoms and molecules. Ionizing radiations like alpha and beta particles are produced during nuclear explosions, scientific experiments where radioisotopes are used and atomic weapon testing. Radiations are also caused by naturally occurring radioactive substances.

7.1.7 Cosmic Radiation

Cosmic rays are high-energy charged particles (mostly protons) of extraterrestrial origin, which are important source of natural radiations. The Earth's atmosphere is under the continuous bombardment of high-energy particles coming from outer space. These particles are the primary cosmic rays. As they penetrate into the atmosphere, primary cosmic rays loss energy and gradually disappear on colliding against the oxygen and nitrogen atoms of the air. At the same time, however, they give rise to secondary rays mostly different in nature from the primary particles and, of course, of lower average energy. Thus, at each point with the atmosphere, one finds a radiation consisting partly of primary particles and partly of secondary particles. This radiation may be denoted as local cosmic radiation.

7.1.8 Suspended Particulate Matter (SPM)

SPM is a major air pollutant. Dust is generated from sources such as coal dust (from oil refineries, power plants, etc.), cement dust and silica dust (from stone crushing). A huge amount of dust is also blown by transport vehicles. A new study by the Massachusetts Institute of Technology (MIT) has found that the concentration of hazardous respirable suspended particulate matter (RSPM) in Delhi's air is double the national limit and six times the threshold set by the World Health Organization (WHO).

RSPMs are more dangerous to health because they are much smaller than suspended particulate matter (SPM), an umbrella term for all such substances with deleterious consequences that are less than 100 micrometres in diameter. These superfine particles can penetrate deeper into the lungs, making those who breathe Delhi's poisonous air more vulnerable to heart diseases, chronic bronchitis, asthma and even lung cancer. The RSPM concentration in Delhi is about 120 micrograms per cubic metre, which is twice the prescribed National Ambient Air Quality Standard of 60 micrograms per cubic metre, the study points out (https://www.indiatoday.in/india/north/story/mit-study-high-rspm-in-delhi-air-pollution-138144-2011-07-24).

Incidences of high lead dust were recorded in some pockets of the city of Kolkata. In Picnic Garden area of Kolkata, there are a number of lead factories producing lead ingots and lead alloys. In the dining table of some local houses, the concentration of lead in the dust exceeded 5000 μg/gm.

7.2 Effects of Air Pollution

Dust, smoke and other suspended particulate matter reduce the visibility. Some of the aerosols as H_2SO_4 mist, $(NH_4)_2SO_4$ mist and water vapour influence the vertical temperature profile in the atmosphere. It affects the thermal mixing and also leads to greenhouse effects and depletion of ozone layer. Air pollution also reduces the aesthetic value of places and exerts considerable adverse effects on the health of biota. The various effects of air pollution are discussed separately.

7.2.1 Effects of Air Pollution on Climate

In urban areas air pollution causes several effects like increase of air temperature, lowering of wind speed, attenuation of solar radiation, etc. The particulate matter in the atmosphere, whose urban concentrations are typically 10 times those of rural areas, is capable of reducing the amount of solar radiation falling on the city by 15–20%.

Greenhouse effect is also a direct consequence of air pollution. A large amount of CO_2 gets introduced into the atmosphere due to burning of fossil fuels, furnaces, etc. The gas gets confined exclusively to troposphere. In dense concentration, it can act as a serious pollutant. The temperature at the surface of the Earth is being maintained by the energy balance between the sun's rays that strike the planet and the heat that gets radiated back into the space. Some of the sun's rays that penetrate the thick layer of CO_2 are able to strike the Earth and get converted into heat. The heated Earth is able to reradiate this absorbed energy as radiations of longer wavelengths. Much of this does not pass through CO_2 layer to the outer space but gets absorbed by the CO_2

layer creating an addition of extra heat. An increased heating of Earth may result in melting of the polar ice caps and finally cause a rise in the sea level. This phenomenon is termed as greenhouse effect. Many workers have predicted that there is a high probability of increment of the global temperature by 1.5–4.5 °C by the year 2050.

Sometimes, in the presence of sunlight, atomic oxygen from the photochemical reduction of NO_2 also reacts with a number of hydrocarbons (such as methane, ethane, toluene, etc., all of which originate from burning of fossil fuels or directly from plants) to form reactive intermediates called radicals. These radicals then take part in a series of reactions to form still more radicals that combine with oxygen, hydrocarbons and NO_2. This results in the regeneration of NO_2, disappearance of nitric oxide, accumulations of ozone and formation of a number of secondary pollutants like formaldehyde, aldehydes and peroxyacetyl nitrate or PAN ($C_2H_3O_5N$). All these collectively form photochemical smog.

The increased use of CFCs by human beings in recent times has another cause in air pollution. These obnoxious compounds emitted mainly by aerosols escape to stratosphere and react with ozone. A sharp drop in the protective ozone layer (about 40%) over Antarctica could have been caused by human-produced pollutants. This may ultimately pave the way of global temperature rise.

7.2.2 *Effects of Air Pollution on Human Health*

Human health is seriously affected by air pollution, and in extreme cases many diseases may be fatal. Diseases like asthma, bronchitis, pollen allergies and lung cancer are all associated with air pollution.

Air pollution can also be the cause of cancer. Higher benzene soluble organic matter (BSOM) in city air means that the air is more polluted and usually automobiles and coal burning are responsible for higher BSOM value. In BSOM, there lie a group of compounds known as polynuclear aromatic hydrocarbon (PAH), and many of them are suspected carcinogens (like benzapyrene). Evidences from epidemiological studies suggest an increase of one benzapyrene unit corresponds to an increase of 5% in the lung cancer death rate.

The pathological effects of various air pollutants along with their sources are given in Table 7.2.

7.2.3 *Effects of Air Pollution on Aesthetic Value*

1. Coal dusts are a source of nuisance for household articles and impart a dirty look.
2. The stone in Parthenon in Athens has deteriorated in the past 50 years for air pollution. Similarly Statue of Liberty is corroded from SO_2 and NO_2 and Taj Mahal from SO_2 emitted from Mathura refineries. Soot and tar stick to the building stones and painted surface, which are very difficult to remove completely.

Table 7.2 Pathological effects of some common air pollutants

Pollutants	Source	Pathological effect
Aldehydes	Thermal decomposition of fats, oil and glycerol	Irritate nasal and respiratory tracts
Ammonia	Chemical processes, dye making, explosives, fertilizer	Inflames upper respiratory passages
Arsine	Processes involving metals or acids containing arsenic soldering	Breaks down red cells in blood, damages kidneys and causes jaundice
Carbon monoxide	Gasoline motor exhausts, burning of coal	Reduces oxygen carrying capacity of blood
Chlorine	Bleaching cotton and flour, many other chemical processes	Attacks entire respiratory tract and mucous membranes of eyes, cause pulmonary oedema
Hydrogen cyanide	Fumigation, blast furnaces, chemical manufacturing, metal plating	Interferes with nerve cells, produces dry throat, indistinct vision and headache
Hydrogen fluoride	Petroleum refining, glass etching, Al and fertilizer production	Irritates and corrodes all body passages
Hydrogen sulphide	Refineries and chemical industries, bituminous fuels	Causes nausea, irritates eyes and throats
Nitrogen oxides	Motor vehicles exhausts, soft coal	Inhibits cilia action so that soot and dusts penetrates far into the lungs
Phosgene (carbonyl chloride)	Chemical and dye manufacturing	Induces coughing irritation and sometimes fatal pulmonary oedema
Sulphur dioxide	Coal and oil combustion	Causes chest constriction, headache, vomiting and respiratory ailments
Suspended particles	Incinerators, any manufacturing units	Causes emphysema, eye irritations and possibly cancer

After Southwick (1976)

7.3 Mitigation of Air Pollution

Air pollution is regarded as a grave danger for the healthy sustenance of all forms of life on this Earth. The various measures to control air pollution are given below:

1. Pollution caused by two-stroke engines can only be minimized by using the technology for four-stroke engines. In case of three wheelers, the Government can insist to design their engine in such a way so that they can run on LPG. Sulphur-free and lead-free fuel should be used for motor vehicles.
2. The height of the chimneys of factory should be tall enough to reduce the rate of pollution at the ground level.
3. To remove the particulate matter in the smoke, it should be filtered before releasing into the air.
4. Air pollution can be reduced by careful planning and locating industries after conducting EIA, designing better equipment, better operation of the equipment, fuel selection and utilization and site selection cum- zoning.

5. Use of clean and non-combustive means of generating energy like solar power, tidal power and geothermal power can reduce air pollution to a great extent (Fig. 7.1).
6. Plantation should be done on a large scale. Plants like *Ficus variegata*, *Phaseolus vulgaris* and *Coleus blumei* can store carbon easily depending on their biomass. Some other plants like *Pinus* sp., *Juniperus* sp., *Quercus* sp. and *Vitis* sp. can use

Fig. 7.1 Wind energy is a possible road map to combat air pollution

oxides of nitrogen. To reduce the effects of automobile pollution, the free space of the city area may be given a vegetation cover on the basis of the air pollution tolerance index (APTI) of trees. Some city plants have been identified to store carbon in considerable amount in their above ground biomass that can be planted in large scale to reduce the level of carbon dioxide in the atmosphere. A detailed study carried out in Kestopur (in the city of Kolkata) and two more sampling stations Barasat in North 24 Parganas and Konnagar in the west bank of River Hooghly generated unique findings (*vide* Annexure 7A) based on which the plantation programme can be carried out to reduce the atmospheric carbon dioxide at the local level.

7. Control equipment like gravity settling tanks porous filters and electrostatic precipitators should be installed in factories to minimize air pollution. The method used for pollutant separation is dependent on the size of the pollutant particles. Gravity settling chambers are used for particles having size more than 50 μm. Electrostatic precipitators or cyclone collectors are used for small-sized particles.

   Cyclone collectors contain a camber in which gas stream with particulate matter is whorled round through a tight circular spiral. Thus particulate pollutants are centrifuged, collected and removed. In electrostatic precipitators, precipitation is due to electric charging of particles and collecting surface bears an opposite charge to attract them. Through this technique, particles with size of 5–20 μm can be removed to the extent of 95%.
8. Permanent air monitoring stations should be set up in the core area of the city to monitor the air quality throughout the year.
9. **Air** (Prevention **and Control of Pollution) Act, 1981, Amended in 1987,** already exists in the legal frame of Indian sub-continent. However, strict implementation of the existing law can reduce the pollution to a considerable extent. This Act was promulgated by the Parliament of India to implement the decisions taken at the United Nations Conference on the Human Environment held at Stockholm in June, 1972. The jurisdiction of the Air Act is the whole of India.

The main objectives are:

1. To provide for the prevention, control and abatement of air pollution
2. To provide for the establishment of Central and State Boards
3. To provide for conferring on such boards powers and assigning functions

The composition of the board, qualification and disqualification of the members are similar to those relating to the board constituted under Water Act. Central Board exercises powers and performs the functions of State Board for Union Territories (UT) either directly or by delegating powers. The functions and powers under Air Act are same as under the Water Act except that it relates to the prevention and control of air pollution instead of prevention and control of water pollution.

7.3.1 Functions of Central Board

(i) To plan and oversee execution of nation-wide programme for prevention, control or abatement of air pollution
(ii) To advise Central Government and State Boards about improvement in the quality of air and issues related to air pollution
(iii) To establish or identify laboratories for sample analysis
(iv) To provide technical assistance and guidance to State Boards and industries
(v) To lay down standards for quality of air
(vi) To utilize the services of mass media in educating public about causes, effects, prevention and control of air pollution
(vii) To collect, compile and publish technical and statistical data relating to air pollution
(viii) To plan and organize training of persons in the field of air pollution

7.3.2 Functions of State Board

(i) To plan programmes for prevention and control or abatement of air pollution
(ii) To advise State Government on matters concerning air pollution, area to be declared as air pollution control area and location of different industries likely to cause air pollution
(iii) To inspect air quality in air control area from time to time and take steps to reduce the pollution
(iv) To lay down standards for air quality in consultation with Central Board
(v) To collect and disseminate information related to causes, prevention and control of air pollution
(vi) To seek guidance and training of persons from Central Board

7.3.3 Powers of State Boards

(i) The Board can restrict industrial operations in air pollution control areas.
(ii) To recommend to State Government to declare any area within the state as air pollution control area or areas. The whole of Punjab has been declared as air pollution control area.
(iii) On the recommendation of State Board, the State Government can direct any industry to use only approved fuel and approved appliances. It can prohibit the burning of any material.
(iv) Every industrial establishment has to move application of consent. After making inquiry and studying details, the State Board shall refuse or grant the consent, subject to such conditions like specifications of control equipment, its periodic checking, alterations, replacements and specifications of chimney.

(v) The Board can take air or emission samples for analysis.
(vi) The Board can withdraw consent to any industry and can impose new conditions.
(vii) The Board can call any industry for information with regard to pollutants emitted into atmosphere.
(viii) The Board can issue directions to authorities under Motor Vehicle Act regarding standards for automobile emissions.
(ix) Board officials have the power to enter any premises at all reasonable times and inspect compliance with requirements of the Act.
(x) Board can take remedial action in case of accidents or unforeseen events causing pollution in pollution control area.

7.3.4 Section 19

Under Section 19 of the Air Act, the State Government in consultation with the State Board can declare any area or areas within the state as air pollution control area for the purposes of this Act. It may be noted that the whole of the Union Territory of Puducherry and the Union Territory of Chandigarh have been declared as air pollution control areas with effect from January 25, 1988 and February 2, 1998, respectively.

7.3.5 Section 20

With a view to ensure that the standards for emission of air pollutants from automobiles, laid down by the State Board are complied with, the State Government in consultation with the State Board may give instructions to the registration authority under the Motor Vehicle Act for ensuring standards for emission from automobiles and the registration authority shall be bound to it.

7.3.6 Section 37

Section 37 of the Air Act provides that:

1. For operating or establishing any industrial plant in an air pollution control area without the consent of the Stare Board
2. For allowing emission of air pollutants in excess of the standards laid down by the Board in exercise of its powers and performance of its functions under the Act. The punishment shall be imprisonment for a term of 1.5 years to 6 years and fine. In case the failure continues, an additional fine of 5000 rupees per day after first conviction.

7.3.7 *Section 38*

Provides penalties for certain other acts such as destroying the property of the board, obstructing the person from excising his functions under the Act, failure to furnish information, giving false information, etc. The punishment for such acts shall be imprisonment for a term which may extend to 3 months or fine which may extend to 10,000 rupees or both.

7.3.8 *Section 39*

Under this Act, punishment for contravention of certain other provisions of the Act is provided which is the same as under Section 38. The only difference is that here an additional fine of 5000 rupees per day may be imposed for continued contravention.

7.4 Take Home Messages

(A) According to WHO, air pollution is defined as '…Phenomenon in which substances put into air by the activity of mankind into concentration sufficient to cause harmful effect to his health, vegetables, property or interfere with the enjoyment of his property…' The main sources of air pollution are industries, automobiles, burning of fossil fuels, aircraft emissions, agricultural activities, SPM, nuclear weapon testing and explosions.

(B) In urban areas air pollution causes several effects like increase of air temperature, lowering of wind speed, attenuation of solar radiation, etc. The particulate matter in the atmosphere, whose urban concentrations are typically 10 times those of rural areas, is capable of reducing the amount of solar radiation falling on the city by 15–20%.

(C) Human health is seriously affected by air pollution, and in extreme cases many diseases may be fatal. Diseases like asthma, bronchitis, pollen allergies and lung cancer are all associated with air pollution. Air pollution can also be the cause of cancer. Higher benzene soluble organic matter (BSOM) in city air means that the air is more polluted and usually automobiles and coal burning are responsible for higher BSOM value.

(D) Air pollution can be reduced by careful planning and locating industries after conducting EIA, designing better equipment, better operation of the equipment, fuel selection and utilization and site selection cum- zoning. Use of clean and non-combustive means of generating energy like solar power, tidal power and geothermal power can reduce air pollution to a great extent. Strict enforcement of the existing acts (Prevention and Control of Air Pollution Act 1981, Amended in 1987) can also reduce air pollution to a great extent.

(E) Plants like *Ficus variegata*, *Phaseolus vulgaris* and *Coleus blumei* can reduce air pollution at local scale by carbon sequestration depending on their biomass. Some other plants like *Pinus* sp., *Juniperus* sp., *Quercus* sp. and *Vitis* sp. can use oxides of nitrogen. To reduce the effects of automobile pollution, the free space of the city area may be given a vegetation cover on the basis of the air pollution tolerance index (APTI) of trees.

7.5 Brain Churners

Air pollution is a burning issue in underdeveloped and developing nations of the world. It is intricately related to industrialization and urbanization and has severe adverse effects on human health, biodiversity, properties and climate of the planet Earth. The readers can consider this chapter as a knowledge box on their health security. This section is a sort of '*self-appraisal approach*' as correct answers to all the questions are marked as black in the box at the end of this section.

1. Phosphate fertilizer manufacturing plants produce
   (a) SiF_4 and CO
   (b) SiF_4 and O_2
   (c) SiF_4 and H_2SiF_6
   (d) None of the above
2. The melting point of the mixture of $CaF_2L_2O_3$ and Na_3AlF_6 is lowered by
   (a) N_2
   (b) Dry ice
   (c) CO_2
   (d) Na_3AlF_6
3. The main ingredient of cement production is
   (a) CaO
   (b) SO_2
   (c) Cu
   (d) CaF_2
4. Benzene and toluene are
   (a) Metallic compounds
   (b) Volatile organic compounds
   (c) Inorganic compounds
   (d) None of the above

5. Chlorinated hydrocarbons are emitted into the air by
   (a) Automobiles
   (b) Crop spraying and dusting for pest control
   (c) Livestock
   (d) Burning of coal
6. Ionization radiations like *alpha* and *beta* particles are produced during
   (a) Nuclear explosions
   (b) Metallurgical process
   (c) Burning of leaf litter
   (d) None of the above
7. Cosmic rays are high-energy charged particles mainly composed of
   (a) Electrons
   (b) Neutrons
   (c) Protons
   (d) None of the above
8. Silica dust is generated in the atmosphere by
   (a) Metallurgical process
   (b) Agricultural activities
   (c) Aquacultural farms
   (d) Stone crushing
9. The vertical temperature profile in the atmosphere is influenced by
   (a) H_2SO_4, $(NH_4)_2\ SO_4$, mist and water vapour
   (b) CO and H_2
   (c) CO and S_2
   (d) Water vapour and CO_2
10. Higher benzene soluble organic matter (BSOM) in the city air is caused by
    (a) Industrial activities
    (b) Cooking process
    (c) Automobiles and Coal Burning
    (d) None of the above
11. Benzene pyrene is
    (a) The cause of skin problem
    (b) Carcinogenic
    (c) The cause of gastric problem
    (d) None of the above
12. Statue of liberty is corroded by
    (a) CO_2 and H_2
    (b) CO_2 and O_2

(c) SiO_2 and CO_2
(d) SiO_2 and NO_2

13. Air pollution can be reduced by using

(a) Four-stroke engine
(b) Two-stroke engine
(c) Petrol with lead
(d) All of the above

14. Air pollution can be reduced by using

(a) Solar power
(b) Tidal power
(c) Geothermal power
(d) All of the above

15. Air Pollution Prevention and Control Act of 1981 has been amended in

(a) 1990
(b) 1987
(c) 2000
(d) 2018

16. The State Government in consultation with the State Pollution Control Board can declare any area within the State as air pollution control area by using

(a) Section 20
(b) Section 19
(c) Section 37
(d) Section 38

17. The difference between Section 38 and Section 39 in context to Air Pollution Prevention and Control Act is

(a) An addition fine of INR 1 lakh/day may be used for continued contravention.
(b) An addition fine of INR 50,000/day may be used for continued contravention.
(c) An addition fine of INR 5 lakhs/day may be used for continued contravention.
(d) An addition fine of INR 5,000/day may be used for continued contravention.

18. Taj Mahal has lost its beauty due to

(a) Emission of NO_2
(b) Emission of CO_2 from coal burning
(c) Emission of SO_2 from Mathura refineries
(d) None of the above

19. SO_2 causes

    (a) Chest constriction, headache, vomiting and respiratory ailments
    (b) Gastric problem and diarrhoea
    (c) Cholera
    (d) Influenza

20. Phosgene (carbonyl chloride) may cause

    (a) Gastric problem
    (b) Vision problem
    (c) Pulmonary oedema
    (d) None of the above

Answer Sheet

S. No.	a	B	c	D
1.			■	
2.				■
3.	■			
4.		■		
5.		■		
6.	■			
7.			■	
8.				■
9.	■			
10.			■	
11.	■			
12.				■
13.	■			
14.				■
15.		■		
16.		■		
17.				■
18.			■	
19.	■			
20.			■	

Annexure 7A: Carbon Storage Potential of Urban Trees

The carbon sequestration of producer community is a function of biomass production capacity, which in turn depends upon interaction between edaphic, climate and topographic factors of an area. Hence, results obtained at one place may not be applicable to another. Therefore, region-based potential of different land types needs to be worked out.

The present research, carried out during 2016, has immense scope as this programme can serve as a road map to minimize the carbon dioxide level and enhance beautification (landscaping) in a cost-effective method.

The objectives of the present research programme are highlighted below:

- Identifying the tree species (in the city of Kolkata and its surrounding areas) having potential for carbon storage
- Selecting tree species with highest efficiency of carbon storage
- Development of allometric equation relating diameter at breast height (DBH) and above ground biomass (AGB) and carbon storage
- Studying the spatial and temporal variations of carbon storage by the selected tree species in the study area.

After a rigorous field work carried out by the present authors, mean values of AGB and AGC data have been presented (Table 7.A.1) for three stations, namely, Kestopur (Station A), Barasat (Station B) and Konnagar (Station C).

The AGB of Station A is in the order *Ficus religiosa* (1143.11) > *Ficus benghalensis* (1095.66) > *Cocos nucifera* (914.9) > *Delonix regia* (854.98) > *Eucalyptus globulus* (853.95) > *Acacia auriculiformis* (661.45) > *Bombax ceiba* (630.03) > *Psidium guajava* (624.92) > *Peltophorum pterocarpum* (604.14) > *Mangifera indica* (474.53) > *Polyalthia longifolia* (341.39) > *Azadirachta indica* (307.89) > *Ziziphus mauritiana* (301.74) > *Terminalia arjuna* (181.24) > *Artocarpus heterophyllus* (128.67) > *Aegle marmelos* (102.00).

Similarly for Station A the order of AGC is *Ficus religiosa* (544.12) > *Ficus benghalensis* (523.73) > *Cocos nucifera* (431.83) > *Eucalyptus globulus* (403.92) > *Delonix regia* (399.26) > *Acacia auriculiformis* (310.88) > *Bombax ceiba* (294.85) > *Psidium guajava* (294.85) > *Peltophorum pterocarpum* (281.53) > *Mangifera indica* (225.40) > *Polyalthia longifolia* (160.11) > *Azadirachta indica* (143.02) > *Ziziphus mauritiana* (143.02) > *Tamarindus indica* (137.45) > *Terminalia arjuna* (85.0) > *Artocarpus heterophyllus* (61.38) > *Aegle marmelos* (48.14).

For Station B the order of AGB is *Ficus benghalensis* (2085.33) > *Eucalyptus globulus* (1765.9) > *Ficus religiosa* (1678.9) > *Delonix regia* (1249.78) > *Cocos nucifera* (998.78) > *Acacia auriculiformis* (985.78) > *Peltophorum pterocarpum* (789.90) > *Mangifera indica* (789.23) > *Psidium guajava* (787.04) > *Bombax ceiba* (786.9) > *Polyalthia longifolia* (785.23) > *Ziziphus mauritiana* (675.78) > *Artocarpus heterophyllus* (654.20) > *Tamarindus indica* (509.33) > *Azadirachta indica* (487.12) > *Terminalia arjuna* (245.89) > *Aegle marmelos* (234.89).

AGC for Station B follows the order *Ficus benghalensis* (975.93) > *Eucalyptus globulus* (831.74) > *Ficus religiosa* (790.76) > *Delonix regia* (573.65) > *Cocos nucifera* (475.42) > *Acacia auriculiformis* (461.35) > *Mangifera indica* (379.62) *Bombax ceiba* (379.29) > *Peltophorum pterocarpum* (368.09) > > *Psidium guajava* (365.97) > *Polyalthia longifolia* (361.21) > *Ziziphus mauritiana* (320.96) > *Artocarpus heterophyllus* (315.98) > *Tamarindus indica* (238.37) > *Azadirachta indica* (230.38) > *Terminalia arjuna* (113.11) > *Aegle marmelos* (112.75).

Table 7.A.1 List of dominant tree species in the selected stations with their respective AGB and AGC

Sl. No.	Species	AGB (tonnes ha^{-1})			AGC (tonnes ha^{-1})		
		A	B	C	A	B	C
1.	 *Cocos nucifera* (coconut)	914.90	998.78	1123.19	431.83	475.42	532.39
2.	 *Azadirachta indica* (neem)	307.89	489.12	875.10	143.48	230.38	415.67
3.	 *Mangifera indica* (mango)	474.53	789.23	985.66	225.40	379.62	461.29
4.	 *Tamarindus indica* (tentul)	295.60	509.33	678.45	137.45	238.37	311.41

(continued)

Table 7.A.1 (continued)

Sl. No.	Species	AGB (tonnes ha^{-1})			AGC (tonnes ha^{-1})		
		A	B	C	A	B	C
5.	 *Bombax ceiba* (shimul)	630.03	786.90	961.23	294.85	379.29	450.82
6.	 *Aegle marmelos* (bel)	102.00	234.89	356.28	48.14	112.75	169.95
7.	 *Terminalia arjuna* (arjun)	181.24	245.89	345.90	85.00	113.11	164.30
8.	 *Delonix regia* (krishnachura)	854.98	1249.78	1987.66	399.26	573.65	940.16

(continued)

Table 7.A.1 (continued)

Sl. No.	Species	AGB (tonnes ha^{-1})			AGC (tonnes ha^{-1})		
		A	B	C	A	B	C
9.	 *Artocarpus heterophyllus* (jackfruit)	128.67	654.20	987.90	61.38	315.98	475.18
10.	 *Psidium guajava* (guava)	624.92	787.04	1230.78	294.34	365.97	577.24
11.	 *Acacia auriculiformis* (akashmoni)	661.45	985.78	1268.89	310.88	461.35	593.84
12.	 *Ziziphus mauritiana* (kul)	301.74	675.78	980.45	143.02	320.96	456.89

(continued)

Table 7.A.1 (continued)

Sl. No.	Species	AGB (tonnes ha^{-1})			AGC (tonnes ha^{-1})		
		A	B	C	A	B	C
13.	 *Eucalyptus globulus* (eucalyptus)	853.95	1765.90	978.45	403.92	831.74	464.76
14.	 *Peltophorum pterocarpum* (radhachura)	604.14	789.90	678.67	281.53	368.09	318.97
15.	 *Polyalthia longifolia* (debdaru)	341.39	785.23	987.34	160.11	361.21	467.01
16.	 *Ficus religiosa* (peepul)	1143.11	1678.90	1989.08	544.12	790.76	954.76

(continued)

Table 7.A.1 (continued)

Sl. No.	Species	AGB (tonnes ha^{-1})			AGC (tonnes ha^{-1})		
		A	B	C	A	B	C
17.	 *Ficus benghalensis* (Banyan)	1095.66	2085.33	2091.95	523.73	975.93	989.49

Note: A, Kestopur; B, Barasat; C, Konnagar

The order of AGB for Station C is *Ficus benghalensis* (2091.95) > *Ficus religiosa* (1989.08) > *Delonix regia* (1987.66) > *Acacia auriculiformis* (1268.89) > *Psidium guajava* (1230.78) > *Cocos nucifera* (1123.19) > *Artocarpus heterophyllus* (987.90) > *Polyalthia longifolia* (987.34) > *Mangifera indica* (985.66) > *Ziziphus mauritiana* (980.45) > *Eucalyptus globulus* (978.45) > *Bombax ceiba* (961.23) > *Azadirachta indica* (875.10) > *Peltophorum pterocarpum* (678.67) > *Tamarindus indica* (678.45) > *Aegle marmelos* (356.28) > *Terminalia arjuna* (345.90).

For Station C the order of AGC is *Ficus benghalensis* (989.49) > *Ficus religiosa* (954.76) > *Delonix regia* (940.16) > *Acacia auriculiformis* (593.84) > *Psidium guajava* (577.24) > *Cocos nucifera* (532.39) > *Artocarpus heterophyllus* (475.18) > *Polyalthia longifolia* (467.76) > *Eucalyptus globulus* (464.76) > *Mangifera indica* (461.29) > *Ziziphus mauritiana* (456.89) > *Bombax ceiba* (450.82) > *Azadirachta indica* (415.67) > *Peltophorum pterocarpum* (318.97) > *Tamarindus indica* (311.41) > *Aegle marmelos* (169.95) > *Terminalia arjuna* (164.30) (Figs. 7.A.1, 7.A.2, 7.A.3, 7.A.4, 7.A.5, 7.A.6, 7.A.7, 7.A.8, 7.A.9, 7.A.10, 7.A.11, 7.A.12, 7.A.13, 7.A.14, 7.A.15, 7.A.16 and 7.A.17).

A simple linear regression equation is the simplest type of representation of the linear models. It is based on the inter-relationship between the response variable (Y) and independent variable (X). It assumes that the relationship is linear in nature and hence can be represented as:

$$Y = a + bX + \varepsilon$$

where a is the y-intercept of the line and b its slope, and that the residuals are of constant variance, $Var(\varepsilon) = \sigma^2$.

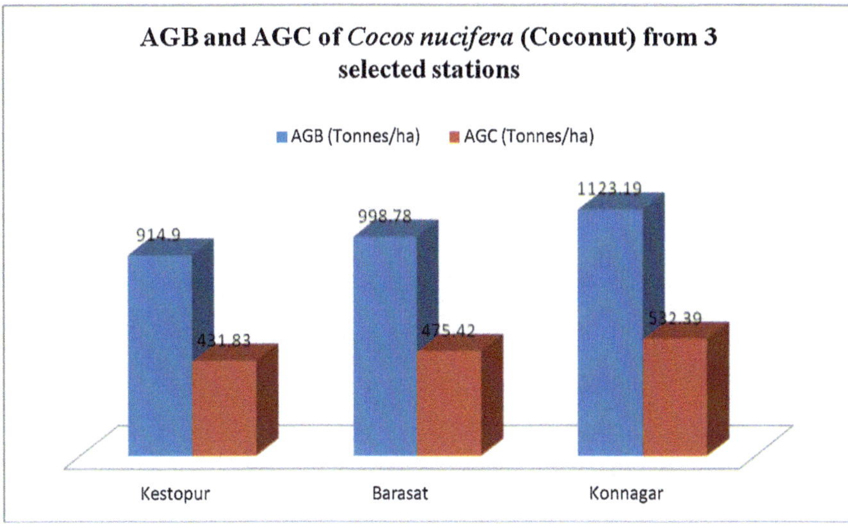

Fig. 7.A.1 AGB and AGC of *Cocos nucifera* (coconut) from three selected stations

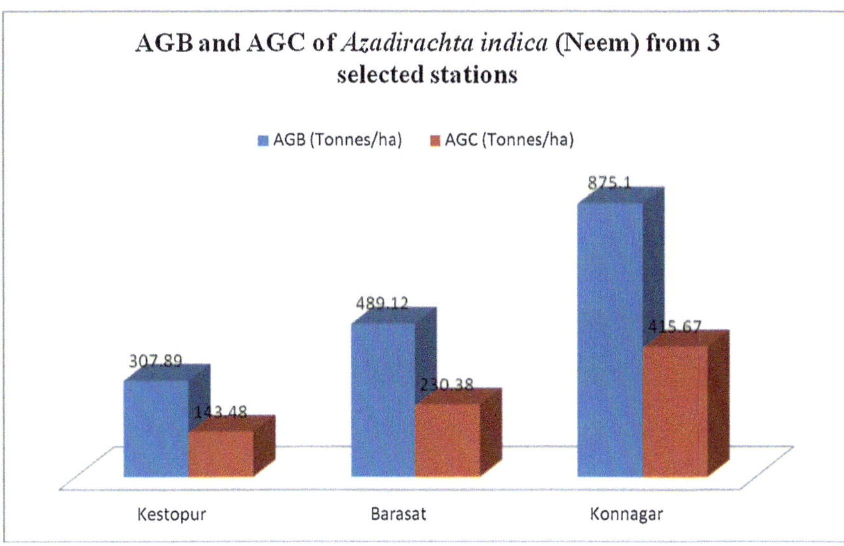

Fig. 7.A.2 AGB and AGC of *Azadirachta indica* (neem) from three selected stations

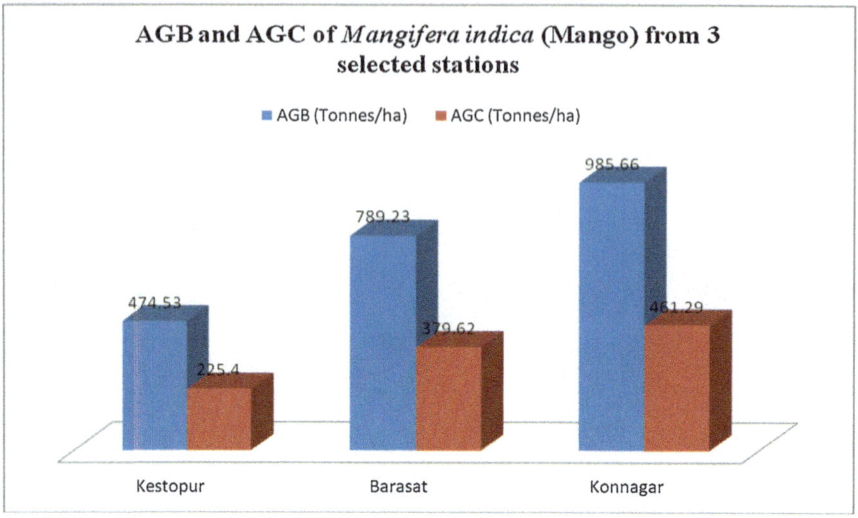

Fig. 7.A.3 AGB and AGC of *Mangifera indica* (mango) from three selected stations

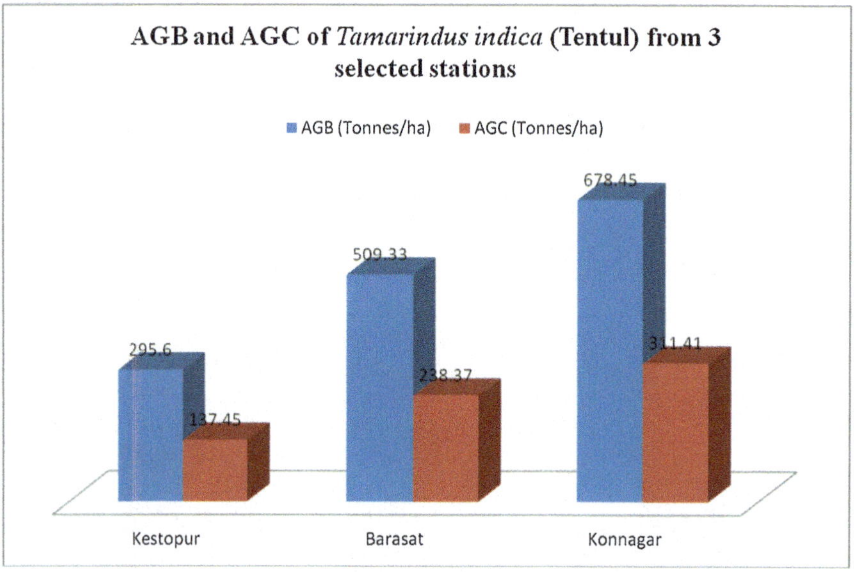

Fig. 7.A.4 AGB and AGC of *Tamarindus indica* (tamarind) from three selected stations

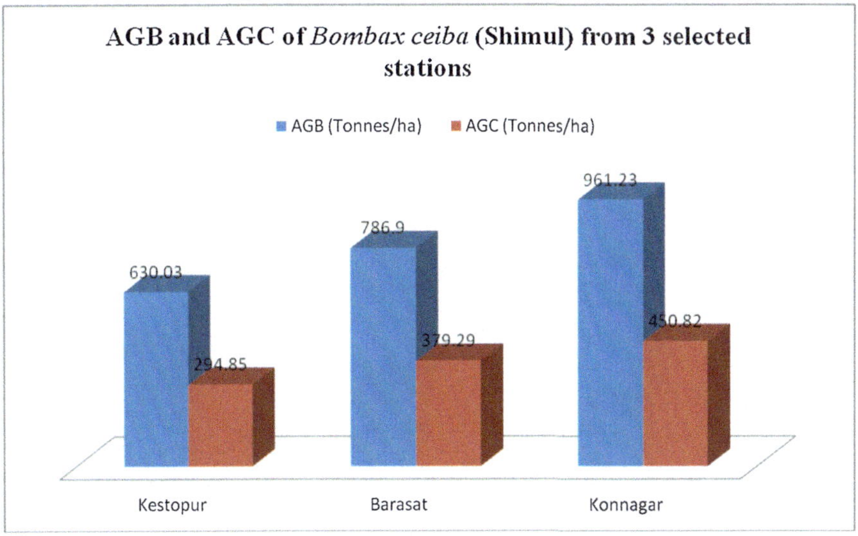

Fig. 7.A.5 AGB and AGC of *Bombax ceiba* (shimul) from three selected stations

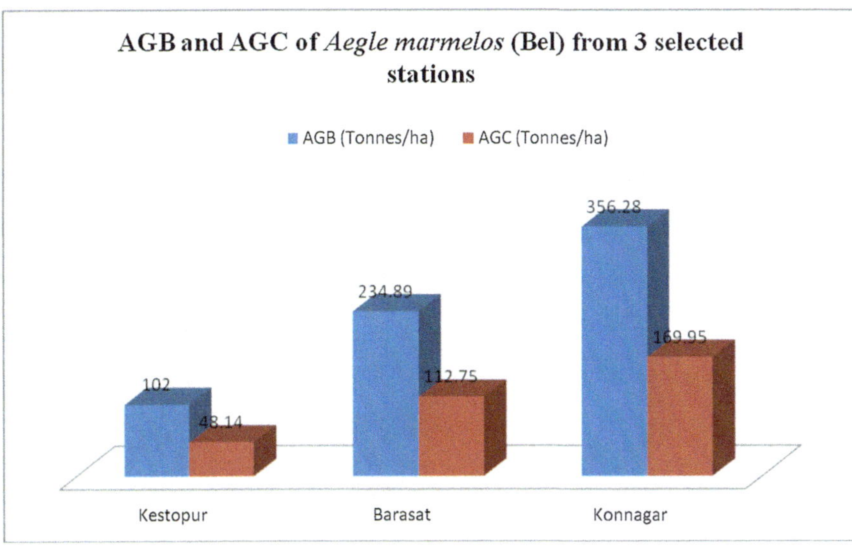

Fig. 7.A.6 AGB and AGC of *Aegle marmelos* (bel) from three selected stations

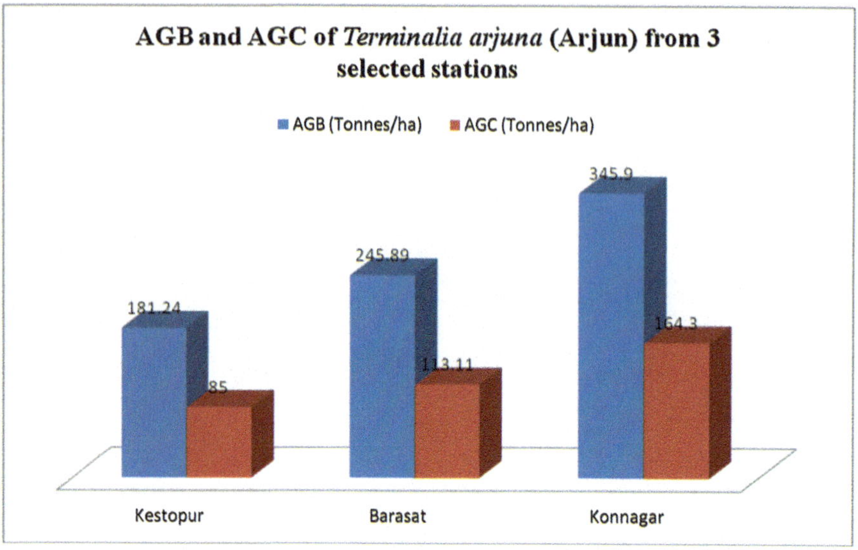

Fig. 7.A.7 AGB and AGC of *Terminalia arjuna* (arjun) from three selected stations

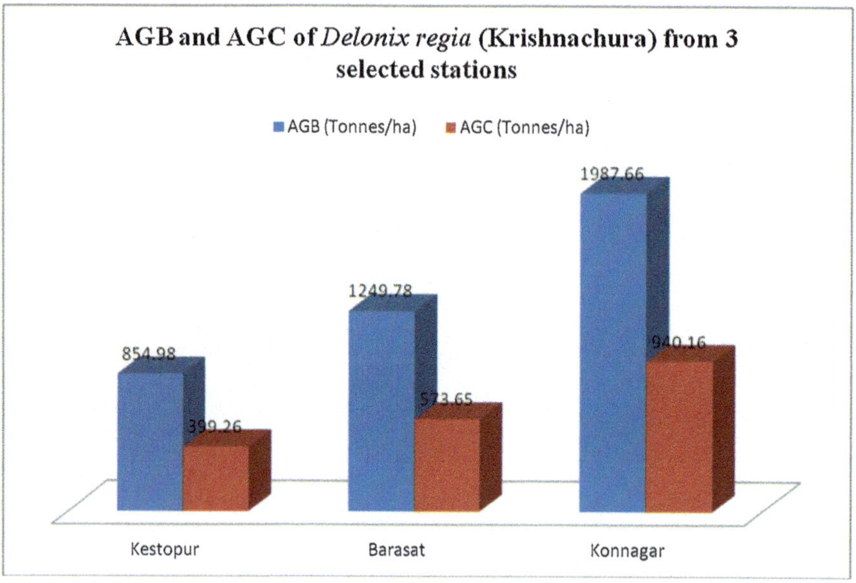

Fig. 7.A.8 AGB and AGC of *Delonix regia* (krishnachura) from three selected stations

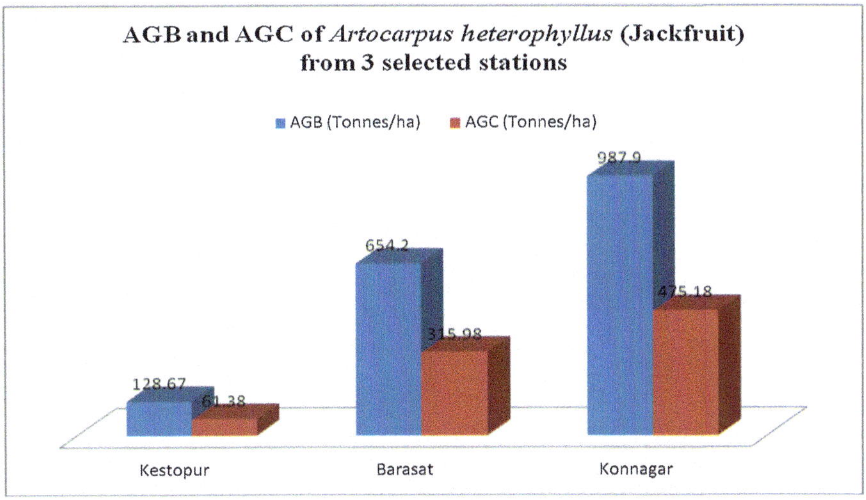

Fig. 7.A.9 AGB and AGC of *Artocarpus heterophyllus* (jackfruit) from three selected stations

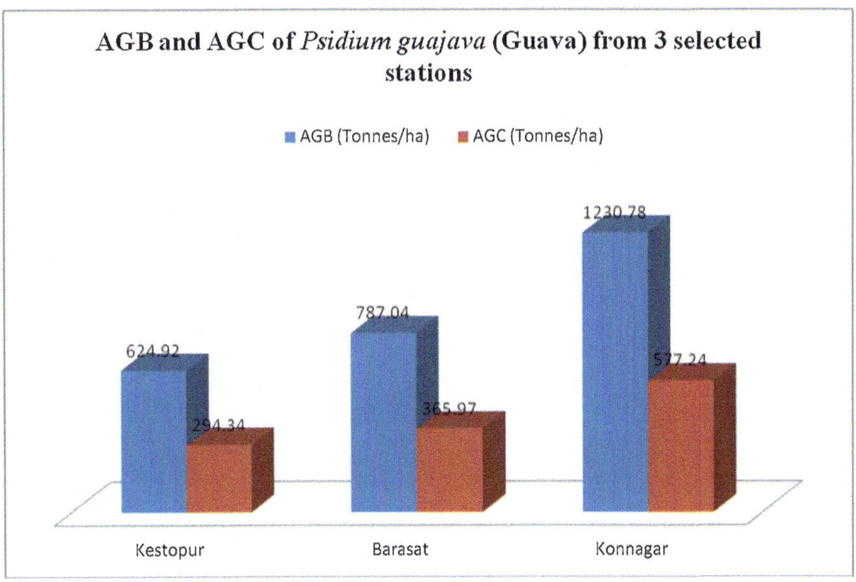

Fig. 7.A.10 AGB and AGC of *Psidium guajava* (guava) from three selected stations

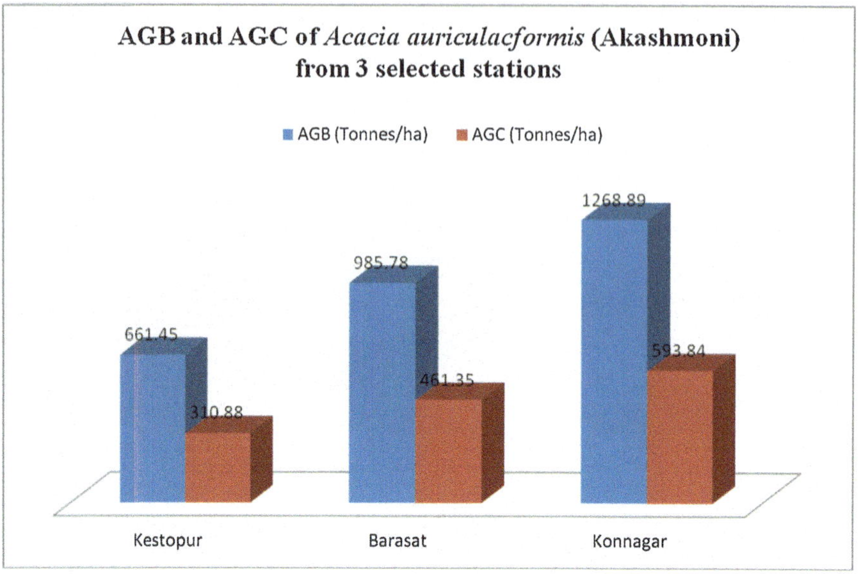

Fig. 7.A.11 AGB and AGC of *Acacia auriculiformis* (akashmoni) from three selected stations

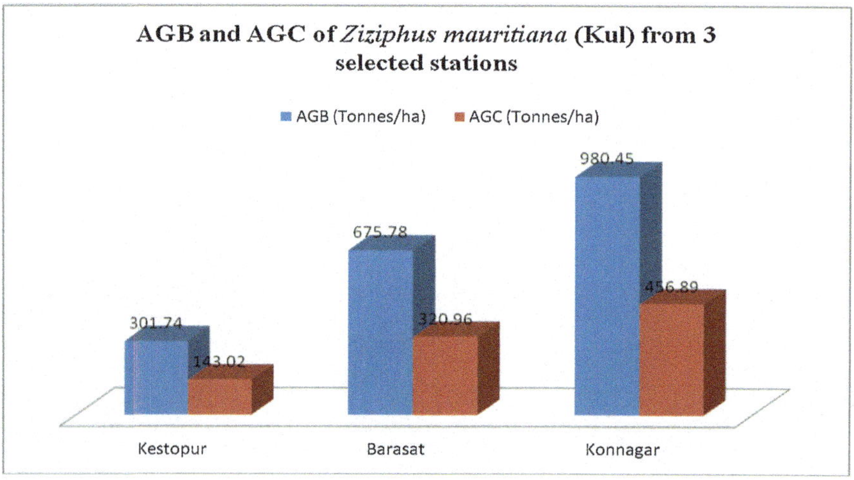

Fig. 7.A.12 AGB and AGC of *Ziziphus mauritiana* (kul) from three selected stations

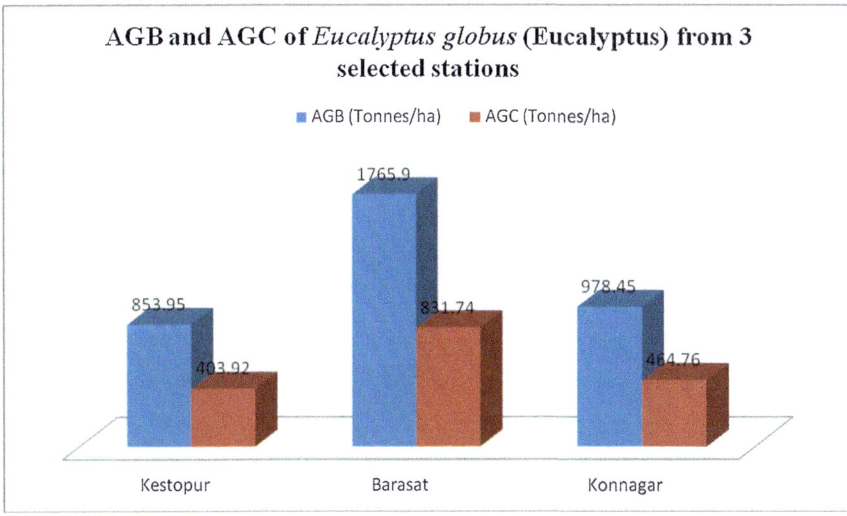

Fig. 7.A.13 AGB and AGC of *Eucalyptus globulus* (eucalyptus) from three selected stations

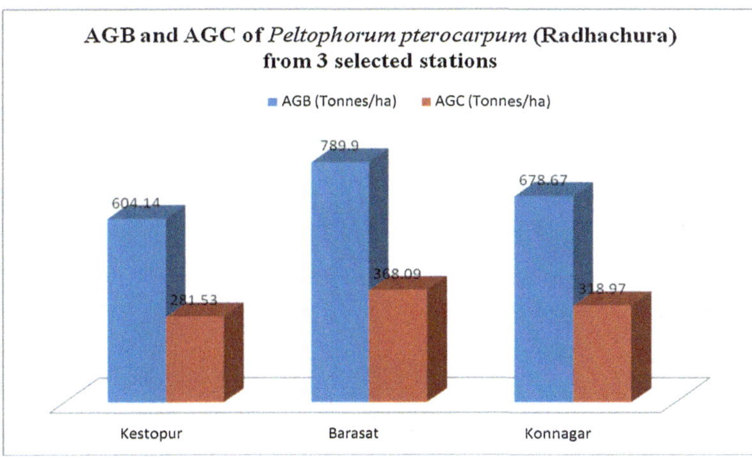

Fig. 7.A.14 AGB and AGC of *Peltophorum pterocarpus* (radhachura) from three selected stations

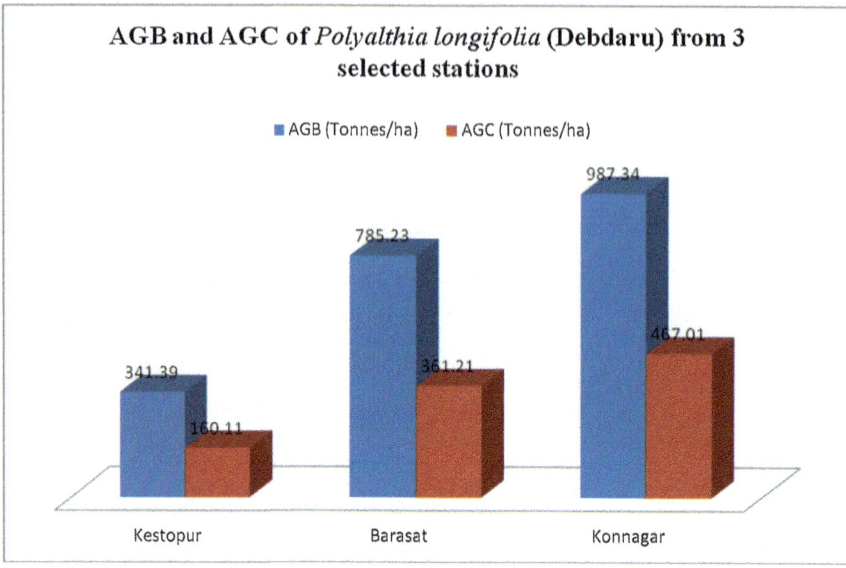

Fig. 7.A.15 AGB and AGC of *Polyalthia longifolia* (debdaru) from three selected stations

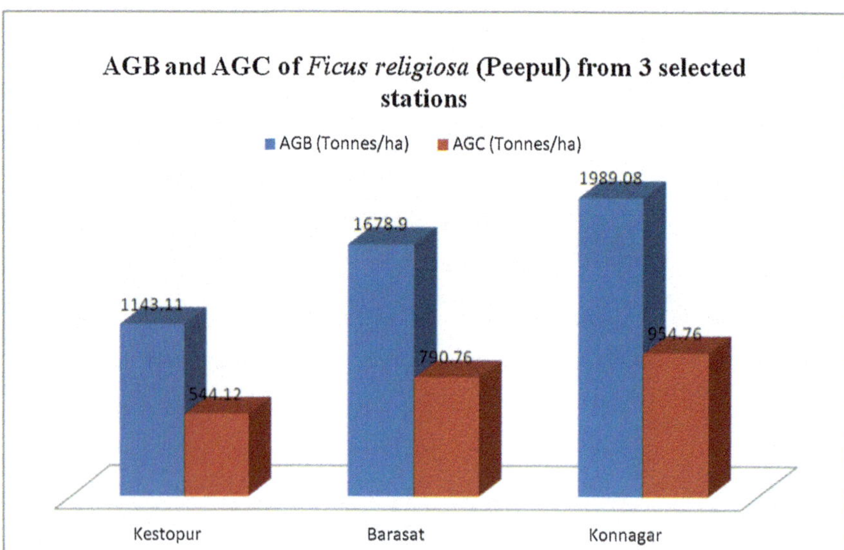

Fig. 7.A.16 AGB and AGC of *Ficus religiosa* (peepul) from three selected stations

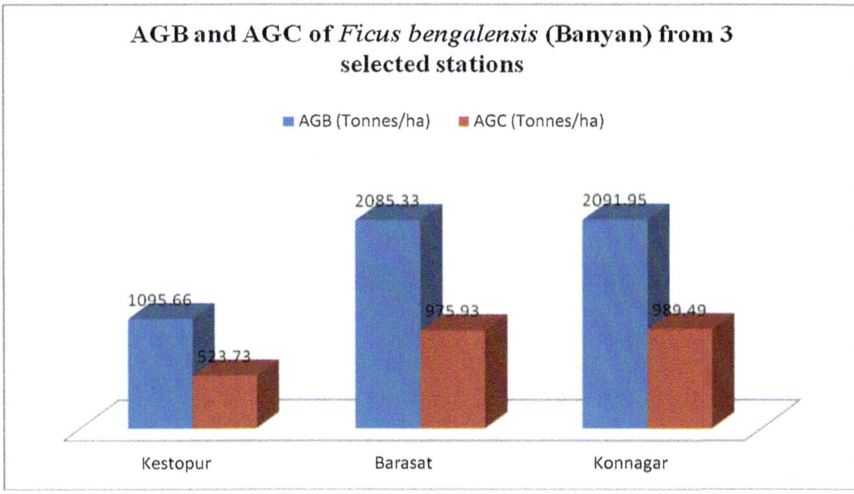

Fig. 7.A.17 AGB and AGC of *Ficus benghalensis* (banyan) from three selected stations

When fitting simple linear regression, several outputs need to be analysed. The determination coefficient, more commonly called R^2, measures the quality of the fit; R^2 is directly related to the residual variance since:

$$R^2 = 1 - \frac{\hat{\sigma}^2 (n-2)/n}{S_Y^2}$$

where $S_Y^2 = \left[\sum_{n}^{ni=1} \left(Y_i - \hat{Y} \right)^2 \right] / \text{n}$ is the empirical variance of Y. The difference $S_Y^2 = \hat{\sigma}\left(\text{n}-2\right)/\text{n}$ between the variance of *Y* and the residual variance corresponds to the variance explained by the model. The determination coefficient R^2 can be interpreted as being the ratio between the variance explained by the model and the total variance. It is between 0 and 1, and the closer it is to 1, the better the quality of the fit.

The linear regression equations in all the cases show that DBH (X) can predict the AGB (Y) in a well-fitted manner in all the selected trees as the R^2 values range between 0.7 and 0.9 in almost all the cases (Table 7.A.2). The scatter plots of all the selected trees are shown in Figs. 7.A.18, 7.A.19, 7.A.20, 7.A.21, 7.A.22, 7.A.23, 7.A.24, 7.A.25, 7.A.26, 7.A.27, 7.A.28, 7.A.29, 7.A.30, 7.A.31, 7.A.32, 7.A.33, 7.A.34, 7.A.35, 7.A.36, 7.A.37, 7.A.38, 7.A.39, 7.A.40, 7.A.41, 7.A.42, 7.A.43, 7.A.44, 7.A.45, 7.A.46, 7.A.47, 7.A.48, 7.A.49, 7.A.50, 7.A.51, 7.A.52, 7.A.53, 7.A.54, 7.A.55, 7.A.56, 7.A.57, 7.A.58, 7.A.59, 7.A.60, 7.A.61, 7.A.62, 7.A.63, 7.A.64, 7.A.65, 7.A.66, 7.A.67 and 7.A.68.

Table 7.A.2 Regression equations with R^2 values representing the inter-relationship between DBH (X) and AGB (Y)

Name of the species	Station	Regression equation	R^2
Cocos nucifera (coconut)	Kestopur	Y = 7.129X − 53.63	0.871
	Barasat	Y = 6.597X − 38.69	0.815
	Konnagar	Y = 6.618X − 34.12	0.716
Azadirachta indica (neem)	Kestopur	Y = 0.348X + 480.5	0.707
	Barasat	Y = 0.452X + 475.2	0.827
	Konnagar	Y = 0.754X + 455.8	0.904
Mangifera indica (mango)	Kestopur	Y = 1.747X + 328.4	0.712
	Barasat	Y = 1.792X + 326.0	0.734
	Konnagar	Y = 1.404X + 341.1	0.748
Tamarindus indica (tentul)	Kestopur	Y = 1.253X + 25.54	0.748
	Barasat	Y = 1.405X + 21.92	0.714
	Konnagar	Y = 1.453X + 21.11	0.840
Bombax ceiba (shimul)	Kestopur	Y = 6.805X − 24.46	0.945
	Barasat	Y = 6.556X − 17.39	0.883
	Konnagar	Y = 5.844X + 4.216	0.858
Aegle marmelos (bel)	Kestopur	Y = 4.396X + 293.5	0.708
	Barasat	Y = 5.242X + 244.2	0.811
	Konnagar	Y = 4.713X + 278.2	0.920
Terminalia arjuna (arjun)	Kestopur	Y = 2.913X + 389.7	0.767
	Barasat	Y = 3.133X + 376.2	0.899
	Konnagar	Y = 3.095X + 378.0	0.909
Delonix regia (krishnachura)	Kestopur	Y = 4.212X + 328.5	0.953
	Barasat	Y = 4.247X + 328.0	0.890
	Konnagar	Y = 3.984X + 342.9	0.903
Artocarpus heterophyllus (jackfruit)	Kestopur	Y = 3.378X + 195.9	0.786
	Barasat	Y = 3.963X + 176.5	0.807
	Konnagar	Y = 3.799X + 180.9	0.903
Psidium guajava (guava)	Kestopur	Y = 1.414X + 20.49	0.886
	Barasat	Y = 1.403X + 20.82	0.854
	Konnagar	Y = 1.381X + 21.32	0.908
Acacia auriculiformis (akashmoni)	Kestopur	Y = 4.952X + 11.84	0.957
	Barasat	Y = 4.574X + 22.00	0.894
	Konnagar	Y = 4.724X + 18.63	0.920
Ziziphus mauritiana (kul)	Kestopur	Y = 2.502X + 3.985	0.920
	Barasat	Y = 2.433X + 5.371	0.867
	Konnagar	Y = 2.567X + 3.259	0.903
Eucalyptus globulus (eucalyptus)	Kestopur	Y = 3.227X + 33.02	0.842
	Barasat	Y = 3.102X + 35.20	0.855
	Konnagar	Y = 3.034X + 36.21	0.912
Peltophorum pterocarpum (radhachura)	Kestopur	Y = 2.563X + 381.6	0.949
	Barasat	Y = 2.229X + 402.7	0.887
	Konnagar	Y = 2.346X + 395.8	0.929

(continued)

Table 7.A.2 (continued)

Name of the species	Station	Regression equation	R^2
Polyalthia longifolia (debdaru)	Kestopur	Y = 0.757X + 67.67	0.848
	Barasat	Y = 0.767X + 67.53	0.864
	Konnagar	Y = 0.702X + 68.31	0.903
Ficus religiosa (peepul)	Kestopur	Y = 8.592X + 248.4	0.712
	Barasat	Y = 8.760X + 238.1	0.806
	Konnagar	Y = 7.752X + 340.7	0.863
Ficus benghalensis (banyan)	Kestopur	Y = 32.55X − 2270.	0.734
	Barasat	Y = 33.95X − 2401.	0.826
	Konnagar	Y = 34.70X − 2467.	0.905
Cocos nucifera (coconut)	Kestopur	Y = 7.129X − 53.63	0.871
	Barasat	Y = 6.597X − 38.69	0.815
	Konnagar	Y = 6.618X − 34.12	0.716
Azadirachta indica (neem)	Kestopur	Y = 0.348X + 480.5	0.707
	Barasat	Y = 0.452X + 475.2	0.827
	Konnagar	Y = 0.754X + 455.8	0.904
Mangifera indica (mango)	Kestopur	Y = 1.747X + 328.4	0.712
	Barasat	Y = 1.792X + 326.0	0.734
	Konnagar	Y = 1.404X + 341.1	0.748
Tamarindus indica (tentul)	Kestopur	Y = 1.253X + 25.54	0.748
	Barasat	Y = 1.405X + 21.92	0.714
	Konnagar	Y = 1.453X + 21.11	0.840
Bombax ceiba (shimul)	Kestopur	Y = 6.805X − 24.46	0.945
	Barasat	Y = 6.556X − 17.39	0.883
	Konnagar	Y = 5.844X + 4.216	0.858
Aegle marmelos (bel)	Kestopur	Y = 4.396X + 293.5	0.708
	Barasat	Y = 5.242X + 244.2	0.811
	Konnagar	Y = 4.713X + 278.2	0.920
Terminalia arjuna (arjun)	Kestopur	Y = 2.913X + 389.7	0.767
	Barasat	Y = 3.133X + 376.2	0.899
	Konnagar	Y = 3.095X + 378.0	0.909
Delonix regia (krishnachura)	Kestopur	Y = 4.212X + 328.5	0.953
	Barasat	Y = 4.247X + 328.0	0.890
	Konnagar	Y = 3.984X + 342.9	0.903
Artocarpus heterophyllus (jackfruit)	Kestopur	Y = 3.378X + 195.9	0.786
	Barasat	Y = 3.963X + 176.5	0.807
	Konnagar	Y = 3.799X + 180.9	0.903
Psidium guajava (guava)	Kestopur	Y = 1.414X + 20.49	0.886
	Barasat	Y = 1.403X + 20.82	0.854
	Konnagar	Y = 1.381X + 21.32	0.908
Acacia auriculiformis (akashmoni)	Kestopur	Y = 4.952X + 11.84	0.957
	Barasat	Y = 4.574X + 22.00	0.894
	Konnagar	Y = 4.724X + 18.63	0.920

(continued)

Table 7.A.2 (continued)

Name of the species	Station	Regression equation	R^2
Ziziphus mauritiana (kul)	Kestopur	Y = 2.502X + 3.985	0.920
	Barasat	Y = 2.433X + 5.371	0.867
	Konnagar	Y = 2.567X + 3.259	0.903
Eucalyptus globulus (eucalyptus)	Kestopur	Y = 3.227X + 33.02	0.842
	Barasat	Y = 3.102X + 35.20	0.855
	Konnagar	Y = 3.034X + 36.21	0.912
Peltophorum pterocarpum (radhachura)	Kestopur	Y = 2.563X + 381.6	0.949
	Barasat	Y = 2.229X + 402.7	0.887
	Konnagar	Y = 2.346X + 395.8	0.929
Polyalthia longifolia (debdaru)	Kestopur	Y = 0.757X + 67.67	0.848
	Barasat	Y = 0.767X + 67.53	0.864
	Konnagar	Y = 0.702X + 68.31	0.903
Ficus religiosa (peepul)	Kestopur	Y = 8.592X + 248.4	0.712
	Barasat	Y = 8.760X + 238.1	0.806
	Konnagar	Y = 7.752X + 340.7	0.863
Ficus benghalensis (banyan)	Kestopur	Y = 32.55X − 2270.	0.734
	Barasat	Y = 33.95X − 2401.	0.826
	Konnagar	Y = 34.70X − 2467.	0.905

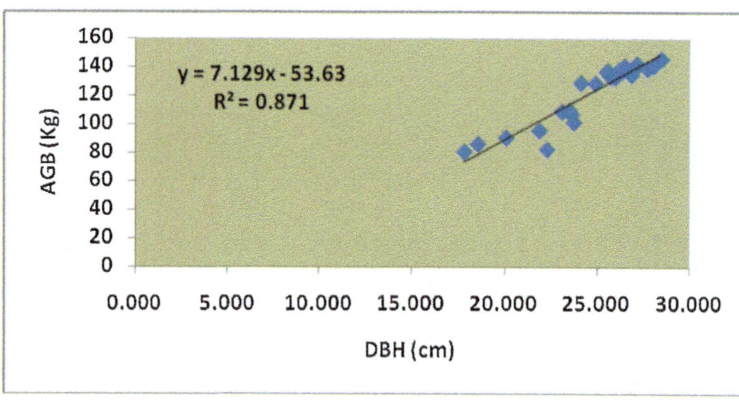

Fig. 7.A.18 Scatter plot showing the inter-relationship between DBH and AGB of *Cocos nucifera* (coconut) at Kestopur

Table 7.A.3 presents significant spatial and species-level variations in AGB and AGC.

Urban, semiurban and rural trees and forests influence climate change but are often disregarded because their ecosystem services are not well-understood or quantified. Trees act as a sink for carbon dioxide by fixing carbon during photosynthesis and storing carbon as biomass. The net long-term carbon dioxide source/sink dynamics of forests change through time as trees grow, die and decay. Trees in

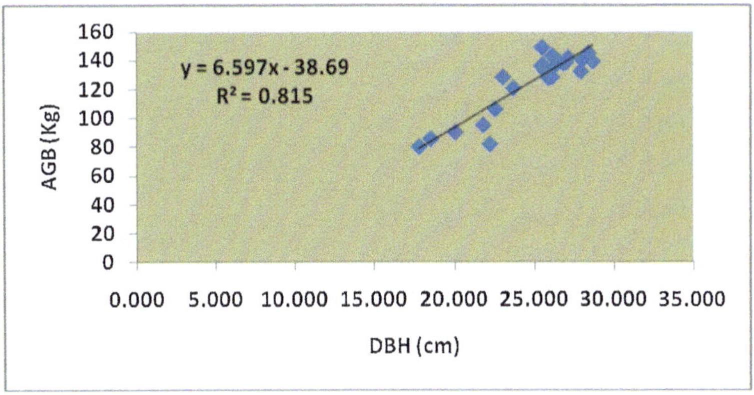

Fig. 7.A.19 Scatter plot showing the inter-relationship between DBH and AGB of *Cocos nucifera* (coconut) at Barasat

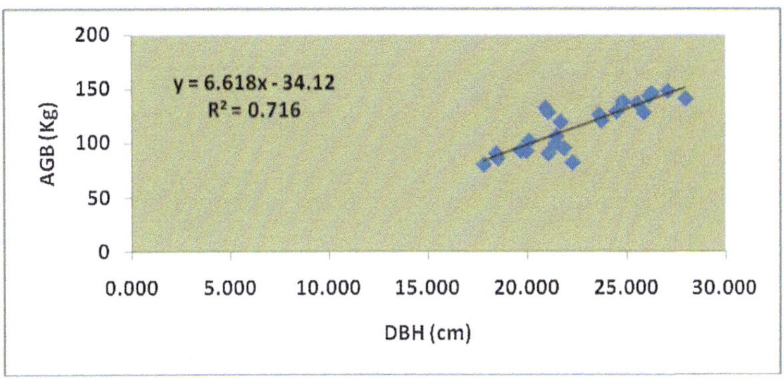

Fig. 7.A.20 Scatter plot showing the inter-relationship between DBH and AGB of *Cocos nucifera* (coconut) at Konnagar

urban and semiurban areas currently store carbon which can be emitted back to the atmosphere after tree death and sequester carbon as they grow. Urban and semiurban trees also influence air temperatures and building energy use and consequently alter carbon emissions from numerous urban sources (e.g. Power Plants) (Nowak 1993). Thus urban trees influence local climate, carbon cycles, energy use and climate change (e.g. Abdollahi et al. 2000; Wilby and Perry 2006; Gill et al. 2007; Nowak 2010; Lal and Augustine 2012).

A study conducted in the United States reveals that urban areas in the conterminous United States have increased from 2.5% of the US land area (19.5 million ha) in 1990 to 3.1% (24.0 million ha) in 2000, an increase in area the size of Vermont and New Hampshire combined. If the growth patterns of 1990s continue, urban land is projected to reach 8.1% by 2050, an increase greater than the area of Montana. Within these urban areas, tree cover is estimated at 35% (Nowak and Greenfield 2012). Given the growing expanse of urban areas, trees within these areas have the

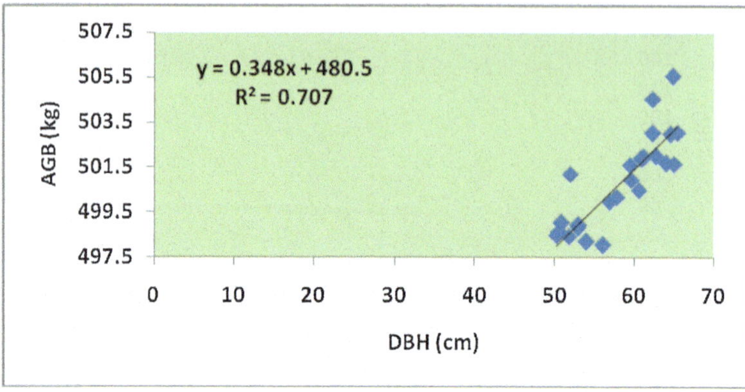

Fig. 7.A.21 Scatter plot showing the inter-relationship between DBH and AGB of *Azadirachta indica* (neem) at Kestopur

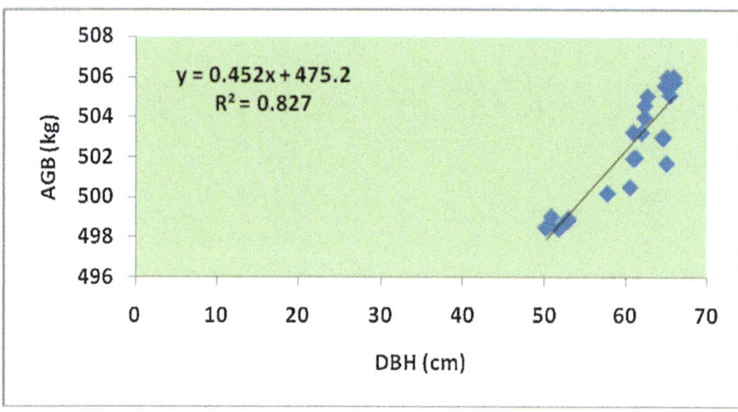

Fig. 7.A.22 Scatter plot showing the inter-relationship between DBH and AGB of *Azadirachta indica* (neem) at Barasat

potential to store and annually sequester substantial amounts of carbon. Understanding this national carbon affect can aid in preparing annual inventories of greenhouse gas (GHG) emissions and sinks (US EPA 2010; Heath et al. 2011). Numerous cities in the United States have analysed carbon storage and sequestration of the trees and forests among various land-use types using the standard methodology (www.itreetools.org) (Hutyra et al. 2011; Chaparro and Terradas 2009; Zhao et al. 2010; Davies et al. 2011; Strohbach and Haase 2012).

In the past city analyses of carbon storage and sequestration have been extrapolated to national estimates using limited data. The first estimate of national carbon storage by urban trees (between 350 and 750 million tonnes, Nowak 1993) was based on an extrapolation of Carbon Data from one city (Oakland, C.A.) and tree cover data from various US cities (e.g. Nowak et al. 1996). A later assessment which

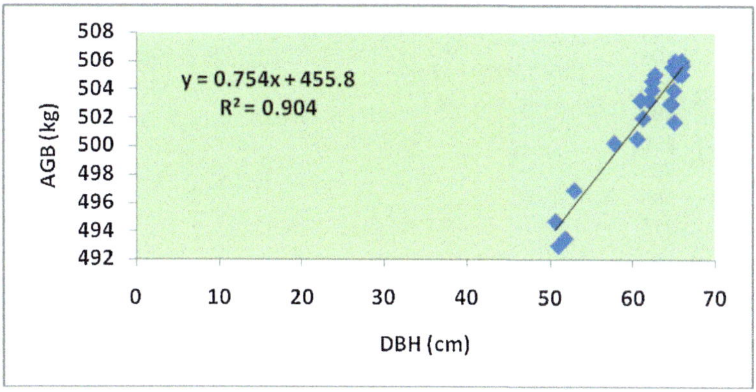

Fig. 7.A.23 Scatter plot showing the inter-relationship between DBH and AGB of *Azadirachta indica* (neem) at Konnagar

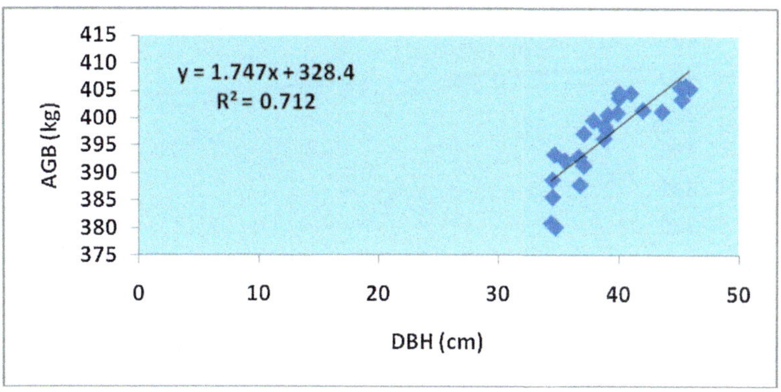

Fig. 7.A.24 Scatter plot showing the inter-relationship between DBH and AGB of *Mangifera indica* (mango) at Kestopur

included data from a second city (Chicago, IL) estimated National Carbon Storage by urban trees between 600 and 900 million tonnes (Nowak 1994). The most recent analysis which uses data from ten cities and urban tree cover estimates (Nowak et al. 2001) derived from 1991 Advanced Very High Resolution Radiometer (AVHRR) data, estimated National Carbon Storage by urban forests at 700 million tonnes (range 335 million to 980 million tonnes) (Nowak and Crane 2002). Above and below ground biomass in all forestland across the United States, which includes forest stands within urban areas, stored approximately 20.2 billion tonnes of carbon in 2008 (Heath et al. 2011).

These carbon storage and sequestration estimates provide better, more up-to-date information for National Carbon estimates (e.g. IPCC 2006) and can be used to help assess the actual and potential role of urban forests in reducing atmospheric carbon dioxide. Although several studies have been carried out on carbon sequestration by

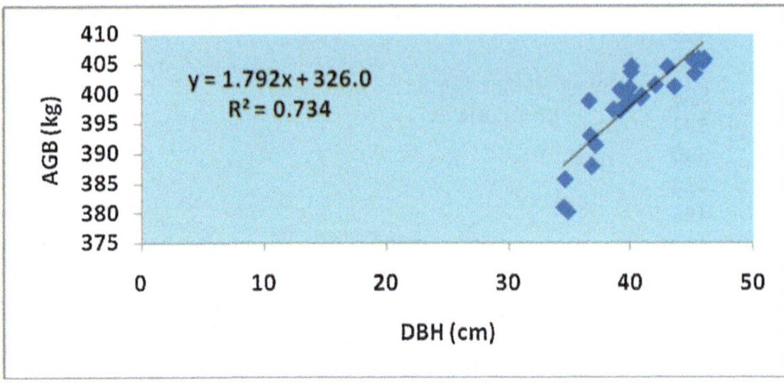

Fig. 7.A.25 Scatter plot showing the inter-relationship between DBH and AGB of *Mangifera indica* (mango) at Barasat

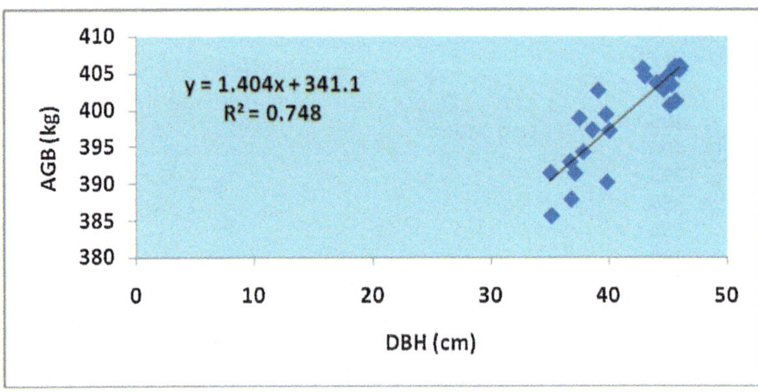

Fig. 7.A.26 Scatter plot showing the inter-relationship between DBH and AGB of *Mangifera indica* (mango) at Konnagar

carbon trees in several cities of the world, very few literatures are available on the potential of trees of Kolkata on carbon sequestration.

There are, however, some studies on Bangalore (Sudha and Ravindranath 2000; Nagendra and Gopal 2010), Visakhapatnam City (Mitra 1993; Madan 1993), Chandigarh (Chaudhary 2006; Chaudhary and Tewari 2010; FSI 2009) and Delhi (FSI 2009). Similar studies such as biodiversity and carbon storage are also available for Bhopal (Dwivedi et al. 2009), Delhi (Khera et al. 2009), Jaipur (Verma 1985; Dubey and Pandey 1993), Mumbai (Zérah 2007) and Pune (Patwardhan et al. 2001). A few studies are also available for specific locations within the urban ecosystems, such as NEERI Campus, Nagpur (Gupta et al. 2008), Indian Institute of Science Campus, Bangalore (Mhatre 2008) and Bangalore University Campus at Jnanabharathi (Nandini et al. 2009). Many policy and robust scientific evidences in last two decades have emphasized the critical necessity of green areas within urban

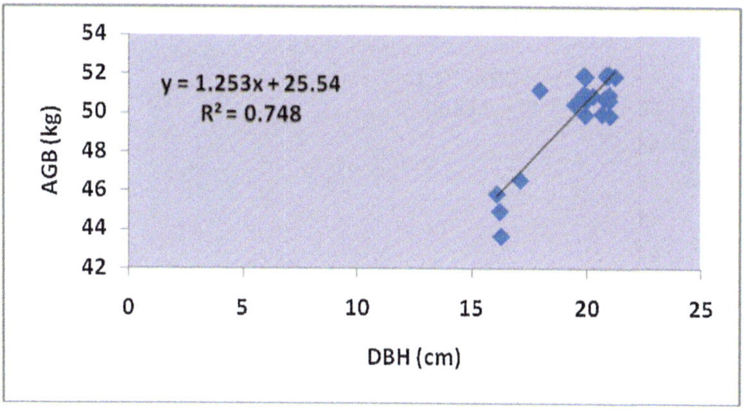

Fig. 7.A.27 Scatter plot showing the inter-relationship between DBH and AGB of *Tamarindus indica* (tentul) at Kestopur

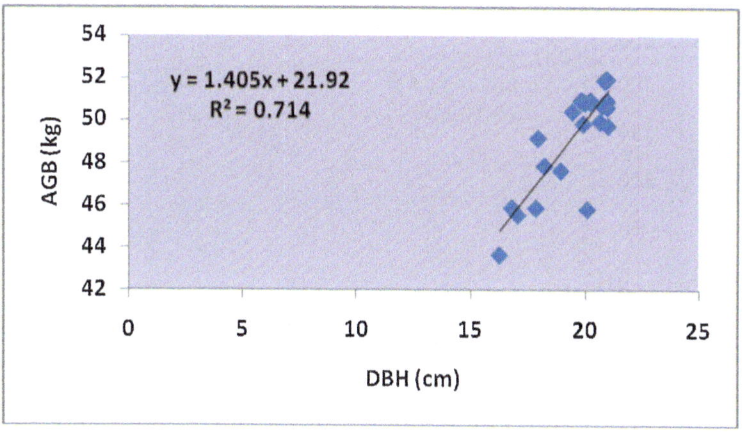

Fig. 7.A.28 Scatter plot showing the inter-relationship between DBH and AGB of *Tamarindus indica* (tentul) at Barasat

social-ecological systems to ameliorate several problems of city-living; however the trend of urban ecology and application of its principles are still lagging behind.

Roadside trees, because of their proximity to the generation of vehicle emissions, are important in reducing pollution. Beckett et al. (2000) found that roadside trees capture more large-size particulate matter than trees not near the road. These effects have implications for air quality standards. Roadside trees additionally have aesthetic value to residents and high ecological value to urban areas as part of our green infrastructure. Urban trees perform important ecological function in cities by sequestering carbon and reducing automobile pollution. The net save in carbon emissions that can be achieved by tree planting can be up to 18 kg CO_2/year per tree,

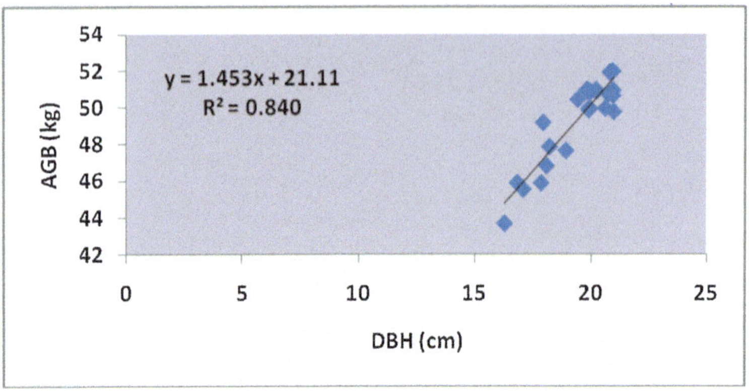

Fig. 7.A.29 Scatter plot showing the inter-relationship between DBH and AGB of *Tamarindus indica* (tentul) at Konnagar

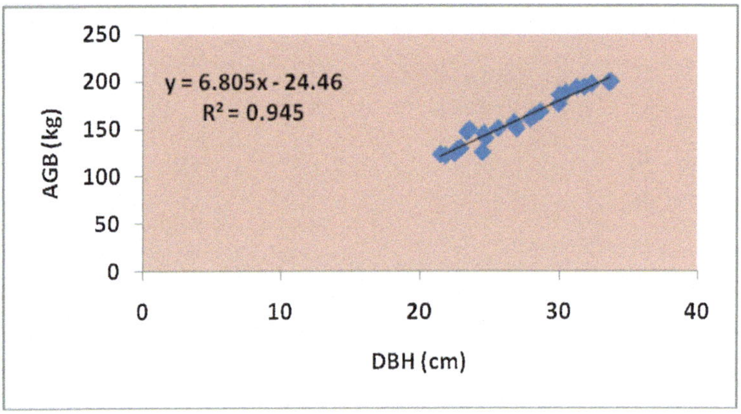

Fig. 7.A.30 Scatter plot showing the inter-relationship between DBH and AGB of *Bombax ceiba* (shimul) at Kestopur

and this benefit corresponds to that provided by 3–5 forest trees of similar size and health (Francesco 2011).

The amount of carbon stored in a tree depends upon its biomass and growth pattern. It is found that fast-growing trees seize more carbon than slow-growing trees (Montagnini and Porras 1998; Redondo-Brenes 2007). In the long term, the amount of carbon accumulated by slow-growing species is larger than by fast-growing species. This indicates faster-growing trees may accumulate larger amount carbon in early stage of their life, while high specific gravity of slow growing trees allow them to accumulate more carbon in longer.

In the present study, *Ficus benghalensis* shows the highest AGB and AGC for all the stations, and *Aegle marmelos* shows the lowest AGB and AGC for Stations A and B. For Station C the lowest AGB and AGC is found in *Terminalia arjuna*.

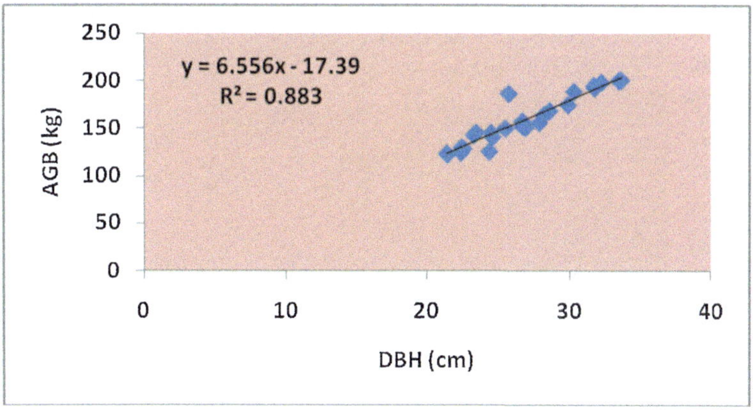

Fig. 7.A.31 Scatter plot showing the inter-relationship between DBH and AGB of *Bombax ceiba* (shimul) at Barasat

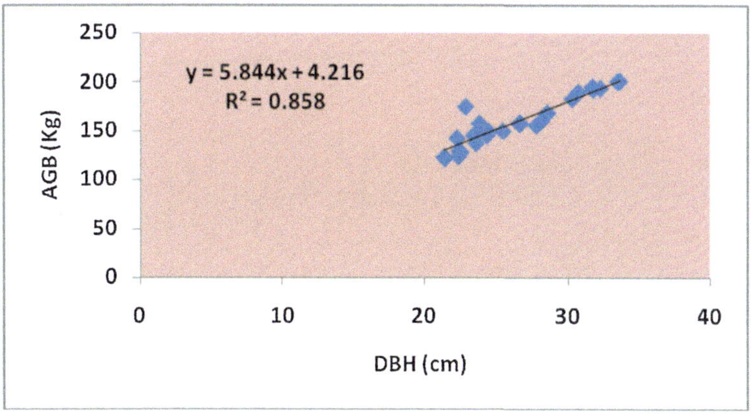

Fig. 7.A.32 Scatter plot showing the inter-relationship between DBH and AGB of *Bombax ceiba* (shimul) at Konnagar

ANOVA results show significant spatial variations of AGB and AGC between the selected stations as well as between the selected tree species due to variation in edaphic factors.

The significant AGB and AGC variations between species and stations indicate that carbon storage potential is regulated not only by edaphic factors (that are different for different locations or stations) but also on the type of the species. Species with C_3 and C_4 modes of photosynthesis have different response to climatic conditions. From the entire result, it can be concluded that urban and semiurban trees are unique store house of carbon. As the diameter of floral species increases, its biomass and carbon storage capacity also increases resulting in the sequestering of more carbon, which removes more carbon dioxide from the atmosphere. Species should be planted considering all environmental parameters (location/biogeographic zone,

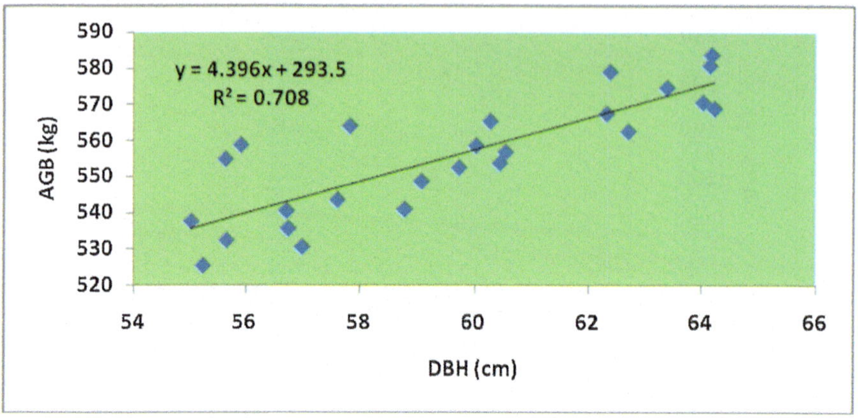

Fig. 7.A.33 Scatter plot showing the inter-relationship between DBH and AGB of *Aegle marmelos* (bel) at Kestopur

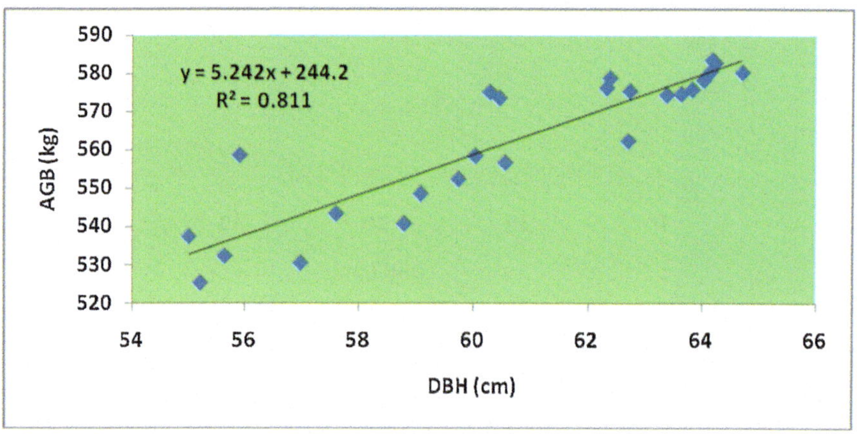

Fig. 7.A.34 Scatter plot showing the inter-relationship between DBH and AGB of *Aegle marmelos* (bel) at Barasat

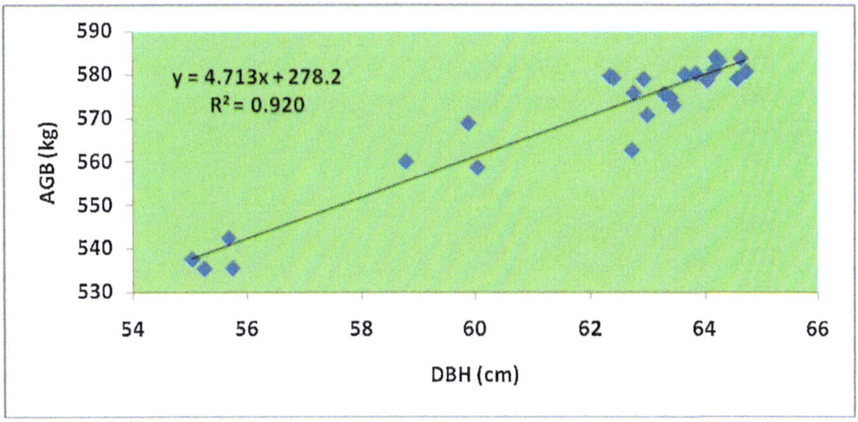

Fig. 7.A.35 Scatter plot showing the inter-relationship between DBH and AGB of *Aegle marmelos* (bel) at Konnagar

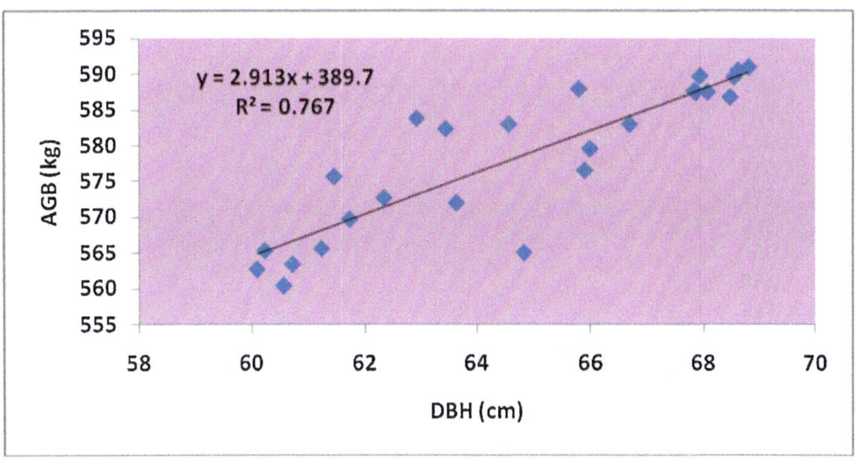

Fig. 7.A.36 Scatter plot showing the inter-relationship between DBH and AGB of *Terminalia arjuna* (arjun) at Kestopur

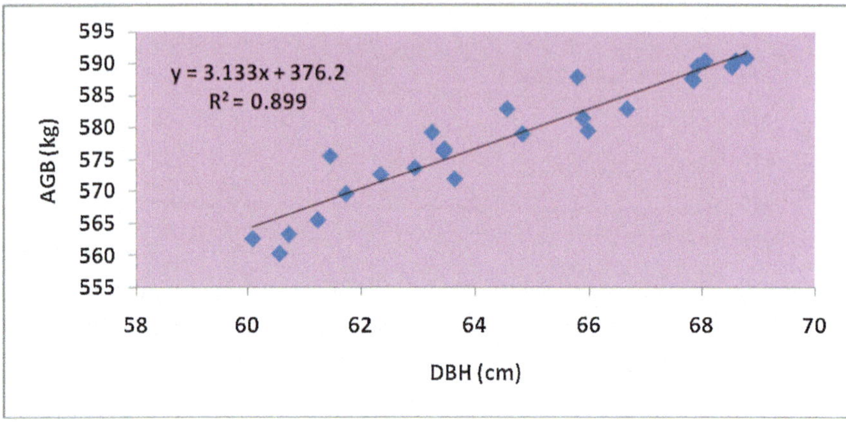

Fig. 7.A.37 Scatter plot showing the inter-relationship between DBH and AGB of *Terminalia arjuna* (arjun) at Barasat

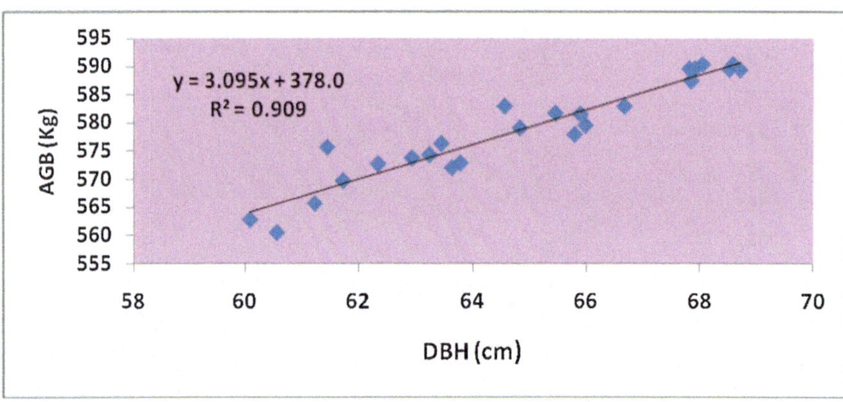

Fig. 7.A.38 Scatter plot showing the inter-relationship between DBH and AGB of *Terminalia arjuna* (arjun) at Konnagar

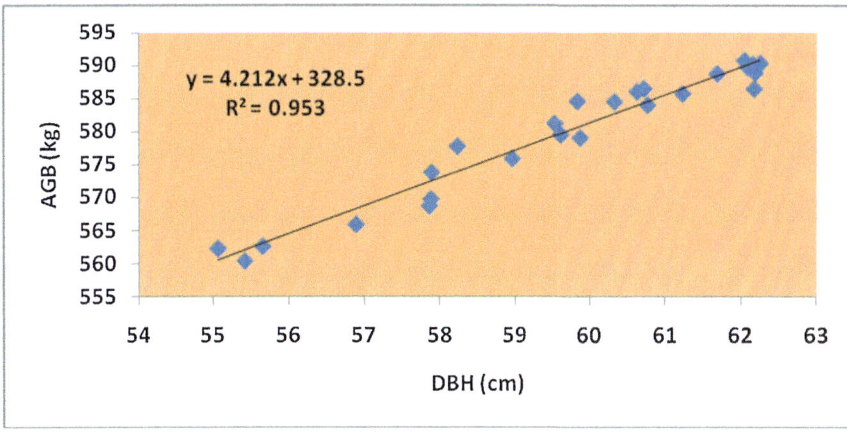

Fig. 7.A.39 Scatter plot showing the inter-relationship between DBH and AGB of *Delonix regia* (krishnachura) at Kestopur

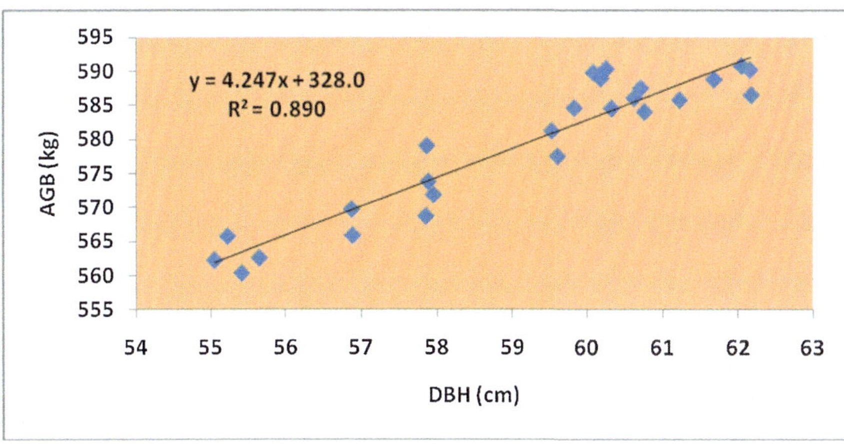

Fig. 7.A.40 Scatter plot showing the inter-relationship between DBH and AGB of *Delonix regia* (krishnachura) at Barasat

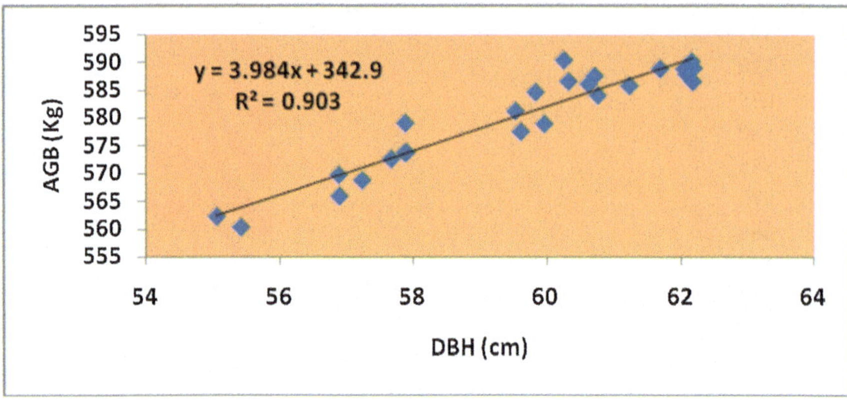

Fig. 7.A.41 Scatter plot showing the inter-relationship between DBH and AGB of *Delonix regia* (krishnachura) at Konnagar

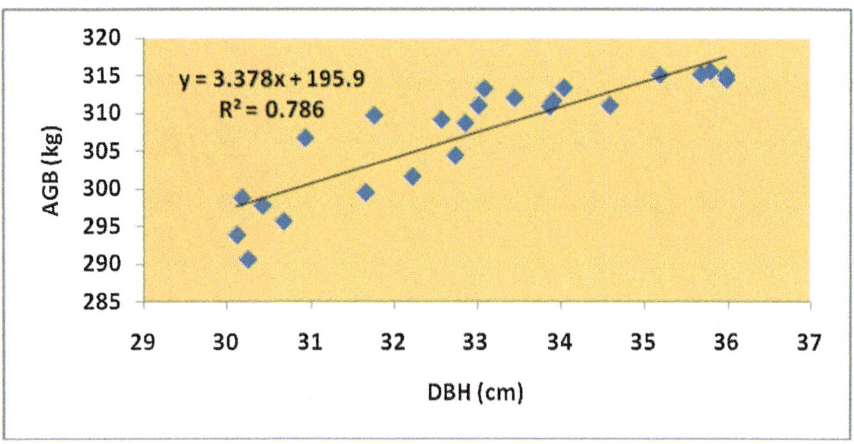

Fig. 7.A.42 Scatter plot showing the inter-relationship between DBH and AGB of *Artocarpus heterophyllus* (jackfruit) at Kestopur

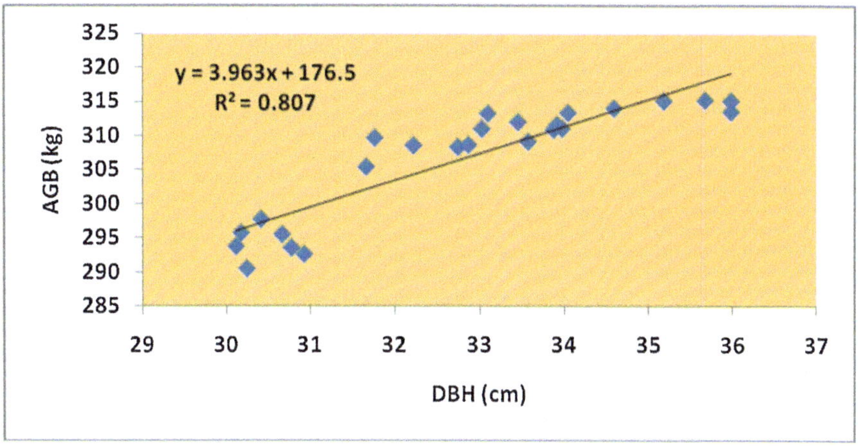

Fig. 7.A.43 Scatter plot showing the inter-relationship between DBH and AGB of *Artocarpus heterophyllus* (jackfruit) at Barasat

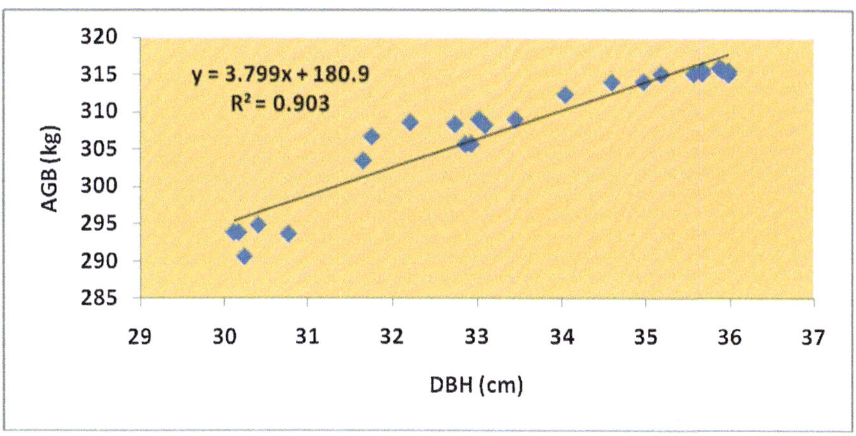

Fig. 7.A.44 Scatter plot showing the inter-relationship between DBH and AGB of *Artocarpus heterophyllus* (jackfruit) at Konnagar

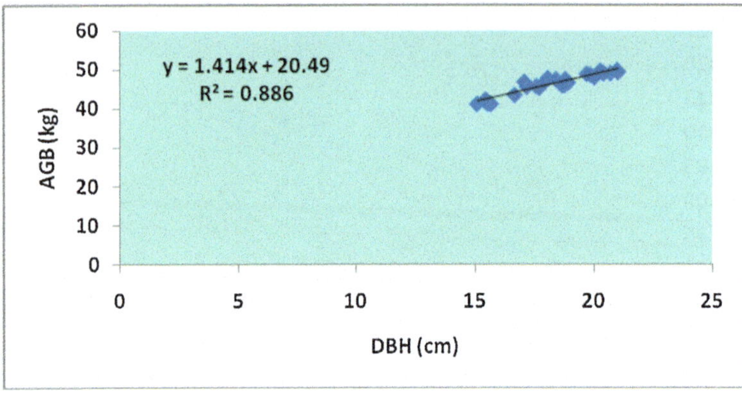

Fig. 7.A.45 Scatter plot showing the inter-relationship between DBH and AGB of *Psidium guajava* (guava) at Kestopur

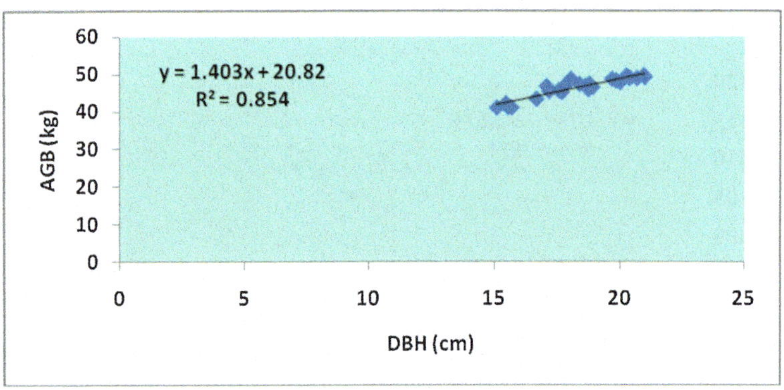

Fig. 7.A.46 Scatter plot showing the inter-relationship between DBH and AGB of *Psidium guajava* (guava) at Barasat

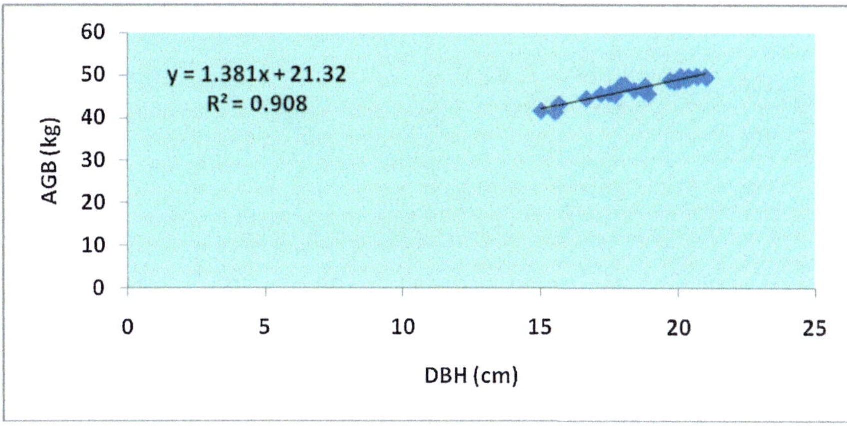

Fig. 7.A.47 Scatter plot showing the inter-relationship between DBH and AGB of *Psidium guajava* (guava) at Konnagar

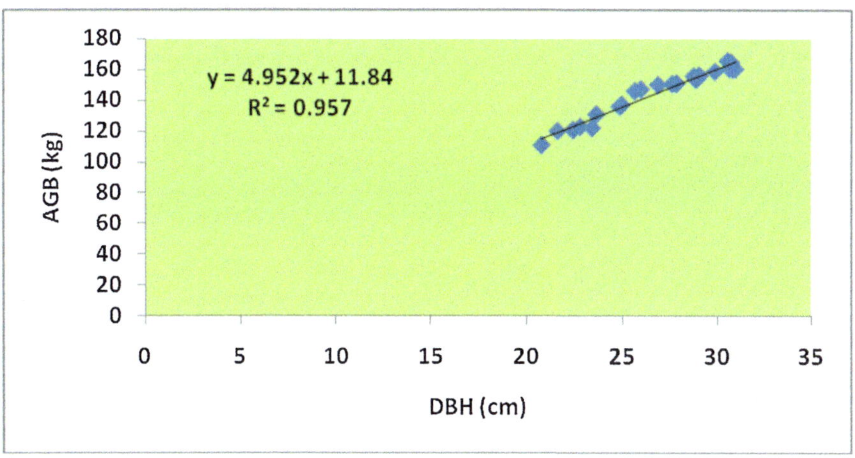

Fig. 7.A.48 Scatter plot showing the inter-relationship between DBH and AGB of *Acacia auriculiformis* (akashmoni) at Kestopur

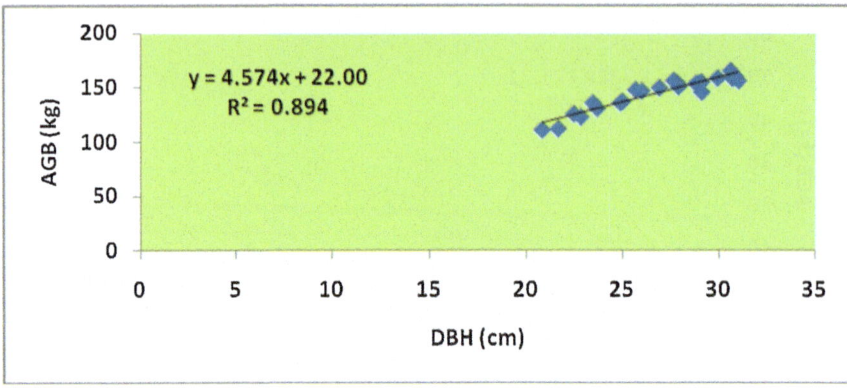

Fig. 7.A.49 Scatter plot showing the inter-relationship between DBH and AGB of *Acacia auriculiformis* (akashmoni) at Barasat

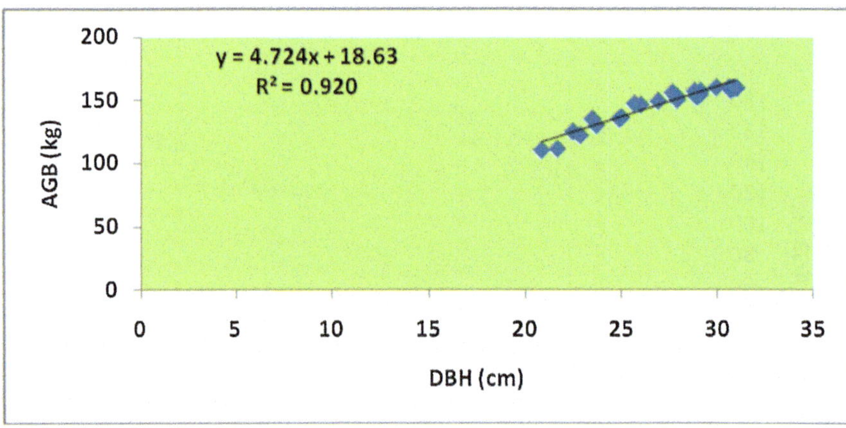

Fig. 7.A.50 Scatter plot showing the inter-relationship between DBH and AGB of *Acacia auriculiformis* (akashmoni) at Konnagar

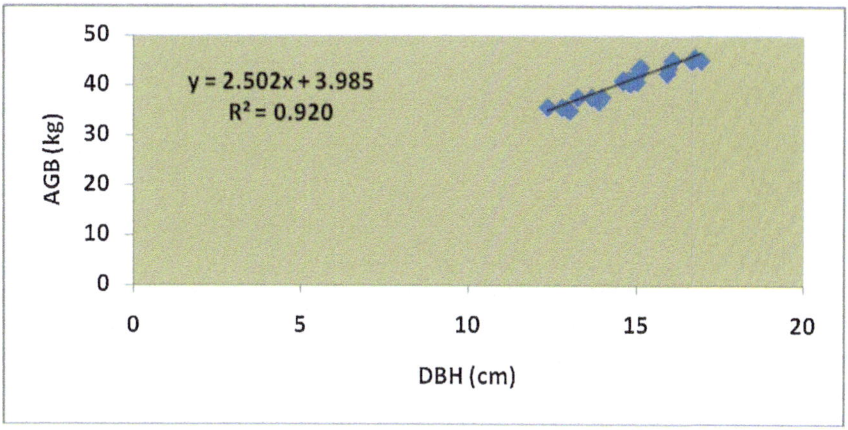

Fig. 7.A.51 Scatter plot showing the inter-relationship between DBH and AGB of *Ziziphus mauritiana* (kul) at Kestopur

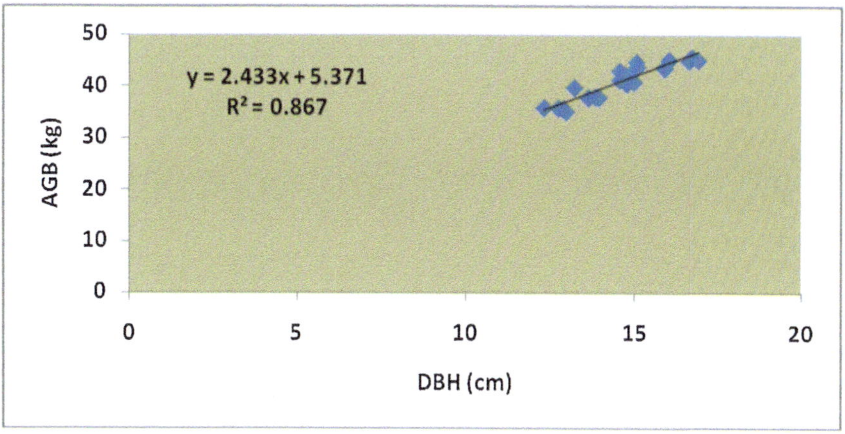

Fig. 7.A.52 Scatter plot showing the inter-relationship between DBH and AGB of *Ziziphus mauritiana* (kul) at Barasat

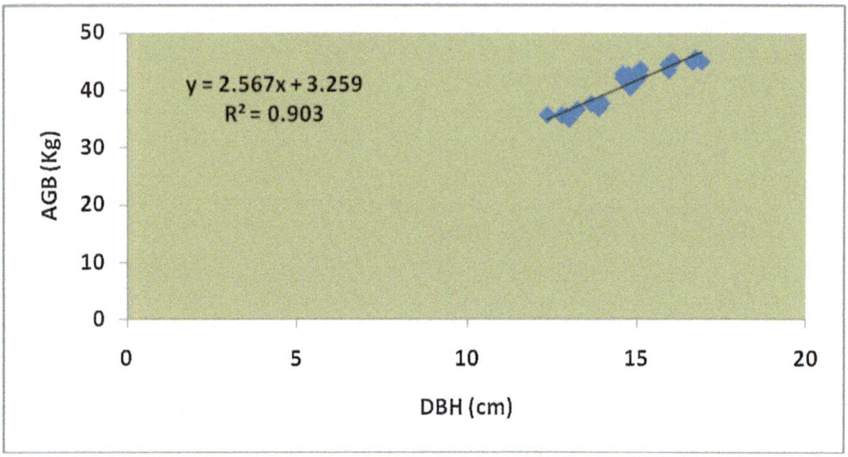

Fig. 7.A.53 Scatter plot showing the inter-relationship between DBH and AGB of *Ziziphus mauritiana* (kul) at Konnagar

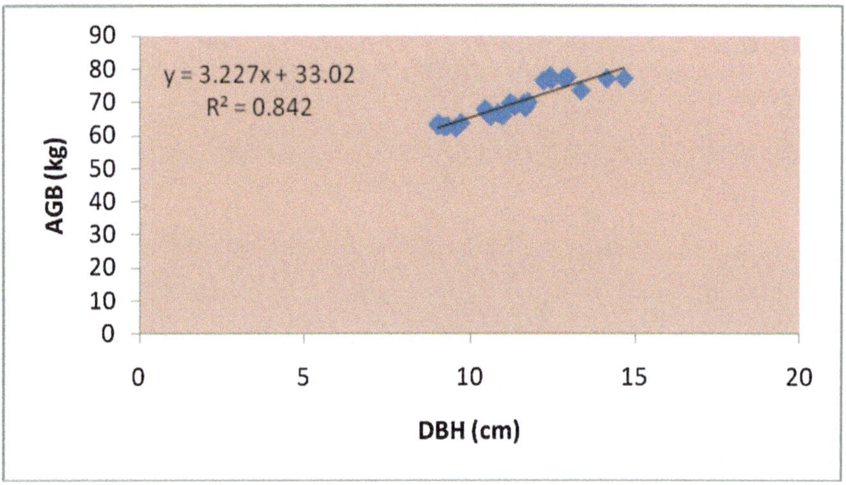

Fig. 7.A.54 Scatter plot showing the inter-relationship between DBH and AGB of *Eucalyptus globulus* (eucalyptus) at Kestopur

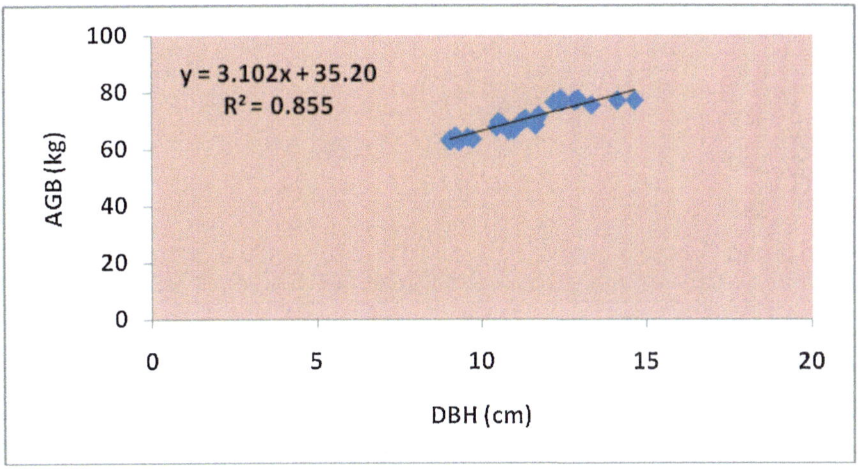

Fig. 7.A.55 Scatter plot showing the inter-relationship between DBH and AGB of *Eucalyptus globulus* (eucalyptus) at Barasat

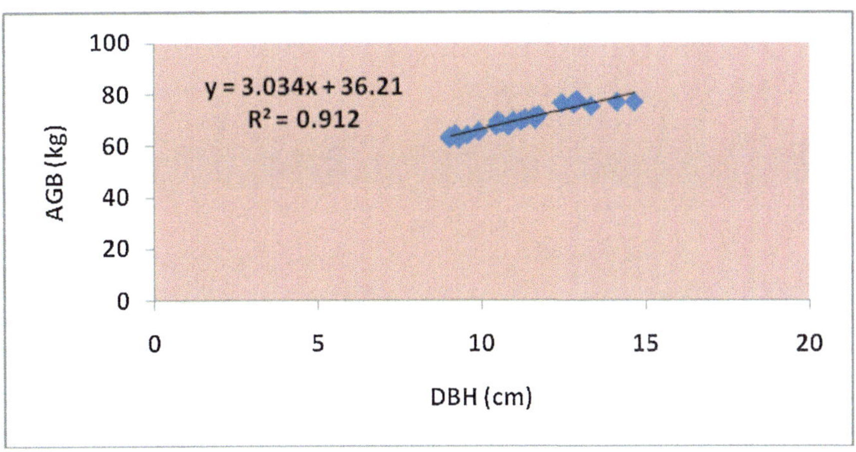

Fig. 7.A.56 Scatter plot showing the inter-relationship between DBH and AGB of *Eucalyptus globulus* (eucalyptus) at Konnagar

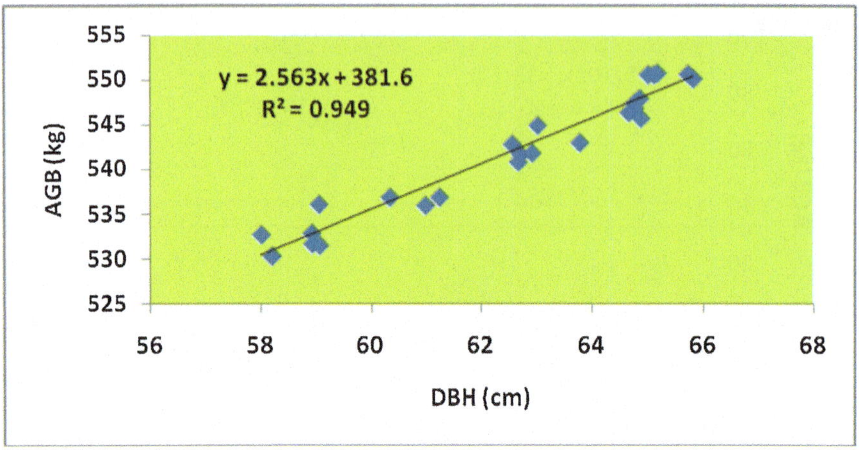

Fig. 7.A.57 Scatter plot showing the inter-relationship between DBH and AGB of *Peltophorum pterocarpum* (radhachura) at Kestopur

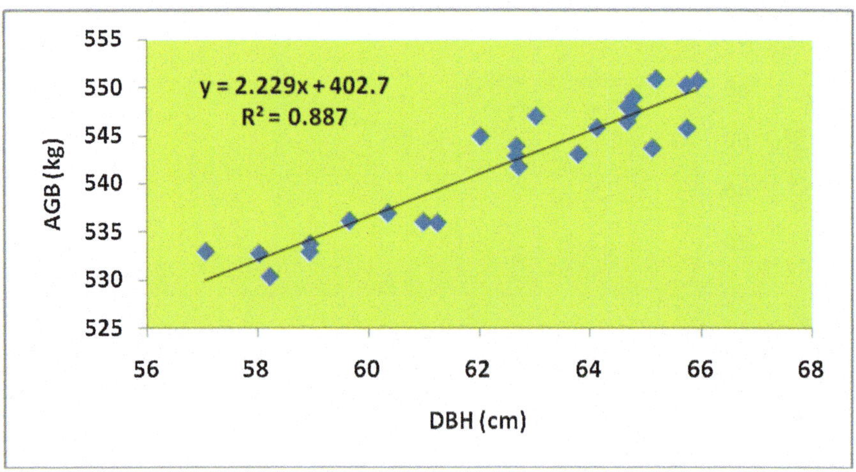

Fig. 7.A.58 Scatter plot showing the inter-relationship between DBH and AGB of *Peltophorum pterocarpum* (radhachura) at Barasat

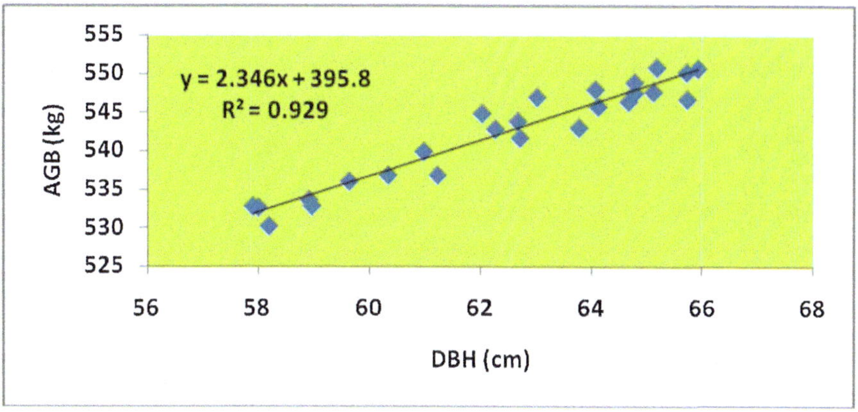

Fig. 7.A.59 Scatter plot showing the inter-relationship between DBH and AGB of *Peltophorum pterocarpum* (radhachura) at Konnagar

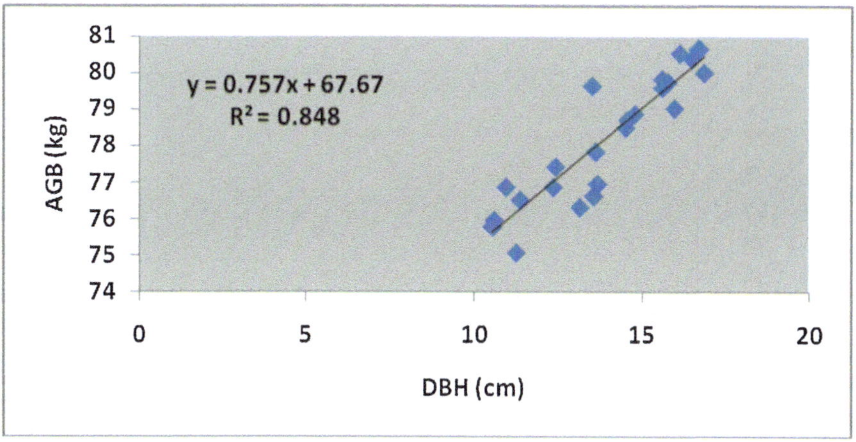

Fig. 7.A.60 Scatter plot showing the inter-relationship between DBH and AGB of *Polyalthia longifolia* (debdaru) at Kestopur

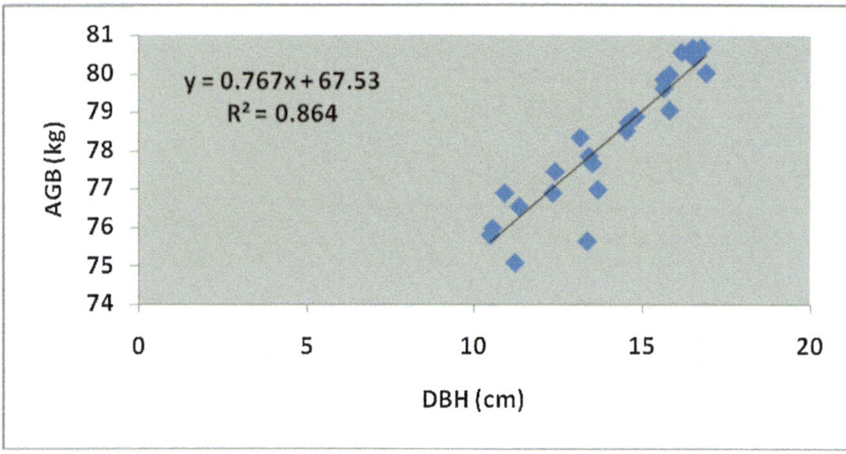

Fig. 7.A.61 Scatter plot showing the inter-relationship between DBH and AGB of *Polyalthia longifolia* (debdaru) at Barasat

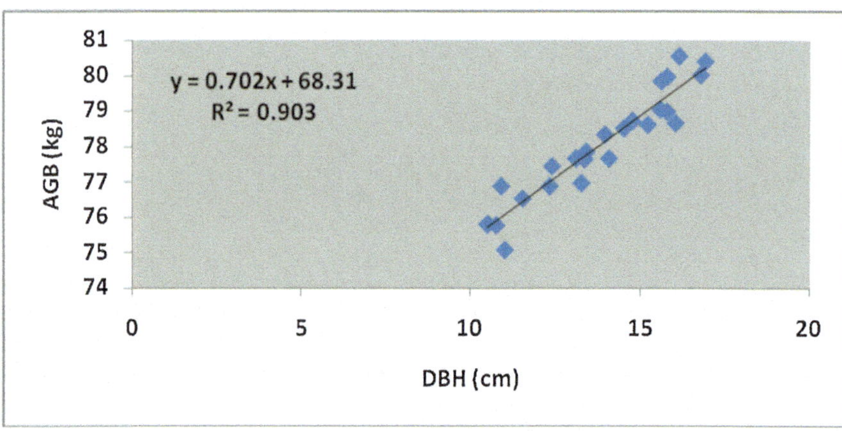

Fig. 7.A.62 Scatter plot showing the inter-relationship between DBH and AGB of *Polyalthia longifolia* (debdaru) at Konnagar

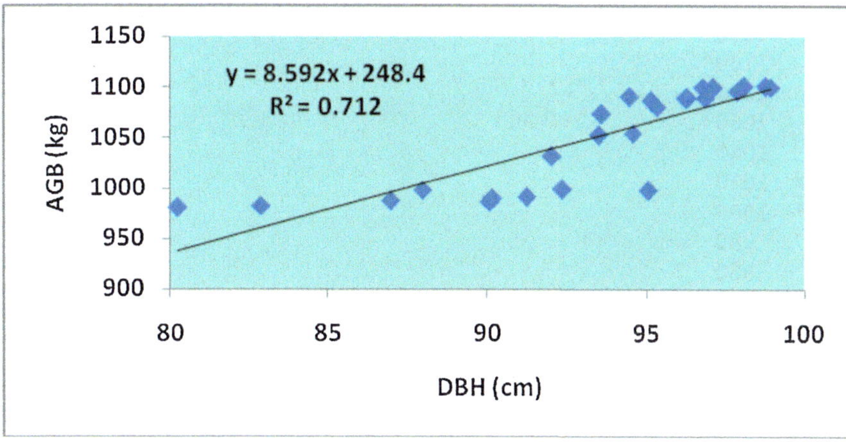

Fig. 7.A.63 Scatter plot showing the inter-relationship between DBH and AGB of *Ficus religiosa* (peepul) at Kestopur

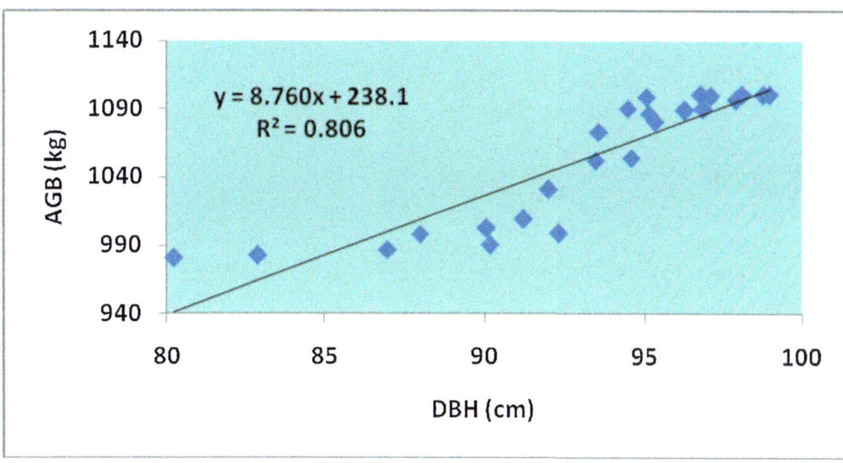

Fig. 7.A.64 Scatter plot showing the inter-relationship between DBH and AGB of *Ficus religiosa* (peepul) at Barasat

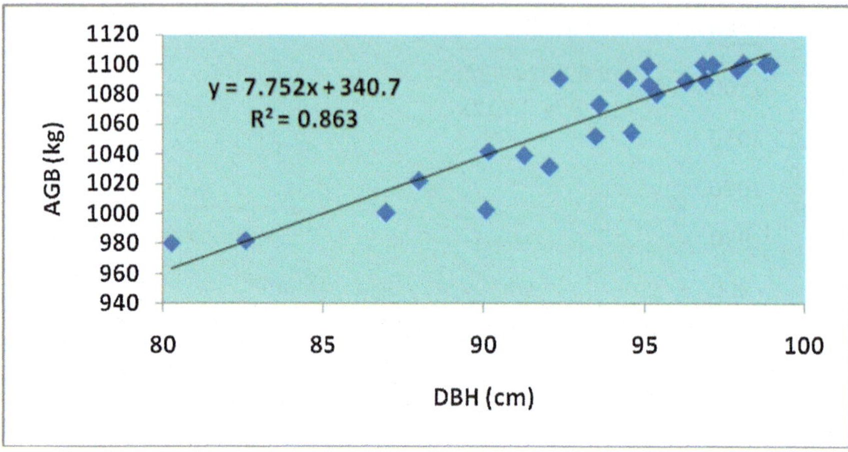

Fig. 7.A.65 Scatter plot showing the inter-relationship between DBH and AGB of *Ficus religiosa* (peepul) at Konnagar

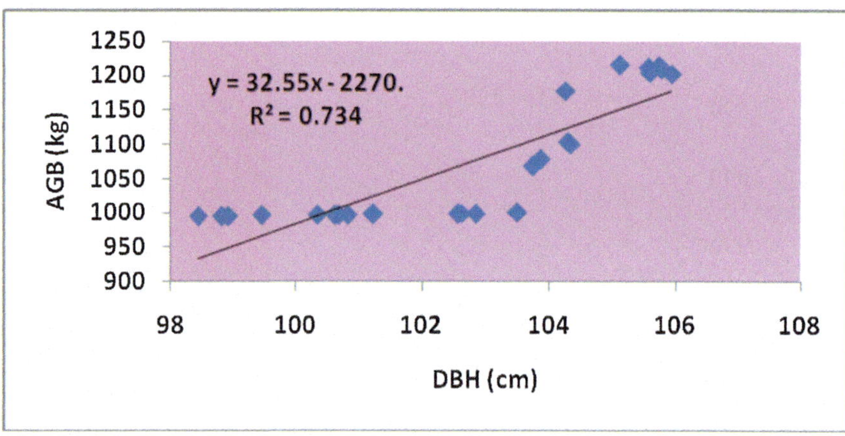

Fig. 7.A.66 Scatter plot showing the inter-relationship between DBH and AGB of *Ficus benghalensis* (banyan) at Kestopur

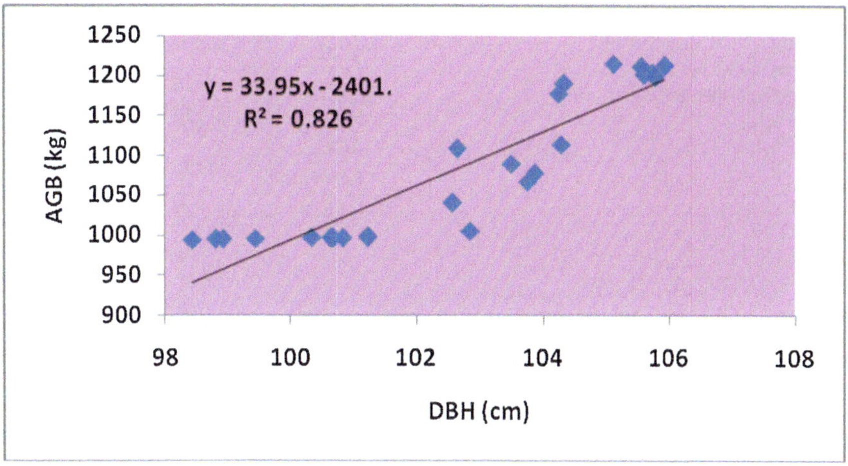

Fig. 7.A.67 Scatter plot showing the inter-relationship between DBH and AGB of *Ficus benghalensis* (banyan) at Barasat

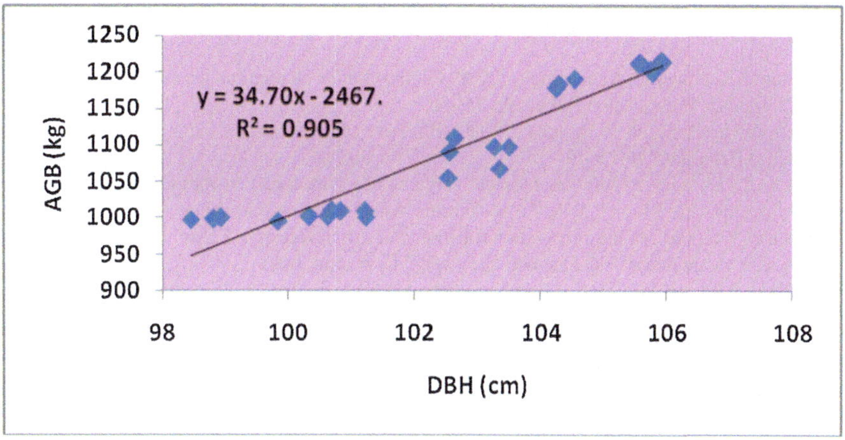

Fig. 7.A.68 Scatter plot showing the inter-relationship between DBH and AGB of *Ficus benghalensis* (banyan) at Konnagar

Table 7.A.3 ANOVA results showing spatial variation of AGB and AGC between selected tree species station wise

Variable	F_{cal}	F_{cric}
AGB		
Between species	12.3102	1.971683
Between stations	27.60294	3.294537
AGC		
Between species	12.38159	1.971683
Between stations	27.72553	3.294537

climate, soil type, annual temperature, groundwater availability, annual rainfall, etc.) in mind. Also the species which harvest more CO_2 from the atmosphere should be planted more.

The author personally feels that plantation of terrestrial plants other than mangroves in the intertidal mudflats of Sundarbans can reduce considerable carbon dioxide level in the atmosphere at local scale. The trees discussed here are abundantly available on the roadside and villages in and around the city of Kolkata. With their considerable biomass, they can serve as potential sink of carbon and reduce the atmospheric carbon dioxide at local scale.

References

Abdollahi KK, Ning ZH, Appeaning A (2000) Global climate change and the urban forest. GCRCC and Franklin Press, Baton Rouge, pp 31–44

Beckett KP, Freer-Smith P, Taylor G (2000) Effective tree species for local air-quality management. J Arboricult 26:12–19

Chaparro L, Terradas J (2009) Ecological services of urban forest in Barcelona. Àrea de Medi Ambient Institut Municipal de Parcs i Jardins, Ajuntament de Barcelona

Chaudhary P (2006). Valuing recreational benefits of urban forestry- a case study of Chandigarh city. Doctorate thesis. FRI Deemed University, Dehradun, India

Chaudhary P, Tewari VP (2010) Managing urban parks and gardens in developing countries: a case from an Indian city. Int J Leis Tour Mark 1(3):248–256

Davies ZG, Edmonson JL, Heinemeyer A, Leake JR, Gaston KJ (2011) Mapping an urban ecosystem service: quantifying above-ground carbon storage at a city wide scale. J Appl Ecol 48:1125–1134

Dubey SK, Pandey DN (1993) The effect of afforestation on the abundance and diversity of birds. In: Dwivedi AP, Gupta GN (eds) Afforestation of arid lands. Scientific Publishers, Jodhpur, pp 313–320

Dwivedi P, Rathore CS, Dubey Y (2009) Ecological benefits of urban forestry: the case of Kerwa Forest Area (KFA), Bhopal, India. Appl Geogr 29(2):194–2009

Francesco F (2011) Sustainable management techniques for trees in the urban areas. J Biodivers Ecol Sci JBES 1(1):1–20

FSI (2009) State of forest report 2009. Forest Survey of India, Ministry of Environment & Forests, Dehradun

Gill SE, Handley JF, Ennos AR, Pauleit S (2007) Adapting cities for climate change: the role of the green infrastructure. Built Environ 33(1):115–133

Gupta RB, Chaudhari PR, Wate SR (2008) Floristic diversity in urban forest area of NEERI campus, Nagpur, Maharashtra (India). J Environ Sci Eng 50(1):55–62

Heath LS, Smith JE, Skog KE, Nowak DJ, Woodall CW (2011) Managed forest carbon estimates for the U.S. Greenhouse Gas Inventory, 1990e 2008. J Forest April/May:167–173

Hutyra LR, Yoon B, Alberti M (2011) Terrestrial carbon stocks across a gradient of urbanization: a study of the Seattle, WA region. Glob Chang Biol 17:783–797

Intergovernmental Panel on Climate Change (IPCC) (2006) 2006 IPCC guidelines for national greenhouse gas inventories. In: Eggleston S, Buendia L, Miwa K, Ngara T, Tanabe K (eds) Agriculture, Forestry and Other Land Use (AFOLU), vol. 4 (last accessed January, 2010). www.ipcc-nggip.iges.or.jp/public2006gl/vol4.html

Khera N, Mehta V, Sabata BC (2009) Interrelationship of birds and habitat features in urban green spaces in Delhi, India. Urban For Urban Green 8(3):187–196

Lal R, Augustine B (2012) Carbon sequestration in urban ecosystems. Springer, New York, p 385

Madan MCS (1993) Composition of the ground vegetation of Visakhapatnam. J Natcon 5:77–82

Mhatre N (2008) Secret lives: biodiversity of the Indian Institute of Science Campus. Indian Institute of Science Press, Bangalore, 229 pp

Mitra S (1993) Some aspects of ecology of walls at Visakhapatnam. Ph.D thesis, Andhra University

Montagnini F, Porras C (1998) Evaluating the role of plantation as carbon sinks; an example of an integrative approach from the humid tropic. Environ Manag 22:459–470

Nagendra H, Gopal D (2010) Street trees in Bangalore: density, diversity, composition and distribution. Urban For Urban Green. https://doi.org/10.1016/j.ufug.2009.1012.1005

Nandini N, Kumar M, Tandon S (2009) Assessment of carbon sequestration in trees of Jnanabharathi Campus – Bangalore University. J Ecol Environ Conserv 15(3):503–508

Nowak DJ (1993) Atmospheric carbon reduction by urban trees. J Environ Manag 37(3):207–217

Nowak DJ (1994) Atmospheric carbon dioxide reduction by Chicago's urban forest. In: McPherson EG, Nowak DJ, Rowntree RA (eds) Chicago's urban forest ecosystem: results of the Chicago urban Forest climate project, general technical report NE-186. USDA Forest Service, Northeastern Forest Experiment Station, Radnor, pp 83–94

Nowak DJ (2010) Urban biodiversity and climate change. In: Muller N, Werner P, Kelcey JG (eds) Urban biodiversity and design. Wiley-Blackwell Publishing, Hoboken, pp 101–117

Nowak DJ, Crane DE (2002) Carbon storage and sequestration by urban trees in the USA. Environ Pollut 116:381–389

Nowak DJ, Greenfield EJ (2012) Tree and impervious cover in the United States. Landsc Urban Plan 107:21–30

Nowak DJ, Rowntree RA, McPherson EG, Sisinni SM, Kerkmann E, Stevens JC (1996) Measuring and analyzing urban tree cover. Landsc Urban Plan 36:49–57

Nowak DJ, Noble MH, Sisinni SM, Dwyer JF (2001) Assessing the U.S. urban forest resource. J Forest 99(3):37–42

Patwardhan AS, Nalavade SK, Utkarsh G (2001) Urban wildlife and protected areas in India. Parks 11(3):28–34

Redondo-Brenes A (2007) Growth, carbon sequestration, and management of negative tree plantation in humid of Costa Rica. New For 34:253–268

Strohbach M, Haase D (2012) The above ground carbon stock of a central European city: patterns of carbon storage in trees in Leipzig, Germany. Landsc Urban Plan 104:95–104

Southwick CH (1976) Pollution, ecology and the quality of our environment. Van Nostrand and Co, New York

Sudha P, Ravindranath NH (2000) A study of Bangalore urban forest. Landsc Urban Plan 47:47–63

U.S. Department of Agriculture Forest Service (2010) Forest inventory and analysis National Core Field Guide: field data collection procedures for phase 2 plots version 5.0, vol. 1. U.S. Department of Agriculture Forest Service, Forest Inventory and Analysis Programme, Arlington, p 361. (last accessed June 2012). www.fia.fs.fed.us/library/field-guides-methods-proc/

Verma SS (1985) Spatio-temporal study of open spaces of part of Jaipur City-Rajasthan. J Ind Soc Remote Sens 13(1):9–16

Wilby RL, Perry GLW (2006) Climate change, biodiversity and the urban environment: a critical review based on London UK. Prog Phys Geogr 30(1):73–98
Zérah MH (2007) Conflict between green space preservation and housing needs: the case of the Sanjay Gandhi National Park in Mumbai. Cities 24(2):122–132
Zhao M, Kong ZH, Escobedo FJ, Gao J (2010) Impacts of urban forest on off-setting carbon emissions from industrial energy use in Hangzhou, China. J Environ Manag 91(4):807–813

Internet References

www.itreetools.org
https://www.indiatoday.in/india/north/story/mit-study-high-rspm-in-delhi-air-pollution-138144-2011-07-24

Chapter 8
Water Pollution and Its Mitigation

Abstract Water is one of the key components to drive the activities of life. It is not only used for drinking, but also for agriculture, aquaculture, industrial activities, running power plants, manufacturing medicines, etc. However, the release of domestic, agricultural, aquacultural and industrial wastes has deteriorated the quality of water to a great extent leading to several ill effects like health disorder, eutrophication, bioaccumulation, alteration of biotic community, etc. The chapter focuses on both the causes and effects of water pollution. The prevention and control of water pollution has also been discussed with special emphasis on cost-effective biological control.

Keywords Bioaccumulation · Goddess Durga · Eutrophication · River Ganga

8.1 Water Pollution: An overview

The term water pollution may be defined as the addition of substances or heat to such a level, which is harmful to humans, animals or any desirable aquatic life, or otherwise causes significant departures from the normal activities of various living communities in or near bodies of water. In practical, the term water pollution refers to several types of aquatic contamination like enrichment of nutrients in ponds, lakes, rivers, bays, estuaries, seas and oceans from sewage and fertilizer, introduction of toxic chemicals (mainly from industries) in water bodies to such a level that the biota are affected to a considerable extent, etc.

The various causes of water pollution may be described on the basis of origin of pollutants that deteriorate the water quality. The important sources of such pollutants are discussed here.

8.1.1 Domestic Wastes and Miscellaneous Factors

These wastes arise from small sources but ultimately spread over a large area. These include wastewater from homes and commercial establishments. The domestic wastes are usually contaminated with nitrates and phosphates and are often responsible for eutrophication – a phenomenon of rise of algal density due

A. Mitra, S. Zaman, *Environmental Science - A Ground Zero Observation on the Indian Subcontinent*, https://doi.org/10.1007/978-3-030-49131-4_8

to enrichment of nutrient in the ambient water. When domestic wastes are dumped into water bodies, the aquatic community in the immediate vicinity does not show any significant alteration. This is because the organic wastes do not get time to decay just at the outfall site, and therefore the dissolved oxygen level does not show any considerable variation. However, somewhat further away, conditions become worse as bacteria and fungi start degrading the organic sewage at the expense of dissolved oxygen. At these sites, the values of dissolved oxygen decrease that has high probability to pose an adverse impact on aquatic lives.

The rising trend of nutrients in the estuaries of lower Gangetic delta has been confirmed by several workers (Mukherjee et al. 2007; Mitra 2013), which is mainly due to some region- specific reasons as highlighted here.

1. Increased industrialization and urbanization
2. Unplanned expansion of shrimp culture units
3. Large-scale use of fertilizers (urea, super phosphate, etc.) for boosting crop production in mono-culturing areas of the islands
4. Mushrooming of tourism and recreational units (as seen in Mandarmani area of Midnapore district of West Bengal)
5. Increased number of fish landing stations
6. Increased number of fishing vessels and trawlers (Table 8.1)
7. Erosion of embankments and mudflats due to wave action
8. Contribution of litters and mangrove detritus from the adjacent landmasses

A study conducted in the three sampling stations in the Hooghly–Matla estuarine complex revealed gradual increase of nitrate, phosphate and silicate with the

Table 8.1 No. of mechanized boat or trawlers during the time period of 1998–2014

Year	No. of mechanized boat/trawlers
1998–1999	3362
1999–2000	3585
2000–2001	3622
2001–2002	3763
2002–2003	4175
2003–2004	4481
2004–2005	4521
2005–2006	4575
2006–2007	4630
2007–2008	5143
2008–2009	5260
2009–2010	5435
2010–2011	5490
2011–2012	5506
2012–2013	5578
2013–2014	5602

passage of time. This study exhibits a long-term data bank of three decades (1984–2014) in the lower Gangetic delta region, and the sharp rise of nutrient level during premonsoon in 2009 in all the three sampling stations (Figs. 8.1, 8.2, 8.3, 8.4, 8.5 and 8.6) is the outcome of Aila, a super cyclone that hit the zone during May 24–25, 2009.

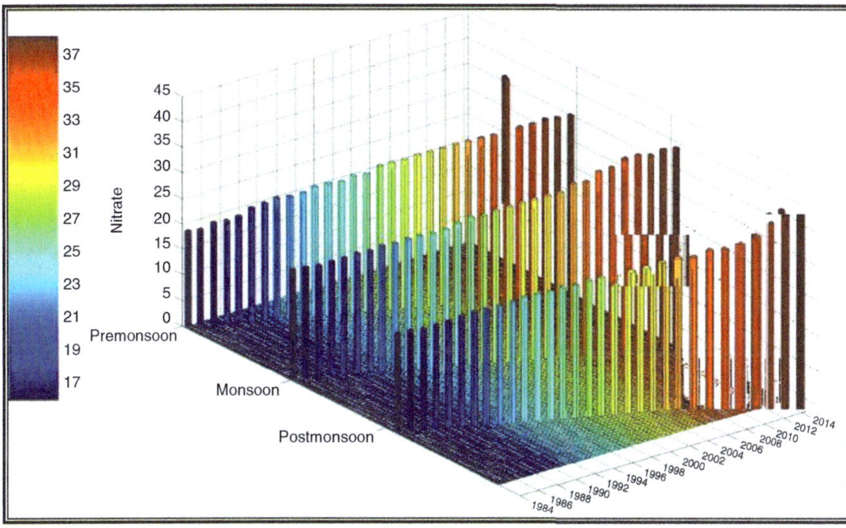

Fig. 8.1 Dissolved nitrate (in µg at l^{-1}) in Diamond Harbour during 1984–2014

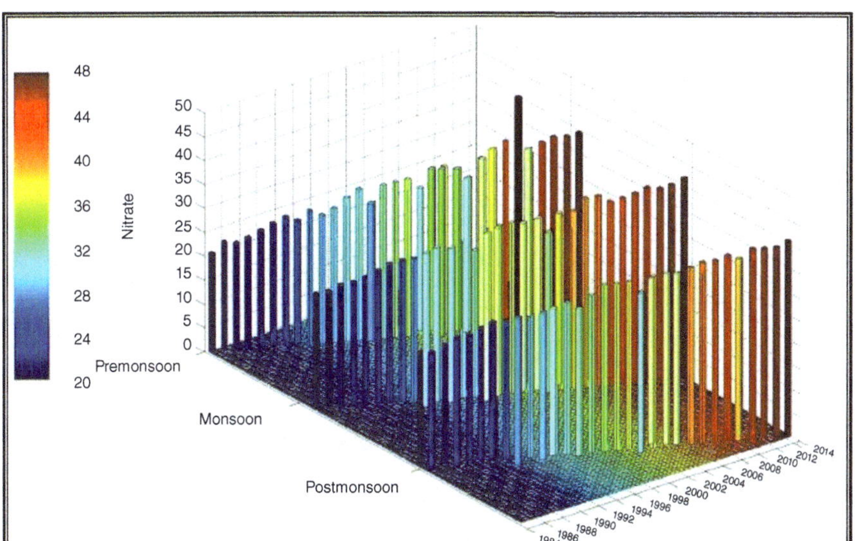

Fig. 8.2 Dissolved nitrate (in µg at l^{-1}) in Namkhana during 1984–2014

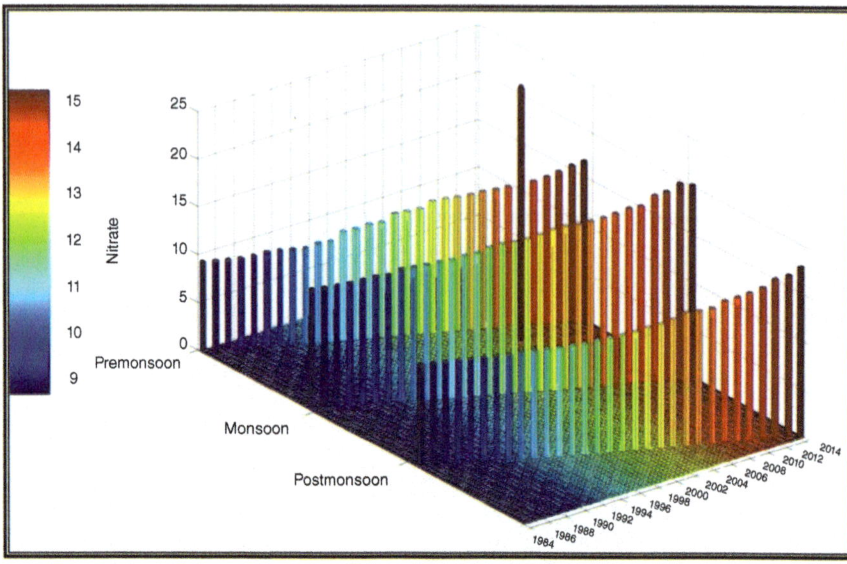

Fig. 8.3 Dissolved nitrate (in μg at l^{-1}) in Ajmalmari during 1984–2014

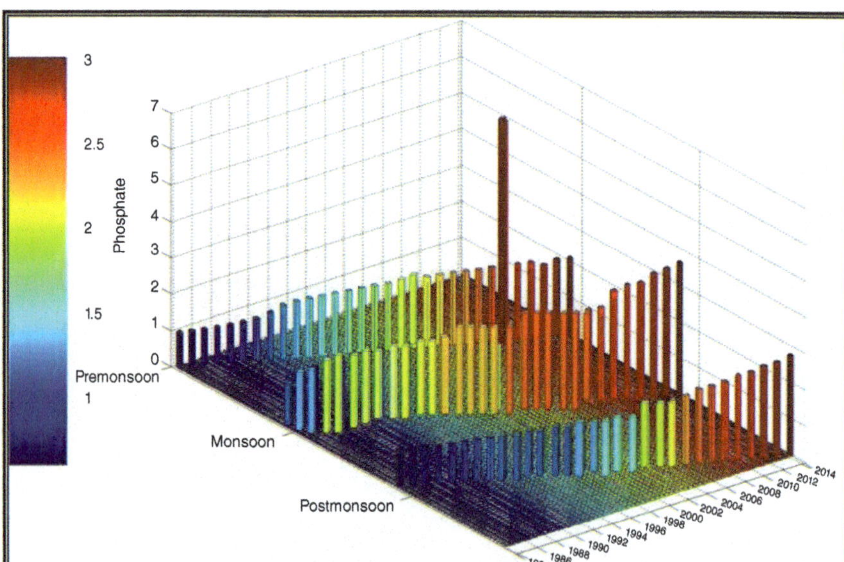

Fig. 8.4 Dissolved phosphate (in μg at l^{-1}) in Diamond Harbour during 1984–2014

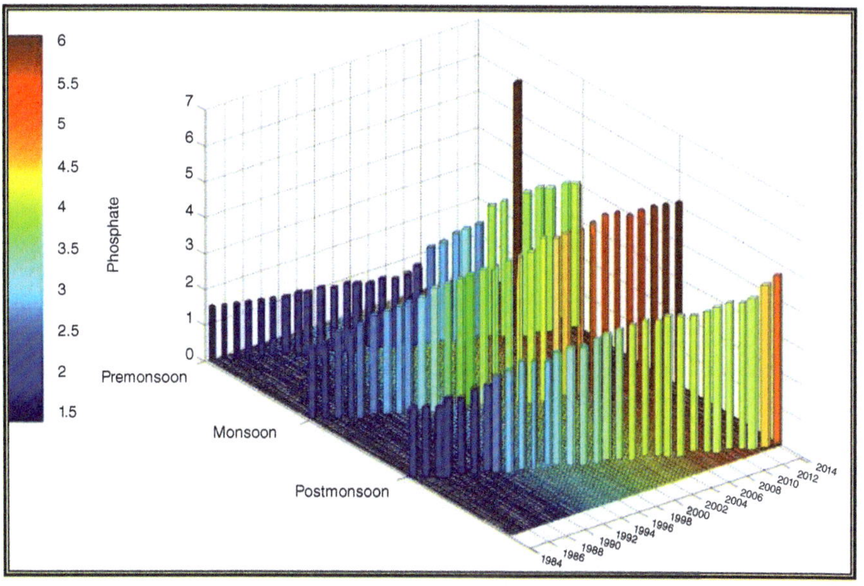

Fig. 8.5 Dissolved phosphate (in µg at l^{-1}) in Namkhana during 1984–2014

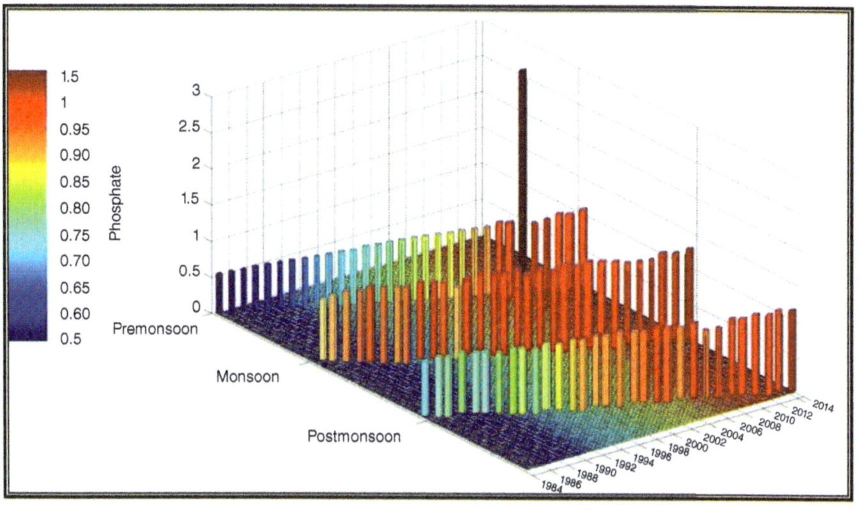

Fig. 8.6 Dissolved phosphate (in µg at l^{-1}) in Ajmalmari during 1984–2014

In addition to the stated factors, aquatic ecosystem also gets polluted during several religious festivals for which India is famous. West Bengal, a maritime state in the northeast part of the country is known for a number of religious festivals of which Durga Puja is the biggest. The tenth day of the Durga Puja, also known as Dashami, is the day when Visarjan (immersion in water) of the idols occur with grand celebrations and processions. The idols are immersed from different banks of the river. Along with the idols, puja articles such as flowers, food offerings, metal polish, plastic sheets, cosmetic items and polythene bags are also thrown into the water. The River Ganga, a sacred river originating from Gangotri in the Himalayas, is the lifeline of the city of Kolkata, and hence any change in the water quality of the river is an issue of concern. A detailed study carried out by the present authors on the alteration of water quality of the mighty River Ganga in terms of heavy metals is presented as Annexure 8A. The study has been discussed in the format of a scientific paper in order to let the budding researches know the art of paper writing.

8.1.2 Industrial Wastes

Several industries contribute heavy metals (Table 8.2) into the aquatic ecosystem either collectively or in an isolated manner depending on the nature and products of the industries.

Industries are major sources of water pollution. A large number of factories and industries located on the banks of rivers and also in the coastal areas discharge their effluents without adequate treatment into the surrounding water bodies. Most of the Indian rivers are polluted by various types of industries (Table 8.3).

Sometimes the industrial wastes carry bulk of conservative wastes (nondegradable pollutants), which accumulate inside the body tissue of flora and fauna inhabiting the aquatic system where the industrial wastes are discharged. Such process is highly dangerous as it can pose threat to human health through food chain. A study conducted by the present authors during 2019 in the inshore region of Bay of Bengal (in Indian Sundarbans) showed considerable accumulation of heavy metals in some edible fishes of coastal waters (Table 8.4).

These results are in accordance with several research works carried out in the present geographical locale (Mitra et al. 2008; Banerjee et al. 2008; Mitra et al. 2009, 2011, 2012; Bhattacharyya et al. 2013; Das et al. 2014; Ray Chaudhuri et al. 2014). Even forecasting based on the available data generated similar trend of heavy metal pollution in the present study area (Zaman et al. 2014).

8.1.3 Agricultural Wastes

The agricultural wastes include sediments, fertilizers, pesticides and farm animal wastes. These pollutants can enter the adjacent water bodies as runoff from agricultural lands. In recent years there has been an increased use of agricultural chemicals,

Table 8.2 Possible sources of Zn, Cu, Mn, Fe, Co, Ni, Pb, Cd, Hg and Cr in aquatic ecosystem

Metals	Possible sources
Zn	(i) Galvanising industry (ii) Antifouling paints used in fishing vessels and trawlers (iii) Painting industry (iv) Pharmaceutical waste (v) Insecticide manufacturing units (vi) Cosmetics manufacturing units (vii) Mining wastes from Pb to Zn mines (viii) Oil sludges (7–80 ppm)
Cu	(i) Pesticide (ii) Algicide (iii) Antifouling paints manufacturing units (vi) Pipe line corrosion (v) Oil sludges (32–120 ppm)
Mn	(i) Pharmaceutical wastes (ii) Galvanising industry (iii) Painting industry
Fe	(i) Acid mine drainage (ii) Corrosion of pipe lines and pumps (iii) Floating old stranding and rusty barges (vi) Explosive manufacturing units
Co	(i) Metal industries
Ni	(i) Electroplating industry (ii) Galvanising industry (iii) Metal industry
Pb	(i) Printing industry (ii) Dyeing industry (iii) Oil refineries
Cd	(i) Electroplating industry (ii) Chemical industry (iii) Acid mine drainage
Cr	(i) Electroplating industry (ii) Painting industry (iii) Dyeing units (iv) Tannery industry (v) Explosive manufacturing units (vi) Ceramics industry (vii) Paper industry (viii) Printing industry (ix) Cr addition to cooling waters for checking corrosion
Hg	(i) Electrical apparatus (ii) Chlor–alkali industry (iii) Paints (iv) Industrial and control instruments (v) Dental care units (vi) Pharmaceuticals units (vii) Paper and pulp industry (viii) Amalgams

Table 8.3 Sources of pollution of some major Indian rivers

River	Sources of pollution
Kali at Meerut	Sugar mills; distilleries; paint, soap, rayon, silk, tin and glycerine industries
Yamuna near Delhi	DDT factory, sewage, Indraprastha Power Station (Delhi)
Ganga at Kanpur	Jute, chemical, metal and surgical industries; tanneries, textile mills and great bulk of domestic sewage having high organic matter
Gomti near Lucknow (U.P.)	Paper and pulp mills; sewage
Dajora in Bareilly (U.P.)	Synthetic rubber factories
Damodar between Bokaro and Panchet	Fertilizers, fly ash from steel mills, suspended coal articles from washeries and thermal power station
Hooghly near Kolkata	Power stations; paper pulp, jute, textiles, chemical mills, paints, varnishes, metal, steel, hydrogenated vegetable oils, rayons, soap, match, shellac and polythene industries and sewage
Sone at Dalmianagar (Bihar)	Cement, pulp and paper mills
Bhadra (Karnataka)	Pulp, paper and steel industries
Coom, Adyar and Buckingham Canal (Madras)	Domestic sewage, automobile workshops
Cauvery (Tamil Nadu)	Sewage, tanneries, distilleries, paper and rayon mills
Godavari	Paper mills
Siwan (Bihar)	Paper, sulphur, cement, sugar mills
Kulu (between Bombay and Kalyan)	Chemical factories, rayon mills and tanneries
Suwao (in Balrampur)	Sugar industries

Table 8.4 Concentration of Pb in the muscle tissue of fish samples collected from Frasergaunge

Finfish species	Common local name	Pb (in μg/gm dry wt.)
Tenualosa ilisha	Ilish	14.84
Pampus argenteus	Pomfret	3.82
Liza parsia	Parse	10.65
Liza tade	Bhangone	19.11
Polynemus paradiseus	Tapsee	7.62
Thunnus albacares	Tuna	3.38
Coilia neglecta	Ruli	2.65
Cynoglossus sp.	Banspata	4.16
Scatophagus argus	Paira chanda	6.43
Ilisha elongata	Dhala	18.08

notably pesticides and fertilizers. The availability of low-priced chemical fertilizers has led to rapid increase in their usage. Agricultural wastes can also lead to eutrophication as plant nutrients and fertilizers contribute appreciable amount of nitrates and phosphates to the ambient media. The greatest agricultural pollution at present is probably due to soil erosion by water and wind, as many agricultural pollutants like pesticides and phosphates are transported by sediments to water bodies.

8.1.4 Shipping Wastes

These include both human sewage and other wastes, the most important of which is oil. About half of the crude oil produced per year is transported by sea. After unloading a cargo of oil from a tanker, it carries seawater as ballast. It is a general practice to fill several of the tanks (25–30% of the total capacity of the tanker) with seawater to ballast the ship for the voyage back to loading terminal. These ballast waters are discharged prior to filling the tanks with new oil. When a tanker is unloaded, it cannot be completely cleaned because some oil remains on the bottom and some cling to the sides of the tank. The ballast water inevitably becomes contaminated with this remaining oil and is responsible for the deterioration of water quality.

8.1.5 Radioactive Wastes

According to the French Environmental Code (Art. L 542.1-1), final radioactive waste means radioactive waste for which no further treatment is possible under existing technical and economic conditions. Treatment particularly entails extracting any part of the waste that can be recycled or reducing any pollutants or hazardous substances it contains.

The radionuclides contained in radioactive waste may be man-made, such as caesium-137, or found in nature, such as radium-226.

Radioactive wastes are classified according to their activity level and the radioactive half-life of the radionuclides they contain. The activity level determines the degree of protection to be provided. Wastes are therefore divided into categories, namely, very low-, low-, intermediate- and high-level waste. Radioactive waste is said to be 'short-lived' if it merely contains radionuclides with a half-life of less than 31 years. It is said to be 'long-lived' if it contains a significant quantity of radionuclides with a half-life of over 31 years (Table 8.5).

Table 8.5 Half-life of radionuclide

Radionuclide	Half-life
Cobalt-60	5.2 years
Tritium	12.2 years
Strontium-90	28.1 years
Caesium-137	30 years
Americium-241	432 years
Radium-226	1600 years
Carbon-14	5730 years
Plutonium-239	24,110 years
Neptunium-237	2,140,000 years
Iodine-129	15,700,000 years
Uranium-238	4,470,000,000 years

The marine and estuarine systems receive radioactive wastes mainly from the nuclear power plants, nuclear power ships and submarines. Also extensive use of radio isotopes as tracers in the fields of medicine, biology, chemistry and biotechnology contributes radioactive substances in the city's drainage system.

Other major sources of radioactive pollutants in the marine and estuarine compartments originate from weapon testing and leaching of radioactive material from soil or sediment which adds to the amount of radiation of the sea.

Radioactive waste is generated not only by the nuclear power industry but also by hospitals, universities and non-nuclear industries. All the regulations applying to waste in general also apply to radioactive waste. However, radioactive waste emits radiation, which has imparted a different status to it in the domain of human health and the environment. It must therefore be managed with special care, from production to final disposal. Finding suitable waste disposal solutions is a major challenge for all stakeholders, industry, regulatory authorities, public authorities, local communities and the population.

8.1.6 Aquaculture Wastes

Aquaculture encompasses the culture of aquatic organisms like finfish, shellfish, seaweeds, algae, etc. in a controlled condition. Although freshwater aquaculture has not flourished yet to the expected target, the coastal aquaculture, particularly the shrimp culture, has achieved a unique position in the Indian sub-continent. A variety of finfishes, oysters, mussels, sea cucumbers, crabs and lobsters are in the export list of India, but shrimp is the single dominant item in the export basket of our marine food which accounts for almost two-thirds of the total export earnings. To enhance the rate of shrimp production (which is usually expressed in tonnes per hectare) from the shrimp culture farms, the stocking density is also significantly increased along with simultaneous input of artificial feed and fertilizers. All these processes increase the nutrient concentrations and organic load per unit area or percent weight of shrimp production, which often exceeds the assimilatory capacity of the system and results in aquatic pollution. The leftover artificial feed supplied for the shrimp growth also generates H_2S in the pond bed through microbial action. The uses of antibiotics like oxytetracycline to prevent the prawn diseases also pollute the water bodies to a great extent.

8.1.7 Heat

A large volume of water is used for cooling purposes by various industries, mainly the steam electric power plants, refrigerator factories, etc. Cooling water is discharged at a raised temperature, and this has profound adverse effect on the growth and physiology of aquatic biota. This is mainly due to the fact that enzymes of organisms have a particular temperature range for their optimum activity, and beyond this range there is high probability of enzyme denaturation.

8.2 Effects of Water Pollution

Water pollution causes several water-borne infectious diseases like cholera, typhoid, etc. Plant nutrients and aquaculture wastes contribute NO_3, PO_4 and NH_3 in the aquatic phase, which trigger the growth of algae leading to eutrophication. Nitrates in drinking water can cause methemoglobinaemia in babies (blue babies), as in an infant's stomach, the nitrate is transformed into nitrite, which acts on the blood haemoglobin to form methemoglobin.

The heavy metals discharged from various industries also accumulate in the body tissues of the aquatic organisms and affect their normal physiological processes and growth. The growth of tiger prawns *Penaeus monodon* cultured with the wastewasters of the Kolkata city has been reduced to a significant level in comparison to their normal growth (Mitra et al. 1999). The diversity of fish juveniles is also reduced by the presence of various pollutants particularly lead, which originate from antifouling paints used for conditioning fishing vessels and trawlers in coastal areas (Mitra et al. 2000). The organic wastes not only reduce the dissolved oxygen concentration to the aquatic phase but also promote the growth of anaerobic bacteria. This ultimately poses a negative stress on the fish population of the water bodies. Oil pollution in coastal zone or estuarine area often affects light penetration and oxygen level in the column water by creating a film on the surface water. This affects the biotic community structure in the oil-spilled area. The important adverse effects caused by water pollutants are shown in Table 8.6.

For convenience of the readers, the effects of water pollution are discussed under two broad headings as discussed here.

1. **Effects of Conservative Wastes**

Conservative wastes are non-biodegradable in nature. Conservative wastes like heavy metals bioaccumulate within the body tissues of organisms depending on the ambient hydrological parameters like concentrations of dissolved heavy metals, pH of the ambient water, etc. A study conducted by the present authors comparing the metal levels in *Channa punctata* collected from the Muriganga River of Indian Sundarbans and Buriganga River of Bangladesh during monsoon 2019 exhibited considerable spatial variation with significantly higher values in the fish tissues sampled from Buriganga. Discharging of sewage and untreated waste containing several heavy metals from various industrial effluents including tannery waste and effluents is the major causes behind the high values of metals in the fish samples of Buriganga River (Table 8.7). The major complaint goes against the Hazaribagh tannery city and other industrial zones for discharging its perilous effluents through untreated waste. The heavy metals emitting through these wastes are obstinate and

Table 8.6 Water pollutants: sources, effects and residence times

Pollutants	Major anthropogenic sources	Adverse effects	Residence time
Oxygen demanding organic wastes (biodegradable)	Domestic sewage, industrial wastes, animal wastes	Depletion of dissolved oxygen, death of fish, damage to plant life, foul smell	Days to weeks
Pathogens	Domestic sewage, animal wastes	Outbreak of water-borne diseases	Days to months
Acids	Industrial wastes, acidic deposition	Death of aquatic organisms, solubility of harmful substances increases	Up to years
Salts	Irrigation, industrial wastes	Increase of salinity, death of aquatic organisms, becomes unfit for drinking purpose, domestic use, industrial use and agricultural use	Variable
Heavy metals, arsenic, etc.	Industrial wastes, leaching from soils by acidic substances	Toxic to humans, animals and aquatic life	Months to years
Plant nutrients (NO_3, PO_4)	Agricultural, domestic and industrial wastes	Eutrophication, fish death, disruption of community structure	Decades
Heat	Cooling water from power plants and other industries	Solubility of oxygen decreases, harmful to aquatic life	Days
Radioactive substances	Mining of radioactive substances, nuclear power generation	Carcinogenic and mutagenic	Days to years
Chlorine and its compounds	Paper and other industries, water disinfection	Fatal to fish possibly carcinogenic	Variable
Oil and grease	Lubricants and solvents, etc., petroleum wastes	Potential damage to ecosystem, taste and odour problems for drinking water	Days to years
Pesticides and herbicides	Agriculture, pesticide and herbicide manufacturing units	Toxic to aquatic life, mammals and humans, some chemicals are carcinogenic and mutagenic	Days to years
Other synthetic organics (phenols ethers, chloroform nitrosamines, etc.)	Industrial effluents	Toxic to aquatic life, mammals and humans, some chemicals are carcinogenic	Months to years

non-biodegradable which are also known for creating detrimental effects on both animal and human health (Javed and Usmani 2016).

2. **Effects of Nutrients**

Nutrient enrichment in the aquatic ecosystem mostly leads to eutrophication, which can change the characteristic form of submerged and shoreline vegetation undesirable, producing swampy tangles of growth through which fish can hardly navigate.

Table 8.7 Selected heavy metal concentrations in *Channa punctata* collected from Muriganga River (A) and Buriganga River (B) during 2019

	Metal concentration (mg/kg) dry wt. basis		
Species	**Zn**	**Cu**	**Pb**
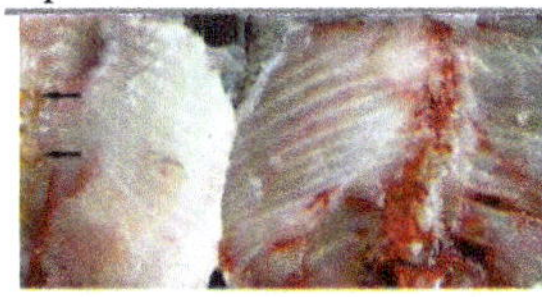 **Muscle**	A = 19.59 B = 35.67	A = 9.63 B = 21.12	A = BDL B = 7.17
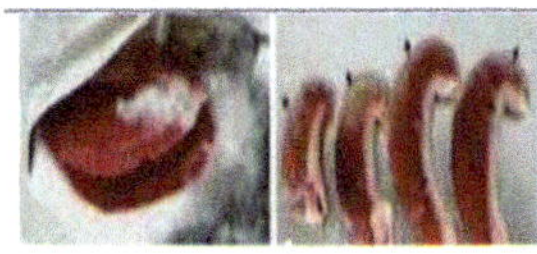 **Gills**	A = 60.12 B = 73.77	A = 13.87 B = 30.10	A = 5.49 B = 12.15
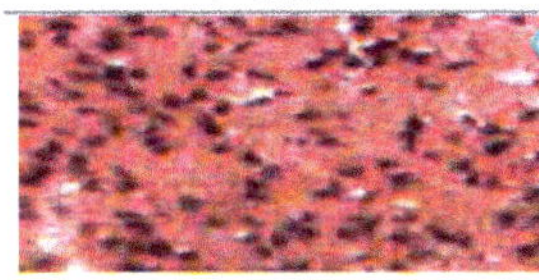 **Hepatopancreas**	A = 63.16 B = 84.58	A = 17.32 B = 41.25	A = 7.05 B = 19.06

Dense blooms of cyanobacteria can occur, crowding out normal and more desirable algae and protists. At night or on cloudy days, the plant life consumes more oxygen than it produces. Finally, the time inevitably comes when a large part of the plant biomass is killed. Its decay then reduces the dissolved oxygen concentration of the water so much that the lake or waterbody cannot support most aquatic animal life and massive fish kills result.

The sequential schematic representation of different events, which take place during eutrophication, is shown here in a flowchart (Fig. 8.7).

The various effects caused by nutrient enrichment are now discussed in points:

(A) Change in the Physico-Chemical Property

A typical eutrophic lake has large surface: volume ratio, that is, the surface area is large in relation to depth. This is because of the accumulation of the organic-rich sediments at the bottom part of the waterbody.

Due to enrichment of nutrients, heavy growth of algae and other aquatic plants occurs. The increased production of phytoplankton increases total biological productivity of the aquatic ecosystem. The phytoplankton becomes concentrated in the upper layers of the water, giving it a musky green cast. The turbidity reduces light penetration and affects the primary production.

The dissolved oxygen concentrations become high during the daytime due to high photosynthetic activity, but during the midnight, the oxygen level drastically reduces as the gas is consumed by the aquatic lives (including the heavily grown floral population) thriving in the waterbody.

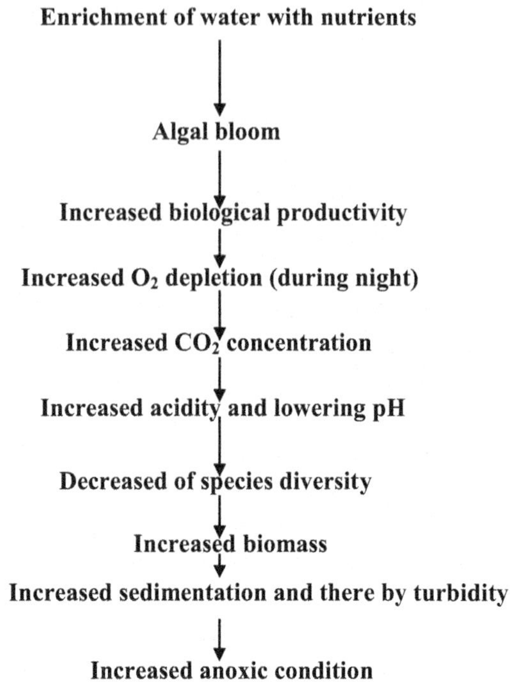

Fig. 8.7 Impact of nutrient enrichment in aquatic ecosystem

(B) **Alteration of Pelagic Community**

Due to enrichment of nutrients in the aquatic phase, the phytoplankton biomass increases greatly which supports diverse type of zooplankton species. Due to increased stressed condition, the biomass of a particular community becomes high due to increased number of certain opportunistic species, but the diversity falls sharply.

(C) **Alteration of Benthic Community**

The herbivores of aquatic ecosystem, mostly grazing zooplankton, are unable to consume the bulk of algae in the eutrophic waterbody. The large biomass of unconsumed algae as well as rooted aquatics dies and sinks to the bottom. On the bottom, the aerobic decomposers are unable to reduce the organic matter to inorganic form and perish due to depletion of oxygen. Gradually the aerobic forms of life at the bottom substratum get replaced by anaerobic organisms that incompletely decompose the organic matter. Partially decomposed bottom sediments build up at the bottom, and sulphate-reducing bacteria release hydrogen sulphide that poses a negative stress on the benthic community of the aquatic ecosystem.

(D) **Toxicity**

Single-celled algae are present in seawater even if the water looks clear. When high concentrations of certain species of dinoflagellates are present, patches of water

Table 8.8 Adverse effects of eutrophication on lakes, reservoirs, rivers and coastal marine waters

Increased biomass of phytoplankton
Toxic or inedible phytoplankton species
Increases in blooms of gelatinous zooplankton
Increased biomass of benthic and epiphytic algae
Changes in macrophyte species composition and biomass
Decreases in water transparency
Taste, odour and water treatment problems
Dissolved oxygen depletion
Increased incidences of fish kills
Loss of desirable fish species
Reductions in harvestable fish and shellfish
Decreases in perceived aesthetic value of the waterbody

From Carpenter et al. (1998); modified from Smith (1998)

look red because these algae contain red pigments – hence the name 'red tide'. High concentrations of other algae may turn seawater orange, yellow, brown or purple. Red tides have been witnessed for centuries and have been seen all over the world. Dense concentrations of algae are referred to as blooms, because the algae have multiplied rapidly to become concentrated in high numbers. In a bloom, there could be tens of millions of cells in a liter of seawater. Most blooms are not harmful, but some have the potential to be harmful, whether by virtue of natural biotoxins (poisons) produced by certain species of algae, or by the oxygen-depleting process initiated upon the death and subsequent decay of large concentrations of algae.

Eutrophication is thus a curse to biodiversity, productivity and optimum hydrological parameters of aquatic system. The adverse effects posed by this event are summarized in Table 8.8.

8.3 Prevention and Control of Water Pollution

Biodegradable pollutants present in water are easy to remove with microbial technology, but conservative pollutants like heavy metals cannot be degraded or decomposed and hence require expensive technologies for their removal.

Water is often referred to as universal solvent. Hence it is practically impossible to get water in pure state in nature. Modern society utilizes water as principle and stable resource for domestic, industrial and agricultural purposes, power supply and aquaculture and even for recreation by enclosing it in artificial reservoirs. Pollution of water will definitely minimize these uses of water and will pose a negative impact on mankind. The various methods for minimizing and controlling water pollution are discussed here.

I. The existing sewage treatment methods (like primary, secondary and tertiary treatment) are very effective in removing selective pollutants from contaminated water.

Primary treatment is a mechanical process which simply removes solids. The relatively larger solid particles are separated by means of metal screens; sands and small stones settle in grit chamber from which the water passes to the sedimentation tank, where the velocity of water is drastically reduced and the small particles settle as sludge. Scum is removed from the uppermost layers.

Secondary treatment is essentially biological process which has the efficiency to remove most of the organic matter. In the activated sludge process of secondary treatment, the incoming sewage is mixed with decomposer bacteria and air or oxygen. Thus the complex organic matter through this treatment is broken down into simpler forms. One important drawback of this method is that if any toxic, industrial chemicals enter the sewage, then the decomposer bacteria may be killed and the entire operation may be inhibited.

Trickling filters are conventional aerobic biological wastewater treatment units, such as active sludge systems or rotating biological contactors. The advantage of all these systems is that they are compact (i.e. applicable in densely populated urban settings) and that they efficiently reduce organic matter (Jenssen et al. 2004). However, they are high-tech and generally require skilled staff for construction as well as for operation.

Trickling filters are a secondary treatment after a primary settling process (Fig. 8.8).

Tertiary treatment is able to remove virtually all the remaining contaminants. Water leaving conventional secondary treatment still has most of the original phosphates and nitrates, many persistent insecticides and herbicides, disease-causing bacteria and viruses and perhaps a number of industrial, organic compounds. Wastewater that is not subjected to tertiary treatment contains the nutrients on which algae thrive.

II. Contaminated water is also treated by adsorption, electrodialysis, ion exchange and reverse osmosis.

Of the various techniques, the reverse osmosis technique is based on the removal of salts and other substances from water by forcing the later through a semipermeable membrane under a pressure that exceeds the osmotic pressure so that the flow is in the reverse direction to the normal osmosis flow. In practice, this involves a

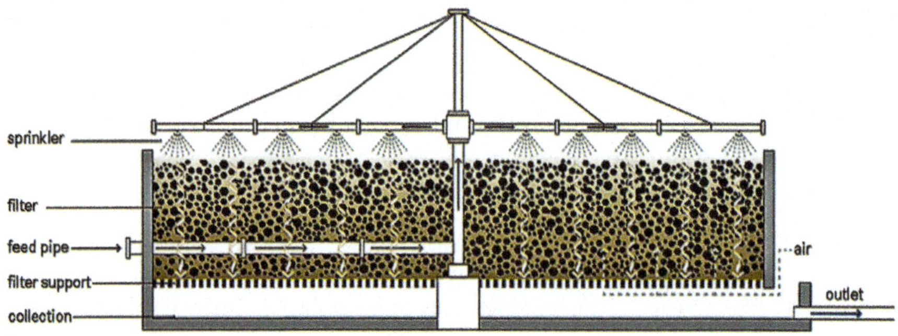

Fig. 8.8 Schematic cross-section of a trickling filter. (*Source*: Tilley et al. 2014)

porous membrane whose chemical nature has been such that it has a preferential attraction for solvent while repulsion for the solute. Reverse osmosis is commonly used to desalinate brackish water and also finds suitable, effective and economical method for the purification of water polluted by sewage effluents.

Electrodialysis is one of the membrane processes developed commercially for desalting saline water to get potable water. This technique can be applied for treatment of wash effluents, especially in electroplating industries (Mayr et al. 1992). The method is used to separate and recover the valuable metals in the form of ions in solution also enable low ion concentration to be brought up to higher level, which can be reused in the process.

Mercury discharge from chlor-alkali plants can be removed and recovered by mercury-selective ion exchange resin. Phenolics in wastewater produced from industries such as pulp and paper mills, petroleum refineries, tanning industries and resin manufacturing units are removed by the use of polymeric adsorbents.

III. The polluted water can also be treated biologically by reusing it for fish culture, oyster culture, algal culture and for plant growth.

A research work done by Niyogi et al. (1997) in the Haldia industrial zone of West Bengal has revealed the excellent growth of some mangrove species in the estuarine water contaminated with heavy metals. In spite of considerable concentrations of zinc, copper and lead in the aquatic phase and vegetative parts (root, stem and leaves) of the mangrove species *Sonneratia apetala*, the growth rate has been accelerated, proving its excellent absorbing capacity of heavy metals. These plants (the woody part) can later be used as source of wood for fencing, boat manufacturing, etc. The water hyacinth (menace in the freshwater aquatic ecosystem) is also very effective in absorbing heavy metals in freshwater systems. A detailed study on the bioaccumulation potential of a mangrove associate species is highlighted as Annexure 8B.

The nutrient enrichment can also be controlled biologically. This method is very effective for shrimp farms, which discharge lot of nutrients in the ambient aquatic phase through water exchange, pond preparation, etc. The nutrients generated through excreta and wasted feed of shrimps/prawns trigger phytoplankton growth (Fig. 8.9), which is the preferred food for filter feeding bivalves.

Also the seaweeds that grow profusely in the shrimp pond absorb considerable load of nutrients (Fig. 8.10)

In the estuaries of Indian Sundarbans, these nutrients from large number shrimp farms are released directly into the water bodies without any treatment, which has resulted in the rise of nutrient level in recent times particularly after early 1990s, when shrimp culture took a boom in the deltaic complex (Mitra 2013).

In order to control the water pollution, strict implementations of the laws like water (Prevention and Control of Water Pollution) Act, 1974, and the Environmental (Protection) Act, 1986, are extremely important in the Indian sub-continent. As a part of protecting and improving the quality of the environment, the standard for emission or discharge of environmental pollutants from the industries, operations or processes has been specified. Any deviation from these specified values must be seriously treated to discourage pollution from point and non-point sources.

Fig. 8.9 Phytoplankton bloom in shrimp ponds due to nutrient loading

Fig. 8.10 Profuse seaweed (*Enteromorpha* spp.) growth in shrimp ponds due to nutrient enrichment

The legal action to ban the semi-intensive shrimp culture farms in the coastal zone of India is a strict step in controlling the environmental pollution due to aquacultural industries. Considering the various negative impacts like salinization of agricultural land (due to saline water intrusion), deterioration of land fertility, release of excessive nutrients and antibiotics in the coastal water, etc., the Supreme Court of India ordered the coastal states on December 11, 1996 that all aquacultural farms other than the traditional and improved traditional ones, operating within 500 metres of the high-tide level (HTL) in the coastal zone be demolished by March 1997. The court also directed that no shrimp farm could be set up within 500 metres of HTL in the zone. In accordance with the principles of 'sustainable development' and 'polluter pays', the court also directed the Government to set up a body under cause (3) of Section 3 of the Environmental Protection Act, 1986. The said body will issue permits to the farmers of traditional aquaculture to adopt improved traditional farming system. Aquafarming in mangrove areas, estuaries, wetlands salt pans and public land is also prohibited. Furthermore, the court ruled that the owners of the closed aquaculture units would be liable to pay 6 years wages as compensation to the workers as well as for eco-restoration of the affected areas by way of creating an Environmental Protection Fund.

8.4 Take Home Messages

(A) The term water pollution may be defined as the addition of substances or heat to such a level, which is harmful to humans, animals or any desirable aquatic life, or otherwise causes significant departures from the normal activities of various living communities in or near bodies of water. In practical, the term water pollution refers to several types of aquatic contamination like enrichment of nutrients in ponds, lakes, rivers, bays, estuaries, seas and oceans from sewage and fertilizer, introduction of toxic chemicals (mainly from industries) in water bodies to such a level that the biota are affected to a considerable extent, etc.

(B) The main sources of water pollution are domestic wastes, industrial wastes, agricultural wastes, aquacultural wastes, shipping wastes, radioactive wastes and explosion.

(C) Water pollution causes several water-borne infectious diseases like cholera, typhoid, etc. Plant nutrients and aquaculture wastes contribute NO_3, PO_4 and NH_3 in the aquatic phase, which trigger the growth of algae leading to eutrophication. Nitrates in drinking water can cause methemoglobinaemia in babies (blue babies), as in an infant's stomach, the nitrate is transformed into nitrite, which acts on the blood haemoglobin to form methemoglobin. The heavy metals discharged from various industries also accumulate in the body tissues of the aquatic organisms and affect their normal physiological processes and growth. Through food chain they also accumulate in the human body.

(D) The existing sewage treatment methods (like primary, secondary and tertiary treatment) are very effective in removing selective pollutants from contaminated water. Primary treatment is a mechanical process which simply removes solids. The relatively larger solid particles are separated by means of metal screens; sands and small stones settle in grit chamber from which the water passes to the sedimentation tank, where the velocity of water is drastically reduced and the small particles settle as sludge. Scum is removed from the uppermost layers. Secondary treatment is essentially biological process which has the efficiency to remove most of the organic matter. In the activated sludge process of secondary treatment, the incoming sewage is mixed with decomposer bacteria and air or oxygen. Tertiary treatment is able to remove virtually all the remaining contaminants. Water leaving conventional secondary treatment still has most of the original phosphates and nitrates, many persistent insecticides and herbicides, disease-causing bacteria and viruses and perhaps a number of industrial, organic compounds. Wastewater that is not subjected to tertiary treatment contains the nutrients on which algae thrive. Contaminated water is also treated by adsorption, electrodialysis, ion exchange and reverse osmosis.

(E) In order to control the water pollution, strict implementations of the laws like water (Prevention and Control of Water Pollution) Act, 1974, and the Environmental (Protection) Act, 1986, are extremely important in the Indian sub-continent. As a part of protecting and improving the quality of the environment, the standard for emission or discharge of environmental pollutants from the industries, operations or processes has been specified. Any deviation from these specified values must be seriously treated to discourage pollution from point and non-point sources.

8.5 Brain Churners

> Water is one of the vital resources of the planet Earth that supports and sustains life. Deterioration of this vital resource would not only affect our health but will also affect all other life forms. Several categories of treatments are now available through which water pollution can be controlled. The readers may go deep into this chapter and consider this section as a sort of 'self-appraisal approach' as correct answers to all the questions are marked as black in the box at the end of this section.

1. Shrimp culture produces

   (a) Organic wastes
   (b) Heavy metals
   (c) Arsenic
   (d) None of the above

2. Antifouling paints used for condition fishing vessels and trawlers contain

   (a) Volatile organic compounds
   (b) Chlorine
   (c) Heavy metals
   (d) Antibiotics

3. Copper is a heavy metal which is used as

   (a) Antibiotics
   (b) Pesticide and algaecide
   (c) As water purifier
   (d) None of the above

4. Lead is produced from

   (a) Printing industry
   (b) Tannery
   (c) Dental care units
   (d) Shrimp culture farms

5. Conservative wastes are

   (a) Biodegradable in nature
   (b) Volatile in nature
   (c) Non-biodegradable in nature
   (d) None of the above

6. Agricultural wastes may lead to

   (a) Oil pollution
   (b) Increase of temperature
   (c) Skin disease
   (d) Eutrophication

7. A ship uses ballast to move safely which is

   (a) 50–60% of the total capacity of the tanker
   (b) 10–20% of the total capacity of the tanker
   (c) 25–30% of the total capacity of the tanker
   (d) 70–80% of the total capacity of the tanker

8. Carbon-14 has a half-life of

   (a) 5730 years
   (b) 6000 years
   (c) 30 years
   (d) 435 years

9. Radioactive waste is generated from

   (a) Painting factory
   (b) Nuclear power plant

   (c) Tourism units
   (d) Nuclear power plants and hospitals

10. Oyster is a
    (a) Shellfish
    (b) Finfish
    (c) Cartilaginous fish
    (d) None of the above

11. Oxytetracycline is a pesticide
    (a) Pesticide
    (b) Antibiotics
    (c) Fungicide
    (d) Radioactive compound

12. Estuaries adjacent to shrimp culture farms are polluted by
    (a) Nitrate
    (b) Phosphate
    (c) Ammonia
    (d) All of the above

13. In eutrophic ponds the dissolved oxygen becomes
    (a) High during the daytime
    (b) Low during the daytime
    (c) Same during the day- and nighttime
    (d) High during the nighttime

14. Red tide is caused by
    (a) Certain species of *Amoeba* sp.
    (b) Zooplankton
    (c) Certain species of dinoflagellates
    (d) None of the above

15. Saline water is converted to potable water by
    (a) Bioremediation
    (b) Primary treatment
    (c) Electrodialysis
    (d) None of the above

16. The Supreme Court of India has banned the
    (a) Semi-intensive method of shrimp culture
    (b) Biofloc system of fish culture
    (c) Traditional method of shrimp culture
    (d) All of the above

17. Trickling filters are conventional
    (a) Aerobic biological wastewater treatment units
    (b) Anaerobic biological wastewater treatment units
    (c) Primary settling process
    (d) None of the above

18. Oil and grease are released from
    (a) Agricultural fields
    (b) Pharmaceutical industries
    (c) Paper industries
    (d) Lubricants and solvents

19. Chromium is released from
    (a) Printing industry
    (b) Tannery industry
    (c) Pesticides
    (d) Oil sludges

20. Radioactive waste is said to be 'short-lived' if it has a half-life of
    (a) Less than 31 years
    (b) More than 31 years
    (c) 300 years
    (d) None of the above

Answer Sheet

S. No.	a	b	c	d
1.	■			
2.			■	
3.		■		
4.	■			
5.			■	
6.				■
7.			■	
8.	■			
9.				■
10.	■			
11.		■		
12.				■
13.	■			
14.			■	
15.			■	
16.	■			
17.	■			
18.				■
19.		■		
20.	■			

Annexure 8A: Water Quality of the River Ganga in and Around the City of Kolkata During and After Goddess Durga Immersion

Introduction

Durga Puja is one of the most famous festivals celebrated in West Bengal and particularly in Kolkata, for worshipping Goddess Durga during the period of Navaratri. It is celebrated for 10 days. However starting from the sixth day until the ninth day, the pandals (structures where the Goddesses are kept) with grand idols of Goddess Durga are open for the visitors. The tenth day, also known as Dashami, marks the Visarjan (immersion in water) of the idol with grand celebrations and processions. Durga Puja in 2018 started on October 15 and continued till October 19. However, Durga Puja, 2018, celebrations in Kolkata started early, with Durga Mahalaya on October 9 and Visarjan on October 19.

According to the Hindu Mythology, Goddess Durga emerged from the collective energy of all Gods/Goddesses as an embodiment of *Shakti* or divine feminine power, to destroy demon *Mahishasura*, who was blessed with the power of not getting defeated by any man or God (Fig. 8.A.1). This powerful form of Goddess Durga is highly revered in the Indian sub-continent, and her return for destroying the evil forces is celebrated with much grandeur and ceremonies.

Fig. 8.A.1 Goddess Durga with ten arms (Dashabhuja)

The 10th day of the Durga Puja festival is called Dashami; it is believed that on this day, Goddess Durga gained victory over the Demon and thus restored the balance on the Earth. It is also known as Vijayadashami. On this day, Goddess Durga is worshipped and offered many things as she is prepared to leave. Highly enthusiastic devotees gather in large numbers to join the procession that carries the Goddess to the ghats (the banks of the River Ganga with bathing facilities and where several religious rituals are performed) to be immersed in water. Women, especially married woman, initiate the procession by first applying red sindoor or vermillion powder on the Goddess and then to each other. In 2018, the date of immersion was October 20.

In this study we made a comparative assessment of the water quality, considering three toxic heavy metals lead, chromium and cadmium leached from idols in six major banks along the stretch of the River Ganga flowing through the mega city of Kolkata during October 10, 2018 (pre-puja/festival period) and October 22, 2018 (post-puja/festival period).

Materials and Methods

Site Selection

Six banks or ghats were selected for the present study, which are basically the zones for immersion. Coordinates of all the study sites in the River Ganga are highlighted in Table 8.A.1.

Analysis of Selective Heavy Metals

Surface water samples were collected from all the six ghats during high tide using 10-l Teflon-lined Go-Flo bottles, fitted with Teflon taps and deployed on a rosette or on Kevlar line, with additional surface sampling carried out by hand. Shortly after collection, samples were filtered through nuclepore filters (0.4 μm pore diameter), and aliquots of the filters were acidified with sub-boiling distilled nitric acid to a pH of about 2 and stored in cleaned low-density polyethylene bottles. Dissolved heavy metals were separated and pre-concentrated from the collected water using dithiocarbamate complexation and subsequent extraction into Freon TF, followed by back extraction into HNO_3. Extracts were analysed for Pb, Cr and Cd by atomic absorption spectrophotometer (Perkin Elmer: Model 3030).

Results

The levels of dissolved Pb, Cr and Cd during the pre-immersion and post-immersion phases in the selected sites of the River Ganga are presented in Figs. 8.A.2, 8.A.3 and 8.A.4. The order of dissolved metals in the study area is Pb > Cr > Cd irrespective of time periods and sites.

Table 8.A.1 Coordinates of all the selected sites/ghats along the River Ganga

Site name	Coordinates	View
Ramkrishna Ghat	**22°34′19.8″ N 88°20′17.0″E**	
Shibpur Ghat	**22°33′41.2″ N 88°19′40.4″E**	
Princep Ghat	**22°33′30.9″ N 88°19′52.5″E**	
Botanical garden	**22°33′06.4″ N 88°18′06.6″E**	
Babughat	**22°34′10.3″ N 88°20′28.5″E**	
2nd Hooghly bridge	**22°33′31.4″ N 88°19′38.5″E**	

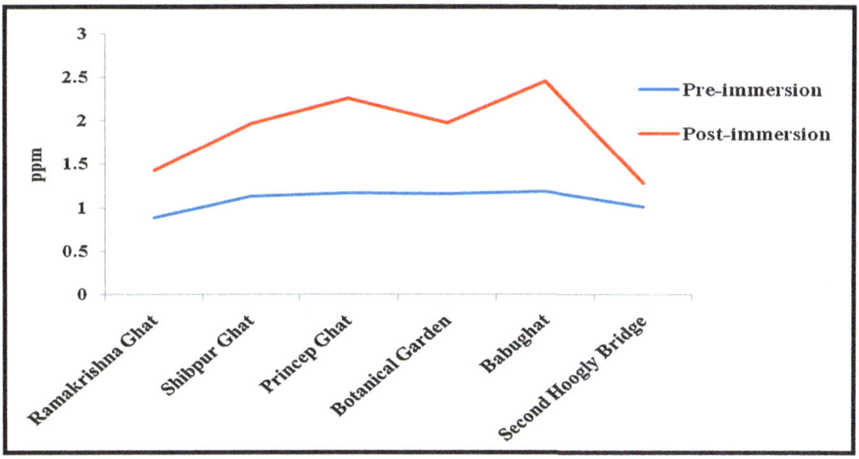

Fig. 8.A.2 Dissolved Pb concentrations in (ppm) during pre-immersion and post-immersion phases

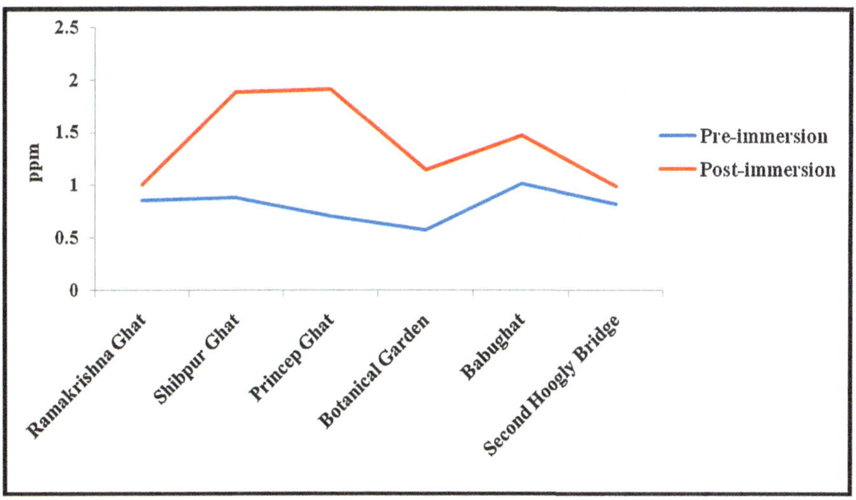

Fig. 8.A.3 Dissolved Cr concentrations in (ppm) during pre-immersion and post-immersion phases

Discussion

It is clear from the data sets that there has been considerable increase in the levels of dissolved heavy metals after the process of immersion of idols. For Ramkrishna Ghat, the levels of increase of Pb, Cr and Cd are 61%, 17% and 125%, respectively. For Shibpur Ghat, the levels of increase of Pb, Cr and Cd are 74%, 111% and 86%, respectively. For Princep Ghat, the levels of increase of Pb, Cr and Cd are 93%,

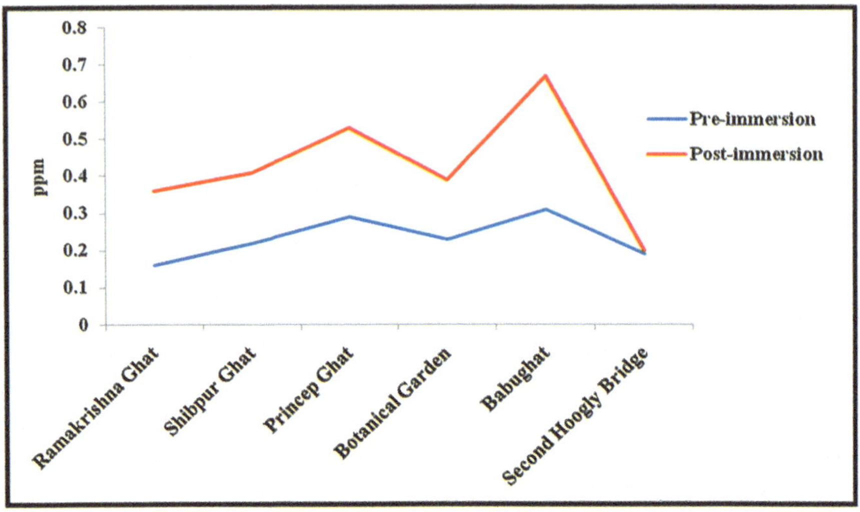

Fig. 8.A.4 Dissolved Cd concentrations in (ppm) during pre-immersion and post-immersion phases

Fig. 8.A.5 Straws and skeletons of idols. (Source: Mitra and Ray Chaudhuri 2017)

169%, and 83% respectively. For Botanical Garden, the levels of increase of Pb, Cr and Cd are 71%, 98% and 70%, respectively. For Babughat, the levels of increase of Pb, Cr and Cd are 107%, 44% and 116%, respectively, and for second Hooghly Bridge, the levels of increase of Pb, Cr and Cd are 27%, 19% and 5%, respectively. The leftover straws and skeletons of the idols are shown in Fig. 8.A.5.

Table 8.A.2 ANOVA for Pb in 6 sample sites during pre-immersion and post-immersion

Source of variation	*SS*	*df*	*MS*	*F*	*P-value*	*F crit*
Between stations	0.801342	5	0.160268	2.445538	0.174379	5.050329
Between periods	1.944075	1	1.944075	29.66468	0.002834	6.607891
Error	0.327675	5	0.065535			
Total	3.073092	11				

Table 8.A.3 ANOVA for Cr in six sample sites during pre-immersion and post-immersion

Source of Variation	*SS*	*df*	*MS*	*F*	*P-value*	*F crit*
Between stations	0.528967	5	0.105793	1.136911	0.445728	5.050329
Between periods	1.032533	1	1.032533	11.09615	0.020754	6.607891
Error	0.465267	5	0.093053			
Total	2.026767	11				

Table 8.A.4 ANOVA for Cd in six sample sites during pre-immersion and post-immersion

Source of variation	*SS*	*df*	*MS*	*F*	*P-value*	*F crit*
Between stations	0.1115	5	0.0223	3.444902	0.100403	5.050329
Between periods	0.112133	1	0.112133	17.32235	0.008807	6.607891
Error	0.032367	5	0.006473			
Total	0.256	11				

ANOVA carried out with the data showed significant variations between two periods (pre-immersion and post-immersion periods) at $p < 0.05$, but no spatial variation was observed (Tables 8.A.2, 8.A.3 and 8.A.4). This may be attributed to proximity of the sites and regular tidal actions in the study stretch that has spatially homogenized the dissolved heavy metals. However, concern is there on the point of significant variation in water quality between the pre-immersion and post-immersion phases.

Lead, a heavy metal that is used in batteries, paints and the oxide, is employed in the production of crystal glass. A higher level of lead results in cognitive impairment in children to peripheral neuropathy in adults.

Intake of chromium in large amounts has severe detrimental health effects like renal, hepatic and gastrointestinal damage.

Cadmium when consumed along with food and water for a longer period of time results in the bioaccumulation in the kidney and liver causing severe damage to these systems.

Way Forward

In India the practice of idol worship is witnessed since ancient times. It has been a tradition to immerse the idols in water bodies after the festival is over. Consequent pollution of such water bodies has been a matter of concern, and there have been public interests and litigations as well. In the present study, also Pb, Cr and Cd increased in the water bodies of six ghats along the stretch of the River Ganga. In this context it is recommended to use traditional clay for idol making rather than baked clay.

Use of painted idols with chemicals should be prohibited as far as possible. In case painted idols are used, water soluble and nontoxic natural dyes may be used. Natural colours used in the food colours and permitted in pharmaceuticals may be preferred. Several dyes can be extracted traditionally from the vegetative and reproductive parts of plants. Red dyes can be sourced from madder, Brazil wood, beetroot, cranberry, safflower and orchid. Orange dye is obtained from the stigmas of the saffron flower. The flowers camomile and milkwort are the sources of yellow colour. The green colour can be obtained from ripe buckthorn berries and ragweed. The blue colour can be sources from wood plant and indigo. It is also recommended that prior to immersion, worship materials like flowers, clothes, decorating materials like paper and plastics should be segregated and collected for recycling or composting.

A revolutionary concept may be a ‘food for thought’ for the policymakers in the domain of immersion of idols. This idea is based on the concept ‘phytoplankton bloom through iron fertilization’. The idols made of iron structures not only can be recycled for making several metallic objects but can also trigger the growth of phytoplankton in the riverine stretch if left for few months in a bounded or closed space. This closed space can be constructed by channelizing the river water in a sloppy area towards the river bank. The phytoplankton being the major food for the fishes and a sink of carbon dioxide would be an innovative spin-off benefit from this festival. Such concept may seem to be highly challenging but if implemented with proper planning may be a footprint for improving fish population and ecological stability of the mighty River Ganga.

Annexure 8B: Bioaccumulation Potential of a Mangrove Associate Species, *Porteresia coarctata*

The wastes of anthropogenic and industrial origin are of complex characters and have considerable percentage of heavy metals. The heavy metals in the saline water phase generally deposit on the sediment bed or remain in dissolved state in the water column, depending on the nature of the chemical species. Sediments are the heavy metal sinks in the marine environment (Govindaswamy et al. 1997). Knowledge of the distribution and concentration of heavy metals in sediments will help to detect the sources of pollution in the aquatic phase (Wittmann 1981).

Indiscriminate releases of untreated or partially treated wastes without considering the assimilative capacity of the waste receiving waterbody have resulted in pockets of polluted environs with depleted coastal resources, public health risks and loss of biodiversity. Over 300 million people living in the coastal zone of India are considered to generate 1.11×10^{10} m^3 of sewage annually, a considerable fraction; particularly from coastal cities and towns where sewage collection network exists, enters the marine water (Zingde 1999). This is a common scenario in all the maritime states of Indian sub-continent.

The rapid industrialization and urbanization of the city of Kolkata, Howrah and the newly emerging Haldia complex has caused considerable ecological imbalance in the adjacent coastal zone. The Hugli estuary, situated on the western sector of coastal West Bengal, receives drainage from several channels, which have sewage outlets into the estuarine system. The chain of factories and industries situated on the western bank of the Hugli estuary is another prominent cause behind the gradual transformation of this beautiful ecotone into stinking cesspools of the megapolis. The lower part of the estuary has multifarious industries such as paper, textiles, chemicals, pharmaceuticals, plastic, shellac, food, leather, jute, tyres and cycle rim (Mitra et al. 2005; Mitra 1998; UNEP 1982). Deadly effluents from these industries might scratch the magnificent network of life spun over a long evolutionary period of time in this biotope. In addition to these, the mushroom growth of hotels, resorts and a number of unofficial fish landing centres at the Canning are some prominent negative pressures on the ecological status of the area. The sewage released by these units into the bay not only alters the water quality in terms of nutrient load but may also pose considerable impact on the biotic community inhabiting the zone.

With this background, the present programme aimed to evaluate the status of heavy metal levels (Zn, Cu and Pb) in the vegetative parts of saltmarsh grass (*Porteresia coarctata*) in the selected stations of mangrove dominated Indian Sundarbans. This saltmarsh grass species is endemic to Sundarban region and is regarded as the pioneer species in island formation (Fig. 8.B.1).

The role of the species in holding the soil particles with long intricate root system is immense to control the erosion of the river banks. The investigation was carried out during 2016–2017 by the present authors in the estuarine zones of Sundarbans at three different stations, namely, Canning, Chotomollakhali and Bali Island with the aim to assess the bioaccumulation pattern of selected heavy metals in the saltmarsh grass inhabiting the intertidal mudflats of the selected stations in Indian Sundarbans.

In *P. coarctata* samples, the heavy metals varied as per the order Zn > Cu > Pb. This sequence is uniform in all the three selected stations. In the present study, the concentration of Zn ranged from 49.80 (at Bali) to 165.48 (at Canning) during 2016, and in 2017 the concentration ranged from 55.78 (at Bali) to 171.46 (at Canning). Cu ranged from 20.19 (at Bali) to 63.21 (at Canning) in 2016 and 23.74 (at Bali) to 66.36 (at Canning) in 2017. Pb ranged from 2.66 (at Bali) to 19.22 (at Canning) in 2016 and from 4.31 (at Bali) to 21.67 (at Canning) in 2017.

The concentrations of heavy metals in the vegetative parts of *P. coarctata* for the present study are tabulated in Tables 8.B.1, 8.B.2 and 8.B.3.

P. coarctata (commonly known as saltmarsh grass) is widely distributed in the mudflats of Sundarbans and is a pioneer species in the process of island ecological succession (Jagtap et al. 2006; Mitra 2013). It is a tetraploid ($2n = 48$) species producing highly recalcitrant seeds (Probert and Longley 1989) and occurs all over the

Fig. 8.B.1 *Porteresia coarctata* in front of the matured mangrove trees in Indian Sundarbans

Table 8.B.1 Monthly variations of heavy metals in *P. coarctata* collected from Canning during 2016 and 2017

	Zn		Cu		Pb	
Month	2016	2017	2016	2017	2016	2017
Jan	100.55	106.53	44.22	46.98	12.99	14.98
Feb	94.69	99.39	41.77	45.94	11.56	13.21
March	92.32	98.3	36.89	39.65	13.78	16.23
April	85.77	90.54	34.33	38.56	10.89	12.54
May	82.33	88.31	32.1	34.86	10.03	10.78
June	78.21	85.26	28.67	31.43	8.64	10.29
July	150.11	156.09	35.89	39.63	11.86	12.99
Aug	152.09	157.21	47.2	49.96	17.33	18.98
Sept	165.48	171.46	63.21	66.36	19.22	21.67
Oct	134.4	141.23	54.9	57.66	15.09	16.74
Nov	112.44	118.42	48.9	51.66	14.33	16.23
Dec	103.66	108.73	41.51	45.23	13.04	15.87

tropics as a mangrove associate, where the soil is inundated (twice a day) with saline river or seawater of 20–40‰. The species can also withstand submergence with saline water for a long period of time due its inherent capacity to tolerate high levels of salinity and submergence (Flowers et al. 1990). The root system of *Porteresia* is well adapted to cope with increasing salinity leading to decreased

Table 8.B.2 Monthly variation of heavy metals in *P. coarctata* collected from Chotomollakhali Island during 2016 and 2017

	Zn		Cu		Pb	
Month	2016	2017	2016	2017	2016	2017
Jan	79.33	85.31	33.29	36.05	8.45	11.23
Feb	73.55	80.67	31.89	35.23	8.78	10.43
March	71.07	77.05	29.87	32.63	7.34	9.91
April	65.45	72.47	26.44	30.61	6.02	7.67
May	62.08	68.06	23.22	25.98	5.89	7.63
June	59.23	64.92	21.67	24.43	5.05	6.7
July	65.23	71.21	30.89	33.65	9.23	10.88
Aug	79.01	84.99	37.34	41.26	9.66	12.96
Sept	102.56	107.98	40.22	42.98	10.21	11.86
Oct	85.22	91.2	38.1	41.98	8.21	10.12
Nov	81.67	88.39	36.3	39.06	5.23	6.88
Dec	78.05	84.03	35.29	37.28	7.12	9.21

Table 8.B.3 Monthly variation of heavy metals in *P. coarctata* collected from Bali Island during 2016 and 2017

	Zn		Cu		Pb	
Month	2016	2017	2016	2017	2016	2017
Jan	66.22	76.55	28	31.67	5.13	6.89
Feb	62.44	68.42	26.23	28.99	4.99	6.64
March	60.28	67.39	24.15	27.37	4.17	5.84
April	58.21	64.19	24.39	28.87	3.89	5.54
May	53.75	60.34	23.08	25.84	2.97	4.34
June	49.8	55.78	20.19	23.74	2.66	4.31
July	70.21	77.45	27.38	30.14	7.51	9.34
Aug	83.65	89.63	31.35	35.26	7.98	9.63
Sept	87.22	92.67	35	37.76	8.2	9.87
Oct	78.28	84.26	33.91	37.88	7.17	8.82
Nov	80.78	86.76	28.68	31.44	6.08	7.21
Dec	67.20	73.18	27.99	30.75	5.11	6.76

water transport. To overcome intertidal strong flow, it often forms pseudo-taproots up to a depth of 1 m and fibrous roots develop from the tip of those pseudo-taproots and internodes of a wide-spreading underground stem called a 'sobole' (Latha et al. 2004). It is reported by some authors that saltmarsh genesis is based on accretion due to sedimentation of suspended matter, supplied by tidal water or flood water of marine and riverine origin.

It is reported by some researchers that the coastal zones and estuaries, which are the primary habitat of this species, are exposed to effluents from chemical industries every day. As a consequence nutrients and contaminants like heavy metals, pesticides and halogenated hydrocarbons are transported into the saltmarsh sediments since they are partly bound to suspended particles. Urban saltmarshes often receive large pollutant loads, including discharges of heavy metals from industry and transportation activities. In *Porteresia* bed, water soluble metals and exchangeable metals are the most available and precipitated inorganic compounds; metal complexes with large molecular weight, humus materials and metals adsorbed to hydrous oxides are also possibly available. *P. coarctata*, being an endemic species of Sundarbans region, is exposed to heavy metal pollution, as the region is contaminated with conservative pollutants (Mitra 1998; Mitra et al. 2011).

Marsh plants are known to absorb and accumulate metals from contaminated sediment (Giblin et al. 1980; Kraus et al. 1986; Kraus 1988; Sanders and Osman 1985). The absorption of contaminants is one reason that wetlands are being used for wastewater treatment. Metals taken up by plants are also capable of re-entering marsh systems through excretion from leaf salt glands (Kraus et al. 1986; Kraus 1988). Metals present within the water column of the plant may be carried through open stoma. This could result in the adhesion of the metals to the outer surface of the leaf. The present study is extremely important for two specific reasons:

1. The saltmarsh grass species *P. coarctata* can act as an agent of bioremediation and can be utilized by coastal industries to biologically treat the effluents contaminated with heavy metals (with a specific retention time, which has not been studied in the present programme).
2. In the mudflats of Sundarbans, cattle from the adjacent island villages graze on *P. coarctata*. Hence, marsh grass contaminated with heavy metals may accumulate in the body tissues of cattle, which may be further transferred to human beings through their milk and meat.

Implementation of water pollution prevention strategies and restoration of ecological systems are integral components of all development plans. To preserve our water and environment, we need to make systematic changes in the way we grow our food, manufacture our goods and dispose of the wastes (Lazaroff 2000). In India, agriculture is the biggest user and polluter of water. If pollution by agriculture is reduced, it would improve water quality and would also eliminate cost incurred for treatment of diseases. Like all other inputs, there is an optimal quantity of fertilizer for given conditions, and excess application does not improve the crop yield. Pricing of fertilizers and pesticides as well as appropriate legislation to regulate their use will also go a long way in stopping indiscriminate use. Industries need to carefully treat their waste discharges. Manufactures may reduce water pollution by reusing materials and chemicals and switching over to less toxic alternatives. Industrial symbiosis, in which the unusable wastes from one product/firm become the input for another, is an attractive solution. Also there is a need to encourage reductions or replacement of toxic chemicals, possibly through fiscal measures. Pollution taxes in the Netherlands, for example, have helped the country slash discharges of heavy metals such as mercury and arsenic into waterways by up to 99% between 1976 and the mid-1990s.

Environmental improvement and restoration should be planned and implemented such that the freshwater and groundwater resources are protected and their quality is maintained and/or enhanced. On the legislative front, laws are required to check littering as well as to implement 'polluter pays' principle. More importantly, these laws should be strictly enforced.

For making the people of various sections of our society aware about the different issues of water resources management, a participatory approach may be adopted. Mass communication programs may be launched using the modern communication means for educating the people (preferably the island dwellers of Sundarbans) about water pollution. Capacity building should be perceived as the process whereby a community equips itself to become an active and well-informed partner in decision-making. The process of capacity building must be aimed at both increasing access to water resources and changing the power relationships between stakeholders. Capacity building not only is limited to officials and technicians but must also include the general awareness of the local population regarding their responsibilities in sustainable management of the water resources. Policy decisions in any water resources projects should be directed to improve knowledge, attitude and practices about the linkages between health and hygiene, provide higher water supply service levels and improve environment through safe disposal of human wastes. Sustainable management of water requires decentralized decisions by giving authority, responsibility and financial support to communities to manage their natural resources and thereby protect the environment. Approaches to water pollution control that focuses on wastewater minimization, in-plant refinement of raw materials and production processes, recycling of waste products, etc., should be given priority over traditional end-of-pipe treatments. An important element in water pollution control strategy is the formulation of realistic standards and regulations. However, the standards must be achievable and the regulations enforceable.

References

Banerjee K, Bhattacharjee R, Amin G, Das S, Mitra A (2008) Concentrations of dissolved Zn, Cu and Pb in and around Indian Sundarbans: A time series analysis. Sci Cult (Indian Science News Association) 74(11–12):476–482

Bhattacharyya SB, Roychowdhury G, Zaman S, Raha AK, Chakraborty S, Bhattacharjee AK, Mitra A (2013) Bioaccumulation of heavy metals in Indian white shrimp (*Fenneropenaeus indicus*: a time series analysis). Int J Life Sci Biot Pharm Res 2(2):97–113. ISSN: 2250-3137

Carpenter SR, Caraco NF, Correll DL, Howarth RW, Sharpley AN, Smith VH (1998) Non-point pollution of surface waters with phosphorus and nitrogen. Ecol Appl 8:559–568

Das S, Mitra A, Zaman S, Pramanick P, Ray Chaudhuri T, Raha AK (2014) Zinc, Copper, Lead and Cadmium levels in edible finfishes from lower Gangetic delta. Am J Biophys Biochem Life Sci 3(1):8–19. ISSN: 2166-126X

Flowers T, Flowers SA, Hajibagheri MA, Yeo AR (1990) Salt tolerance in the halophytic wild rice, *Porteresia coarctata*. New Phytology 114:675–684

Giblin AE, Bourg A, Valiela I, Teal JM (1980) Uptake and losses of heavy metals in sewage sludge by a New England saltmarsh. Am J Bot 67:1059–1068

Govindaswamy C, Kannan L, Azariah J (1997) Trace metal distribution in sediments of the Coromandel Coast, Bay of Bengal. In: Iyer CSP (ed) Advances in environmental science. Educational Publishers and Distributors, New Delhi, pp 53–62

Jagtap TG, Bhosale S, Singh C (2006) Characterization of *Porteresia coarctata* beds along the Goa coast, India. Aqua Bot 84:37–44

Javed M, Usmani N (2016) Accumulation of heavy metals and human health risk assessment via the consumption of freshwater fish *Mastacembelus armatus* inhabiting, thermal power plant effluent loaded canal. Springer Plus 5(5):1

Jenssen PD, Greatorex JM, Warner WS (eds) (2004) Sustainable wastewater management in urban areas. (= Kapitel 4. Kurs WH33, Konzeptionen dezentralisierter Abwasserreinigung und Stoffstrom management). University of Hannover, Hannover

Kraus ML (1988) Accumulation and excretion of five heavy metals by the saltmarsh cordgrass *Spartina alterniflora*. Bull N J Acad Sci 33:39–43

Kraus ML, Weis P, Crow JH (1986) The excretion of heavy metals by the salt marsh cord grass, *Spartina alterniflora*, and *Spartina*'s role in mercury cycling. Mar Environ Res 20:307–316

Latha R, Hosseini Salekdeh G, Bennett J, Swaminathan MS (2004) Molecular analysis of a stress-induced cDNA encoding the translation initiation factor, eIF1, from the salt-tolerant wild relative of rice, *Porteresia coarctata*. Funct Plan Biol 31:1035–1042

Lazaroff C (2000) Hidden groundwater pollution problem runs deep. Environment News Service, Worldwatch Institute, Washington, DC

Mitra A (1998) Status of coastal pollution in West Bengal with special reference to heavy metals. J Indian Ocean Stud 5(2):135–138

Mitra A (2013) Sensitivity of mangrove ecosystem to changing climate. Springer India, New Delhi

Mitra A, Mandal T, Bhattacharyya DP (1999) Concentrations of heavy metals in *Penaeus* spp. of brackishwater wetland ecosystem of West Bengal, India. Indian J Environ Ecopl 2(2):97–106

Mitra A, Das KL, Choudhury A (2000) Ichthyoplankton richness around Haldia port-cum-industrial complex in relation to dissolved Zn and Cu. Res J Chem Environ 4(1):9–11

Mitra A, Banerjee K, Chakraborty R, Banerjee A, Mehta N, Berg H (2005) Study of the water quality of the shrimp culture ponds in Indian Sundarbans. Indian Sci Cruiser 20:34–43

Mitra A, Chakraborty R, Banerjee K (2008) Monthly variation of Zn, Cu and Pb in and around Indian Sundarbans. Proc Nat Acad Sci India 78:234–245

Mitra A, Amin G, Chakrabarti G, Banerjee K (2009) Heavy metals in edible fish muscles from Gangetic delta – how safe for consumption? Nat Acad Sci Lett 32(11 & 12):357–362

Mitra A, Banerjee K, Sinha A (2011) Shrimp tissue quality in the lower Gangetic delta at the apex of Bay of Bengal. Toxicol Environ Chem 93(3):565–574

Mitra A, Choudhury R, Banerjee K (2012) Concentrations of some heavy metals in commercially important finfish and shellfish of the River Ganga. Environ Monitor Assess 184:2219–2230. Springer. https://doi.org/10.1007/s10661-011-2111-x

Mukherjee M, Roy I, Chatterjee P, Basu S, Mitra A (2007) Chapter XVII: Sea shell craft industry and possibilities of edible oyster culture in coastal West Bengal. In: Sunderban wetlands. Department of Fisheries, Aquaculture, Aquatic Resources and Fishing Harbours, Government of West Bengal, pp 271–284

Niyogi S, Mitra A, Aich A, Choudhury A (1997) Sonneratia apetala – an indicator of heavy metal pollution in the coastal zone of West Bengal (India). In: Iyer CSP (ed) Advances in environmental science. Educational Publishers and Distributors, pp 283–287

Probert RJ, Longley PL (1989) Recalcitrant seed storage physiology in three aquatic grasses (*Zizania palustris, Spartina anglica,* and *Porteresia coarctata*). Ann Bot 63:53–63

Ray Chaudhuri T, Fazli P, Zaman S, Pramanick P, Bose R, Mitra (2014) Impact of acidification on heavy metals in Hooghly Estuary. J Harmon Res Appl Sci 2(2):91–97

Sanders JG, Osman RW (1985) Arsenic incorporation in a saltmarsh ecosystem. Estuar Cost Shelf Sci 20:387–392

Smith VH (1998) Cultural eutrophication of inland, estuarine, and coastal waters. In: Pace ML, Groman PM (eds) Successes, limitations and frontiers in ecosystem science. Springer, New York, pp 7–49

Tilley E, Ulrich L, Luethi C, Reymond P, Zurbruegg C (2014) Compendium of sanitation systems and technologies, 2nd rev edn. Swiss Federal Institute of Aquatic Science and Technology (Eawag), Duebendorf. https://www.eawag.ch/de/abteilung/sandec/. Accessed: 28 July 2014

UNEP (1982) Pollution and marine environment in the Indian Ocean, UNEP regional seas report and studies, vol 13. UNEP, Geneva

Wittmann GTW (1981) Toxic metals. In: Forstner U, Wittman GTW (eds) Metal pollution in the aquatic environment. Springer Verlag, Berlin, pp 3–68

Zaman S, De Kumar U, Pramanick P, Thakur S, Amin G, Fazli P, Mitra A (2014) Forecasting heavy metal level on the basis of acidification trend in the major estuarine system of Indian Sundarbans. J Energy Environ Carb Credit 4(2):1–9

Zingde MD (1999) Marine pollution – what we are heading for? In: Somayajulu BLK (ed) Ocean science trends and future directions. Indian National Science Academy, New Delhi, pp 229–246

Chapter 9
Soil Pollution and Its Mitigation

Abstract Soil being an important component of environment supports the pillars of life. Human activities in recent times have led to modifications of this life-supporting matrix through use of chemical fertilizers and pesticides in an uncontrolled manner. The chapter discusses the causes, effects and control of soil pollution along with some innovative recommendations like application of PGPR (plant growth-promoting rhizobacteria) to improve the nutrient reservoir of the soil, planting native species with high carbon storage potential and replacement of chemical fertilizer with low cost organic fertilizer.

Keywords Chemical fertilizer · Pesticide · Crop rotation · Plant growth-promoting rhizobacteria (PGPR) · Organic fertilizer

9.1 Causes of Soil Pollution

Soil is derived from Latin word *solum* meaning floor. The study of soil science is known as edaphology. Soil is an important component of the environment and vital for all living beings, as it produces food on which various life forms depend. Due to the rapid increase in population, the need for food has also increased. To fulfil this basic requirement, there is need to produce more food products with limited land resources. Human activities have led to modification in soil characteristics through use of fertilizers and pesticides to increase the agricultural production. The other activities that have changed natural soil characteristics include deforestation and improper disposal of wastes, especially the garbage refuse and fly ash.

The solid rocks constituting the Earth's crust comprise the lithosphere. Rocks are continuously subjected to weathering either by physical, chemical or biological processes. The top soil consists of organic matter called humus, which is a dark brown decomposed material of dead plants and leaves. Microorganisms are the key players in the process of decomposition.

Soil consists of solid materials, organic matter and air spaces. The four important components of soil are mineral matter (40%), soil water (25%), organic matter (10%) and soil air (25%).

A. Mitra, S. Zaman, *Environmental Science - A Ground Zero Observation on the Indian Subcontinent*, https://doi.org/10.1007/978-3-030-49131-4_9

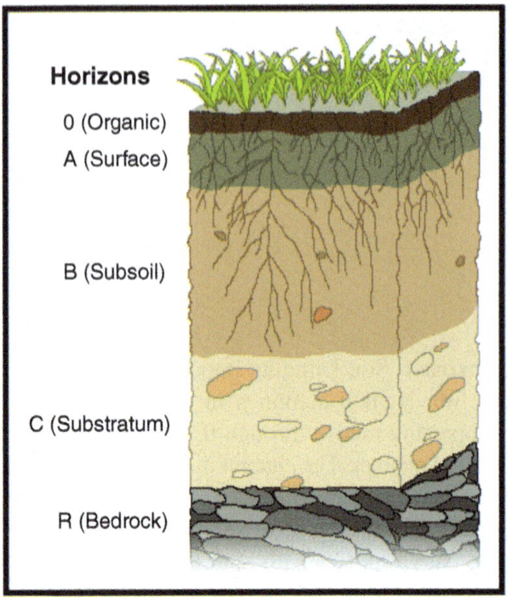

Fig. 9.1 Soil profile

The topmost layer is called A-horizon which is several inches in thickness. The layer has maximum biological activity and has the organic matter humus. This top layer is important for the agricultural crops and vegetation. There is continuous leaching of metal ions. Due to deforestation this topmost layer is prone to soil erosion which in turn reduces the soil fertility and also the production. The next layer beneath to A-horizon is known as B-horizon. The organic matter, minerals and salts leach from A-horizon and reach B-horizon. The next layer is called as C-horizon which consists of weathered parent rocks from which soil originates (Fig. 9.1).

The major causes of soil pollution are due to use of agrochemicals. The other factors causing soil pollution include deforestation, improper waste disposal, etc. All these major causes are discussed separately.

1. **Fertilizers**

Agrochemicals such as fertilizers result in the pollution of both surface and underground soil. Soil nutrients are essential for the growth and development of plants. The plants obtain nutrients from the soil such as potassium, nitrogen, calcium, magnesium and phosphorus and carbon, hydrogen and oxygen from air and water. To overcome the deficiency of the above nutrients, fertilizers are added to the soil. However, along with the nutrients, some impurities also get added to the soil. The impurities come from the raw materials used for the manufacture of fertilizers and mainly consist of arsenic, lead and cadmium which are commonly found in small quantities in rock phosphate. These metals are not degradable and concentrate over the soil surface above toxic levels (thus are poisonous for agricultural crops) due to indiscriminate use of fertilizers by the farmers.

The excess use of fertilizers reduces the protein content of cereal crops and also reduces the carotene as well as vitamin content in fruits and vegetables. The crops grown on excessively fertilized soil are more prone to pests and diseases.

The other important impact of indiscriminate use of fertilizer is eutrophication which triggers the growth of algal species in the adjacent aquatic ecosystem. It is the process of nutrient accumulation mainly of nitrogen and phosphorus in water bodies which occurs through agricultural runoff. This results in increasing the productivity of the lake. The lake on progressive enrichment of nutrients changes from oligotrophic (nil or few nutrients) to mesotrophic (intermediate nutrients) and finally to eutrophic (well nourished) level.

Sediment deposition decreases the depth of the water body and it changes to marsh. Eutrophication causes several chemical, physical and ecological changes including dissolved oxygen depletion and BOD build-up. The depletion of oxygen results in anaerobic condition in which case the water body starts smelling due to release of gases such as methane and hydrogen sulphide.

The fertilizers also contaminate the underground water, e.g. indiscriminate use of nitrogen fertilizer results in leaching of nitrate from soil surface to underground water and pollute it. The nitrate contamination causes blue baby syndrome in infants.

2. **Pesticides**

Farmers use pesticides to protect their crops from the attack by bacteria, fungi, viruses, rodents and weeds. The first widely used pesticide was dichloro trichloroethane (DDT). This pesticide remains in the environment long after its use as a non-biodegradable chemical and gets stored in the fat of organisms. The concentration of these harmful toxic chemicals gets magnified at each successive trophic level in the food chain, with the topmost trophic level having the maximum concentration. This increased accumulation of harmful toxic chemicals at each successive trophic level is known as biological magnification. Due to the presence of such chemicals, the egg shells of osprey, pelicans, etc. become thin and result in premature hatching of these eggs, causing death of the young ones. The other harmful pesticides include benzene hexachloride (BHC), malathion, aldrin and dieldrin. The consumption of pesticide-contaminated crops affects the biological processes, and their use also decreases the soil fertility to large extent.

Few pesticides along with their harmful effects are highlighted here.

(i) **Aldrin**

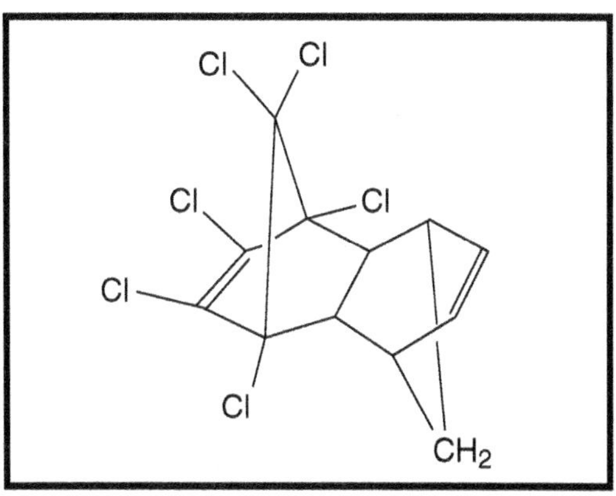

A pesticide applied to soil to kill termites, grasshoppers, corn rootworm and other insect pests; aldrin can also kill birds, fish and humans. In one incident, aldrin-treated rice is believed to have killed hundreds of shorebirds, waterfowl and passerines along the Texas Gulf Coast when these birds ate either the rice or animals that had eaten the rice. In humans, the fatal dose for an adult male is estimated to be about 5 grams. Humans are exposed to aldrin through dairy products and animal meats. Studies in India indicate that the average daily intake of aldrin and its by-product dieldrin is about 19 micrograms per person. The use of aldrin has been banned or severely restricted in many countries.

(ii) **Chlordane**

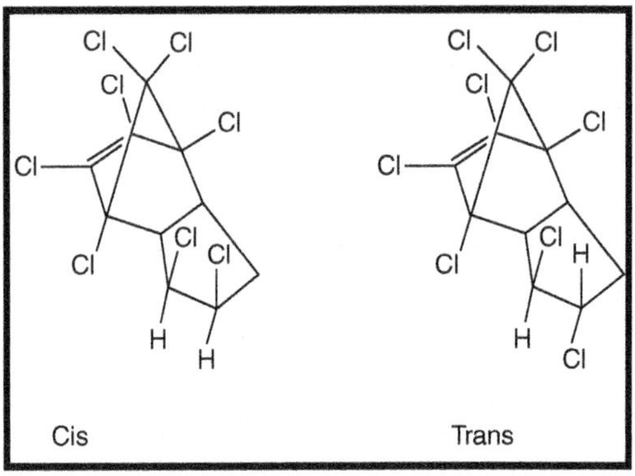

Used extensively to control termites and as a broad-spectrum insecticide on a range of agricultural crops, chlordane remains in the soil for a long time; others severely restrict its use, and it is still detected in food from all over the world. Although residues in domestic animals have declined steadily over the last two decades, food-borne DDT remains the greatest source of exposure for the general population. The short-term acute effects of DDT on humans are limited, but long-term exposures have been associated with chronic health effects. DDT has been detected in breast milk, raising serious concerns about infant health.

(iii) **Dieldrin**

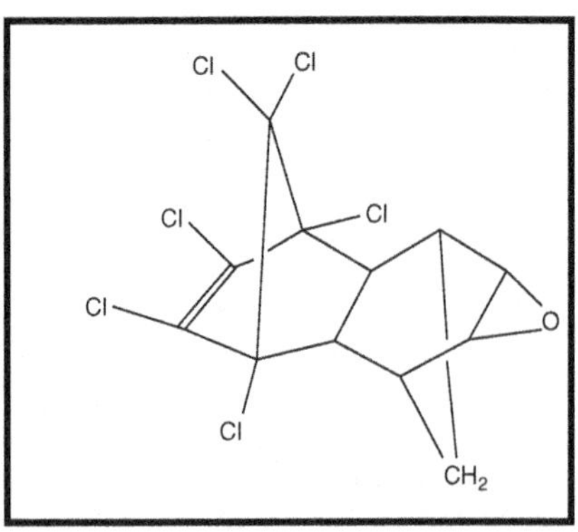

Used principally to control termites and textile pests, dieldrin has also been used to control insect-borne diseases and insects living in agricultural soils. Its half-life in soil is approximately 5 years. The pesticide aldrin rapidly converts to dieldrin, so concentrations of dieldrin in the environment are higher than dieldrin use alone would indicate. Dieldrin is highly toxic to fish and other aquatic animals, particularly frogs, whose embryos can develop spinal deformities after exposure to low levels. Dieldrin residues have been found in air, water, soil, fish, birds and mammals, including humans. Food is the primary source of exposure for the general population; dieldrin was the second most common pesticide detected in a US survey of pasteurized milk.

(iv) **DDT**

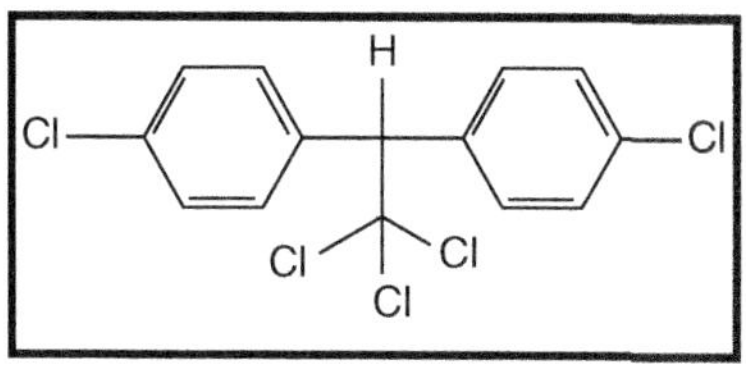

DDT is highly insoluble in water and is soluble in most organic solvents. It is semi-volatile and can be expected to partition into the atmosphere as a result. Its presence is ubiquitous in the environment, and residues have even been detected in the arctic. It is lipophilic and partitions readily into the fat of all living organisms and has been demonstrated to bioconcentrate and biomagnify. The breakdown products of DDT, 1,1-dichloro-2,2-bis(4-chlorophenyl)ethane (DDD or TDE) and 1,1-dichloro-2,2bis(4-chlorophenyl)ethylene) (DDE), are also present virtually everywhere in the environment and are more persistent than the parent compound. DDT is not highly acutely toxic to laboratory animals, with acute oral LD50 values in the range of 100 mg/kg body weight for rats to 1770 mg/kg for rabbits. In a six-generation reproduction study in mice, no effect on fertility, gestation, viability, lactation or survival was observed at a dietary level of 25 ppm. A level of 100 ppm produced a slight reduction in lactation and survival in some generations, but not all, and the effect was not progressive. A level of 250 ppm produced clear adverse reproductive effects. In both these and other studies, no evidence of teratogenicity has been observed.

DDT is highly toxic to fish, with 96-h LC50 values in the range of 0.4 µg/L in shrimp to 42 µg/L in rainbow trout. It also affects fish behaviour. Atlantic salmon exposed to DDT as eggs experienced impaired balance and delayed appearance of normal behaviour patterns. DDT also affects temperature selection in fish.

(v) **Dioxins**

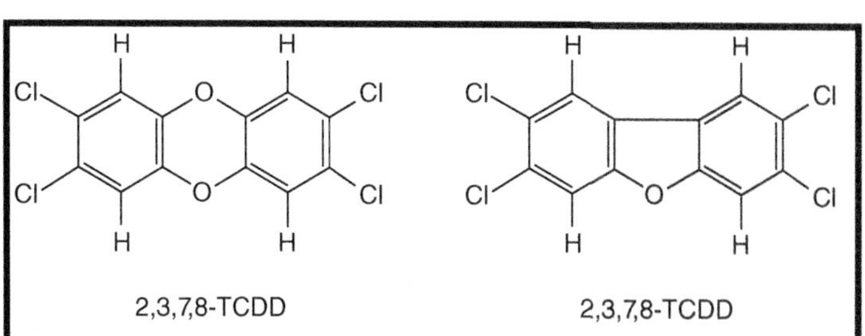

These chemicals are produced unintentionally in incomplete combustion as well as during the manufacture of pesticides and other chlorinated substances. They are emitted mostly in the burning of hospital, municipal and hazardous wastes but also when burning peat, coal and wood and in automobile emissions. Of the 75 different dioxins, 7 are considered to be of concern. One type was found to be present in the soil 10–12 years after the first exposure. Dioxins have been linked to a number of adverse effects in humans, including immune and enzyme disorders and chloracane, and they are classified as possible human carcinogens. In laboratory animals dioxins caused a variety of effects, including an increase in birth defects and stillbirths. Fish exposed to dioxins died shortly after the exposure. Food (particularly from animals) is the major source of exposure for humans.

(vi) **Furans**

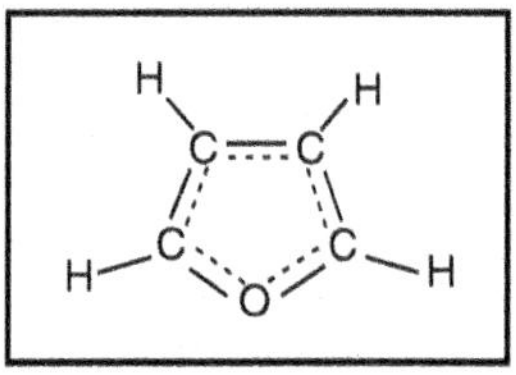

These compounds are produced unintentionally from many of the same processes that produce dioxins and during the production of PCBs. They have been detected in emissions from waste incinerators and automobiles. Furans are structurally similar to dioxins and share many of their toxic effects. The toxicity of the 135 different types varies. Furans persist in the environment for long periods and are classified as possible human carcinogens. Food, particularly animal products, is the major source of exposure for humans. Furans have been detected in breast-fed infants.

(vii) **Endrin**

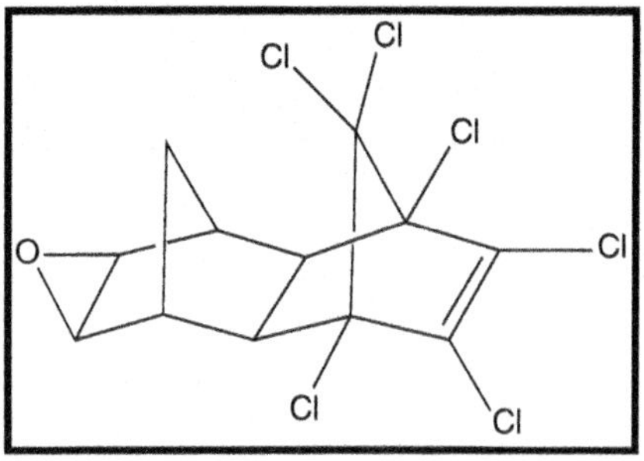

This insecticide is sprayed on the leaves of crops such as cotton and grains and is also used to control rodents such as mice and voles. Animals can metabolize endrin, so it does not accumulate in their fatty tissue to the extent that structurally similar chemicals do. It has a long half-life, however, persisting in the soil for up to 12 years. In addition, endrin is highly toxic to fish. When exposed to high levels of endrin in the water, sheep shed minnows hatched early and died by the ninth day of their exposure. The primary route of exposure for the general human population is through food, although current dietary intake estimates are below the limits deemed safe by world health authorities.

(viii) **Hexachlorobenzene (HCB)**

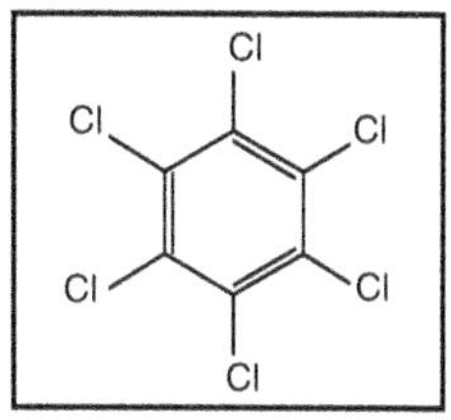

First introduced in 1945 to treat seeds, HCB kills fungi that affect food crops. It was widely used to control wheat bunt. It is also a by-product of the manufacture of certain industrial chemicals and exists as an impurity in several pesticide formulations. When people in eastern Turkey ate HCB-treated seed grain between 1954 and 1959, they developed a variety of symptoms, including photosensitive skin lesions, colic and debilitation; of several thousand who developed a metabolic disorder called *Porphyria turcica*, 14% died. Mothers also passed HCB to their infants through the placenta and through breast milk. In high doses, HCB is lethal to some animals and at lower levels adversely affects their reproductive success. HCB has been found in all food types. A study of Spanish meat found HCB present in all samples. In India, the estimated average daily intake of HCB is 0.13 micrograms per kilogram of body weight.

(ix) **Heptachlor**

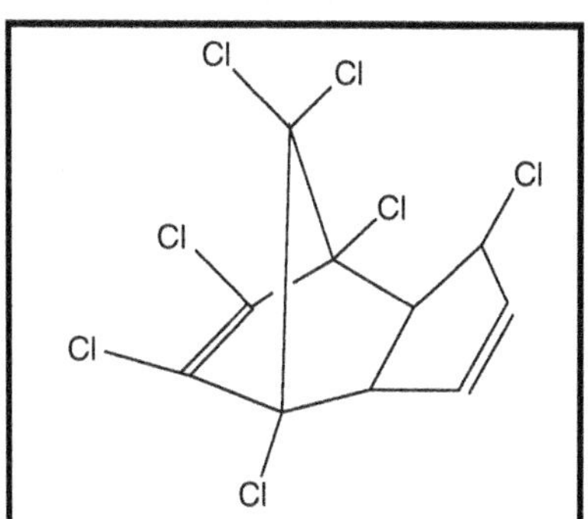

Primarily used to kill soil insects and termites, heptachlor has also been used to kill cotton insects, grasshoppers, other crop pests and malaria-carrying mosquitoes. It is believed to be responsible for the decline of several wild bird populations, including

Canadian geese and American kestrels in the Columbia River basin in the United States. The geese died after eating seeds treated with levels of heptachlor lower than maximum levels recommended by the manufacturer, suggesting that even responsible use of heptachlor may kill wildlife. Laboratory tests have shown high doses of heptachlor to be fatal to mink, rats and rabbits and lower doses to cause adverse behavioural changes and reduced reproductive success. Heptachlor is classified as a possible human carcinogen, and some two dozen countries have either banned it or restricted its use. Food is the major source of exposure for humans, and residues have been detected in the blood of cattle from the United States and Australia.

(x) **Mirex**

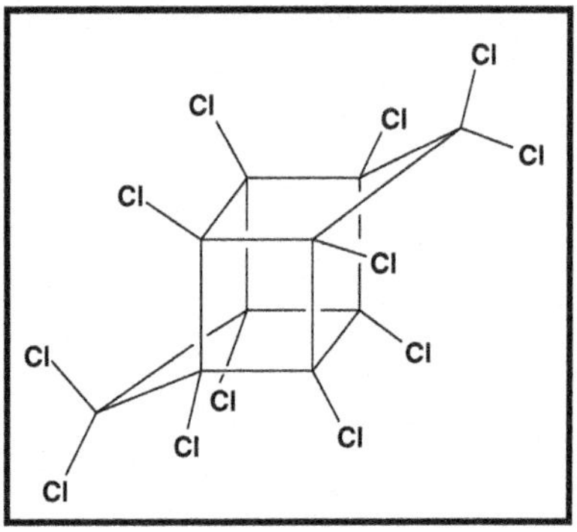

This insecticide is used mainly to combat fire ants, and it has been used against other types of ants and termites. It has also been used as a fire retardant in plastics, rubber and electrical goods. Direct exposure to mirex does not appear to cause injury to humans, but studies on laboratory animals have caused it to be classified as a possible human carcinogen. In studies, mirex proved toxic to several plant species and to fish and crustaceans. It is considered to be one of the most stable and persistent pesticides, with a half-life of as great as 10 years. The main route of human exposure to mirex is through food, particularly meat, fish and wild game.

(xi) **Polychlorinated Biphenyls (PCBs)**

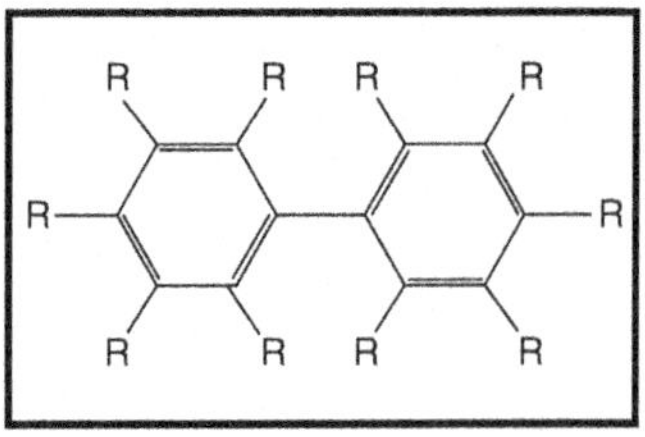

These compounds are used in industry as heat-exchange fluids in electric transformers and capacitors and as additives in paint, carbonless copy paper and plastics. Of the 209 different types of PCBs, 13 exhibit a dioxin-like toxicity. Their persistence in the environment corresponds to the degree of chlorination, and half-lives can vary from 10 days to 1 and a half years. PCBs are toxic to fish, killing them at higher doses and causing spawning failures at lower doses. Research also links PCBs to reproductive failure and suppression of the immune system in various wild animals, such as seals and mink. Large numbers of people have been exposed to PCBs through food contamination. Consumption of PCB contaminated rice oil in Japan in 1968 and in Taiwan in 1979 caused pigmentation of nails and mucous membranes and swelling of the eyelids, along with fatigue, nausea and vomiting. Due to the persistence of PCBs in their mothers' bodies and children born, as many as 70 years after the Taiwan incident showed developmental delays and behavioural problems. Similarly, children of mothers who ate large amounts of contaminated fish from Lake Michigan showed poorer short-term memory function. PCBs also suppress the human immune system and are listed as probable human carcinogens.

(xii) **Toxaphene**

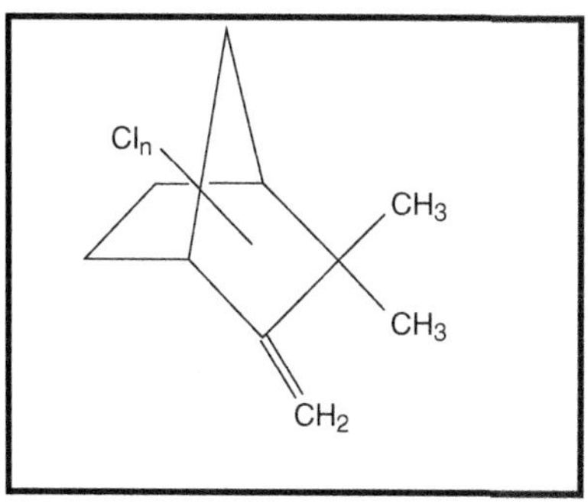

This insecticide is used on cotton, cereal grains, fruits, nuts and vegetables. It has also been used to control ticks and mites in livestock. Toxaphene was the most widely used pesticide in the United States in 1975. As much as 50% of a toxaphene release can persist in the soil for as long as 12 years. For humans, the most likely source of toxaphene exposure is food. While the toxicity to humans of direct exposure is not high, toxaphene has been listed as a possible human carcinogen due to its effects on laboratory animals. It is highly toxic to fish; brook trout exposed to toxaphene for 90 days experienced a 46% reduction in weight and reduced egg viability, and long-term exposure to levels of 0.5 micrograms per litre of water reduced egg viability to zero. Thirty-seven countries have banned toxaphene, and 11 others have severely restricted its use.

3. **Industrial Effluents**

The effluents of different industries such as textile mills, fertilizers, pesticides, refineries, distilleries, pharmaceutical and food-processing industries pollute the soil environment. The human excreta or sewage sludge contain many pathogenic viruses and bacteria which contaminate and disturb the microflora and microfauna of the soil. The industrial waste gets disposed on the soil and changes its physical, chemical and biological properties. They also contaminate drinking water sources, both surface and underground. Many toxic chemicals leached from landfill sites into the soil cause a large number of birth defects, cancers, respiratory, nervous and kidney diseases.

4. **Improper Waste Disposal Techniques**

Human activities generate huge quantities of city wastes including biodegradable materials (vegetables, animal wastes, papers, wooden pieces, carcasses plane parts) and many non-biodegradable materials (plastic bags/bottles/wastes, glass bottles, glass pieces and construction waste materials). The microbial decomposition of organic wastes produces large quantities of methane in addition to other chemicals to pollute the soil and water flowing on its surface. Some of the waste materials release chemicals, such as cadmium, chromium, lead, arsenic and selenium, which get deposited in the underground soil. These chemicals damage the normal activities and ecological balance in the soil.

5. **Deforestation**

The vast clearance of forest land for agricultural, mining and large developmental projects such as road and highways, rail and hydropower leads to changes in the natural soil characteristics. These anthropogenic activities cause deterioration of the top soil. A detailed study conducted by the present authors during 2019 in four iron mines of Jharkhand, India, exhibited significant spatial variations in N, P and K levels among different compartments of the mines (barren land, newly plantation site, forest, buffer and adjacent agricultural land) (Table 9.1). The soil nutrients were extremely low in the barren land compartment. Hence the mine owners have taken great initiative to eco-restore the situation through massive plantation after nourishing the soil with fertilizers from external sources (Figs. 9.2, 9.3, 9.4, 9.5, 9.6, 9.7, 9.8, 9.9, 9.10, 9.11, 9.12 and 9.13).

Table 9.1 Comparative account of soil nutrients in different compartments of the mines

Mine no. 1	**Barren land**	**Newly plantation**	**Forest**	**Buffer**	**Agricultural land**
N	106.34	146.89	494.88	527.69	407.52
P	15.93	28.11	33.92	46.76	35.11
K	143.32	129.49	620.33	716.34	483.29
Mine no. 2	**Barren land**	**Newly plantation**	**Forest**	**Buffer**	**Agricultural land**
N	91.05	130.45	477.32	509.31	–
P	4.22	16.78	20.44	33.15	–
K	118.56	105.49	597.59	691.24	–
Mine no. 3	**Barren land**	**Newly plantation**	**Forest**	**Buffer**	**Agricultural land**
N	83.56	121.66	418.55	469.20	385.57
P	3.57	14.75	17.20	28.19	21.65
K	108.33	98.39	501.76	617.56	452.08
Mine no. 4	**Barren land**	**Newly plantation**	**Forest**	**Buffer**	**Agricultural land**
N	–	110.59	403.87	411.55	472.66
P	–	12.97	15.60	21.90	65.05
K	–	87.44	496.85	567.20	622.43

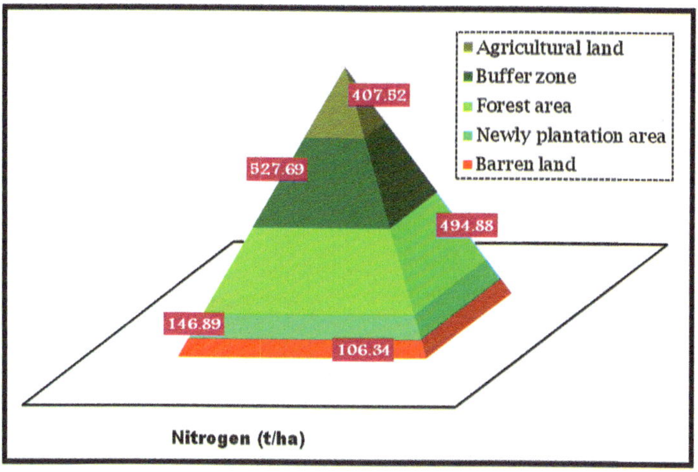

Fig. 9.2 Soil nitrogen (t/ha) in five compartments of Mine no. 1

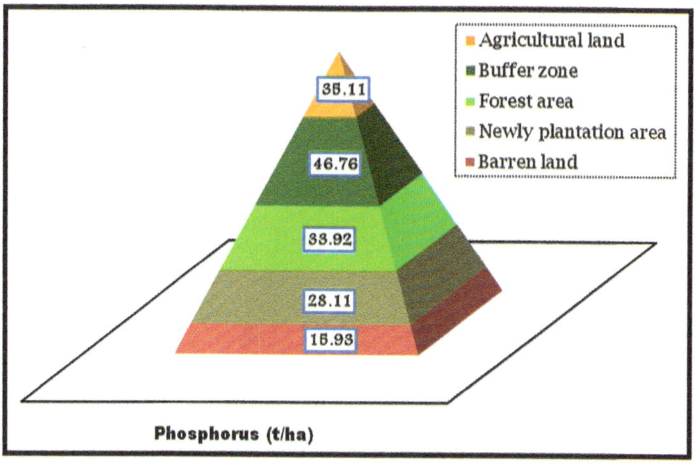

Fig. 9.3 Soil phosphorus (t/ha) in five compartments of Mine no. 1

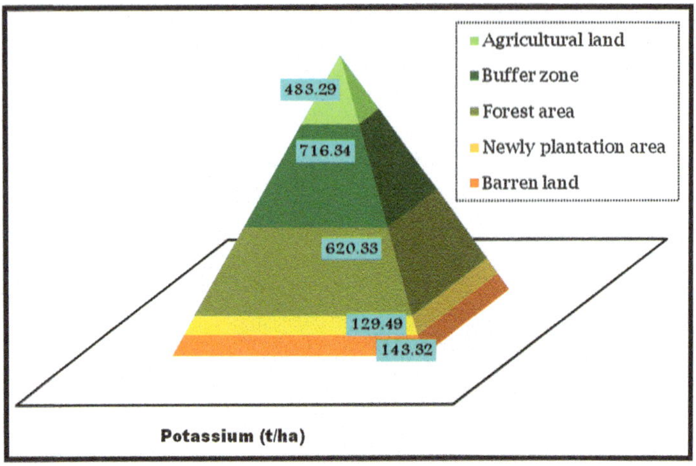

Fig. 9.4 Soil potassium (t/ha) in five compartments of Mine no. 1

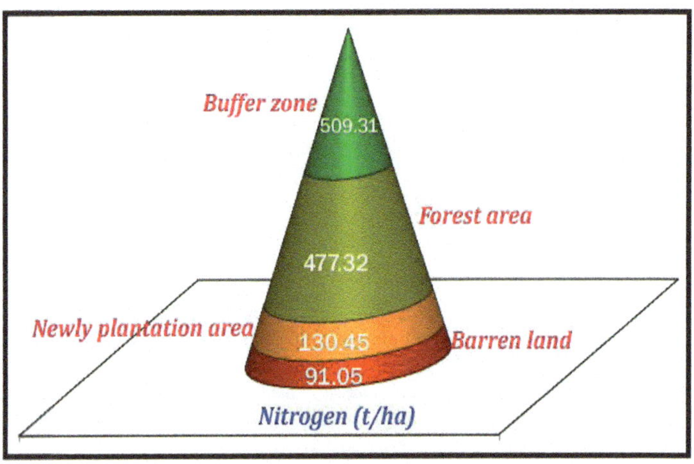

Fig. 9.5 Soil nitrogen (t/ha) in five compartments of Mine no. 2

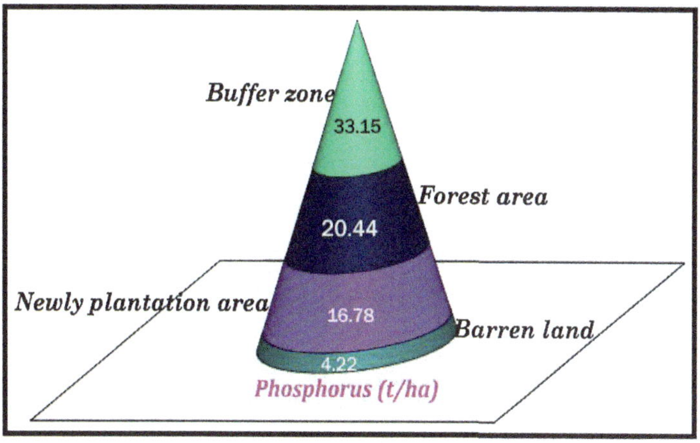

Fig. 9.6 Soil phosphorus (t/ha) in five compartments of Mine no. 2

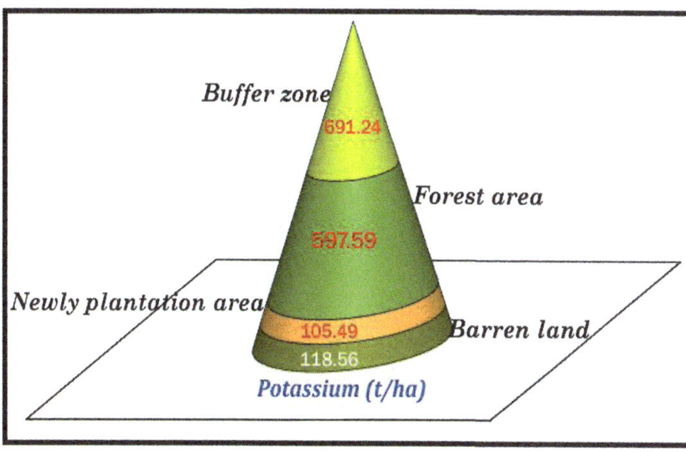

Fig. 9.7 Soil potassium (t/ha) in five compartments of Mine no. 2

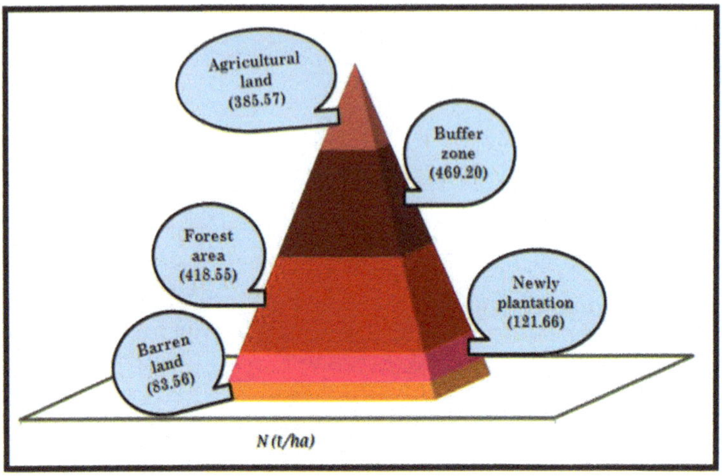

Fig. 9.8 Soil nitrogen (t/ha) in five compartments of Mine no. 3

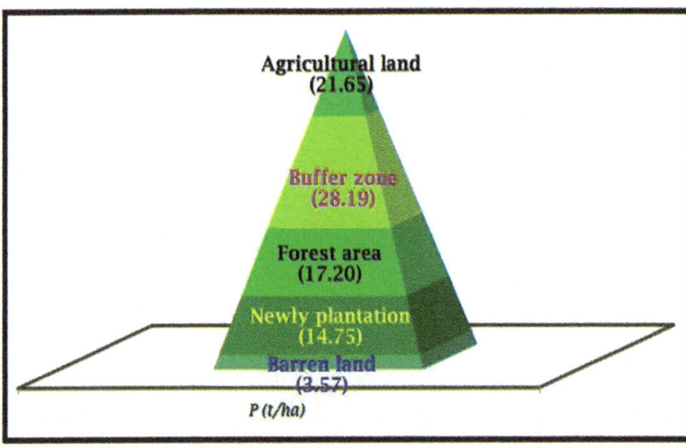

Fig. 9.9 Soil phosphorus (t/ha) in five compartments of Mine no. 3

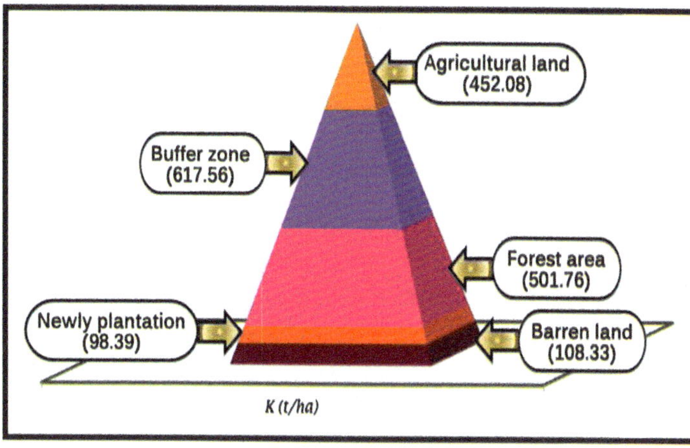

Fig. 9.10 Soil potassium (t/ha) in five compartments of Mine no. 3

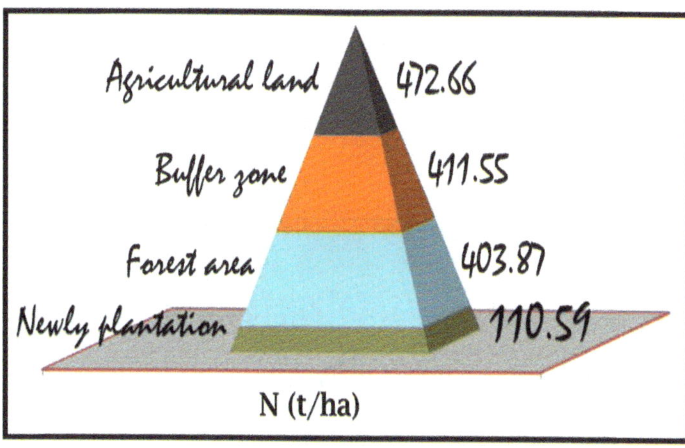

Fig. 9.11 Soil nitrogen (t/ha) in four compartments of Mine no. 4

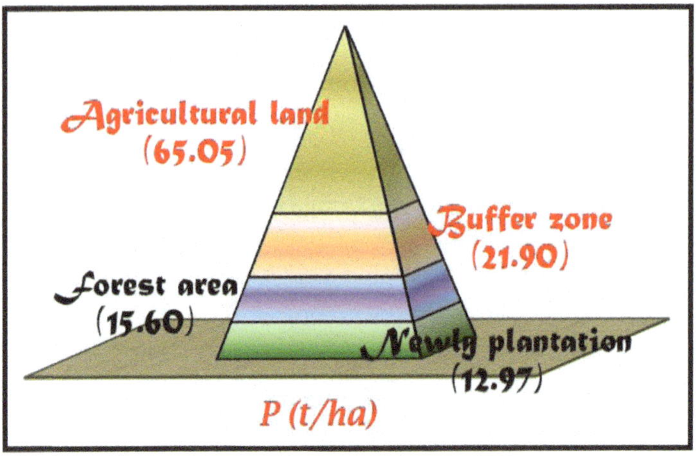

Fig. 9.12 Soil phosphorus (t/ha) in four compartments of Mine No. 4

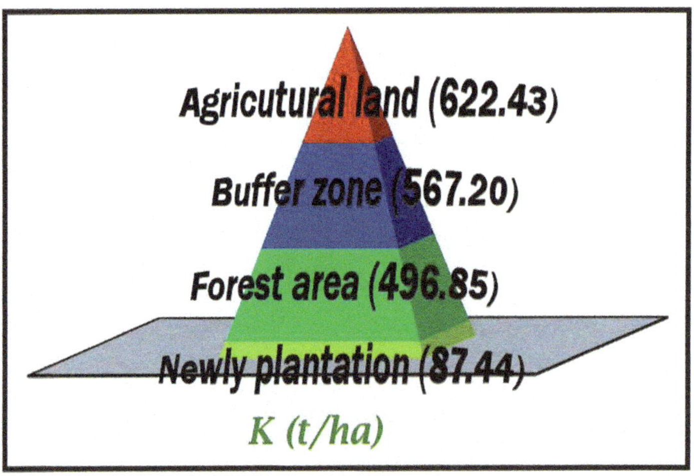

Fig. 9.13 Soil potassium (t/ha) in four compartments of Mine No. 4

9.2 Effects of Soil Pollution

The major effects of soil pollution are discussed here in points:

1. Soil pollution decreases the soil fertility, loss of soil nutrients, thereby reducing the crop yield and imbalance of flora and fauna.
2. Oxides of sulphur and nitrogen present in the atmosphere combine with rain water to form sulphuric acid and nitric acid, which reach the soil and increase its acidity. This increased acidity is unfavourable for plant growth. Mining activities lower the soil pH that results in the loss of biodiversity. We studied the change in

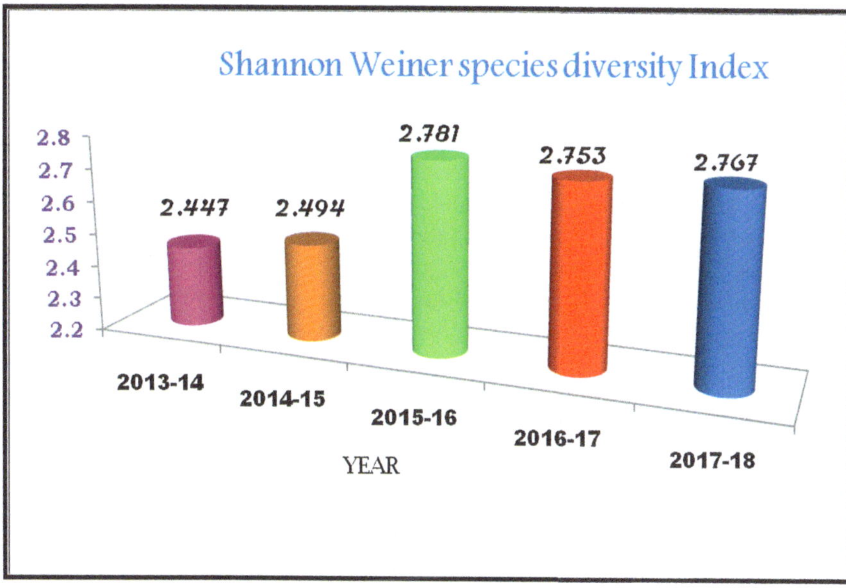

Fig. 9.14 Trend of Shannon–Weiner species diversity index over a period of 5 consecutive years (2013–14, 2014–15, 2015–16, 2016–17, 2017–18)

Fig. 9.15 Soil pH during 2013–14 (~ 3.8) and 2017–18 (~ 6.2) in the same site

biodiversity in an iron mine in the state of Odisha, India, for 5 successive years and observed the gradual increase in biodiversity (Fig. 9.14) with rise of soil pH value from ~4.0 to ~6.5 (Figs. 9.14 and 9.15).

3. Dumping of radioactive wastes is lethal for microorganisms and plants.
4. Processes such as biological magnification and eutrophication occur due to indiscriminate use of pesticides and fertilizers. These agrochemicals are the main culprits of changes in soil characteristics.

9.3 Minimizing Soil Pollution Through Bioremediation

Soil pollution can be controlled in the following ways:

1. The impact of pesticide can be lowered by reducing its use in agriculture and replacing the same with biofertilizer.
2. Biological pest control methods should be implemented. Predators and parasites of pests should be used to control their population.
3. Methods such as crop rotation and mixed cropping need to be promoted so as to reduce the use of agrochemicals. To reduce the impact of fertilizers, practices such as growing of leguminous crops (crops having property of nitrogen fixation), fallowing of land for some time and crop rotation should be followed.
4. Use of organic fertilizers and farm yard manures should be promoted.
5. Agricultural runoff should be controlled before reaching water bodies so as to prevent eutrophication. Check dams can be used for runoff control.

We have few recommendations to eco-restore the already deteriorated soil as discussed here. These will not only upgrade the quality of soil but will also promote the biodiversity as well.

Restoration projects differ in their objectives and their methods of achieving those goals. Many restoration projects aim to establish ecosystems composed of a native species; other projects attempt to restore, improve or create particular ecosystem functions, such as pollination or erosion control. Some examples of different kinds of restoration include the following:

- *Revegetation* – the establishment of vegetation on sites where it has been previously lost, often with erosion control as the primary goal. For example, with vetiver plantation on the slope.
- *Habitat enhancement* – the process of increasing the suitability of a site as habitat for some desired species mostly by upgrading the soil parameters in terms of SOC, N, P, K and pH.
- *Remediation* – improving an existing ecosystem or creating a new one with the aim of replacing another that has deteriorated or been destroyed.
- *Mitigation* – legally mandated remediation for loss of protected species or ecosystems.

At a given restoration site, it may be possible to establish a number of different communities. When choosing a target state for a restoration project, however, restorationists generally select only one (or a small range) of possible community types. Often some sort of 'natural' pre-disturbance condition, or reference state, is selected, along with its presumed properties. This is often represented by a nearby undisturbed reference site. In this context, the nearby/adjacent buffer area may be selected for reference.

Considering the entire scenario of soil disturbance due to mining activities, few core recommendations are suggested:

Recommendation 1

Soil pH is an important parameter to regulate the growth of plants. Soil pH below 5.5 might result in reduced yields and damages to the plants. Under these pH conditions, the availability of micronutrients such as manganese, aluminium and iron increases and toxicity problem of micronutrients might occur. On the other hand, at low soil pH, the availability of other **essential nutrients**, such as N, P, K, Ca and Mg, decreases, and this might result in deficiencies.

(a) **Raising Soil pH Using Lime**

The most commonly used technique to raise the soil pH is applying agricultural lime. The solubility of lime is relatively low, so if it is applied only to the soil surface, it usually affects only the top layer of the soil, not more than a few centimetres deep.

It is recommended that lime should be incorporated into the substratum prior to planting and the process is usually logistically difficult. Waiting until after planting only makes it more complicated, because the lime should then be individually applied to each growing container or each plant. Again, due to its very low solubility, it's impossible to apply it through irrigation. Sprinkling 4% lime at an interval of 3 days is needed for barren area before adding soil for new plantation. Then after soil addition from outside sources, liming is to be carried out in doses only after monitoring the soil dumped from external sources.

The effectiveness of a liming material is directly related to its purity. Purity of a liming material is a measurement of the calcium carbonate equivalent (CCE). It is a measure to the amount of acid that the liming material will neutralize, compared to pure calcium carbonate ($CaCO_3$).

The CCE values of several common liming materials are presented in the table below (Table 9.2).

(b) **Raising Soil pH Using Potassium Carbonate**

Unlike lime, potassium carbonate is highly soluble and therefore can be applied by drip irrigation. Due to its high solubility, potassium carbonate can be easily distributed throughout the root zone together with irrigation water and reach deeper soil profile.

Table 9.2 CCE value of some liming material

Liming material	CCE
Pure calcitic lime ($CaCO_3$)	100
Pure dolomitic lime $CaCO_3 * MgCO_3$	108
Quicklime/burned lime (CaO)	179
Hydrated lime $CaOH_2$	136
Slag ($CaSiO_3$)	86

In both soils and growing media, potassium carbonate can rapidly affect chemical reactions in the root zone, thus elevate root zone pH.

Recommendation 2

PGPR application can be an effective treatment for accelerating the plant growth. The mechanism of rhizobacteria-mediated plant growth promotion is not completely identified, but the so-called plant growth-promoting rhizobacteria have been observed to expedite plant growth. Several researches are available on the beneficial role of rhizobacteria like heavy metal detoxifying potentials (Ma et al. 2011; Wani and Khan 2010), pesticide degradation/tolerance (Ahemad and Khan 2012a, b), salinity tolerance (Tank and Saraf 2010; Mayak et al. 2004) and biological control of phytopathogens and insects (Hynes et al. 2008; Russo et al. 2008; Joo et al. 2005; Murphy et al. 2000) along with the normal plant growth-promoting properties such as phytohormones (Ahemad and Khan 2012c; Tank and Saraf 2010), siderophore (Jahanian et al. 2012; Tian et al. 2009), 1-amino-cyclopropane-1-carboxylate, hydrogen cyanide (HCN) and ammonia production, nitrogenase activity (Glick et al. 1999; Khan 2005) and phosphate solubilization (Ahmed and Khan 2012c) (Table 9.3).

There are certain particular mechanisms through which plant growth promotion occurs. One of such mechanism is the alteration of the whole microbial community in rhizosphere niche through production of various types of plant growth-promoting rhizobacteria (PGPR), which accelerate the growth of plant by two ways: the first way is by facilitating the process of resource acquisition like nitrogen, phosphorous and several essential minerals present in the soil, and the second way is by serving as biocontrol agents and thereby acting against the pathogens (Fig. 9.16).

Recommendation 3

Role of exotic or native species in reclamation needs careful consideration as newly introduced exotic species may become pests in other situations. Therefore, candidate species for plantation should be screened carefully to avoid becoming problematic weeds in relation to local/regional floristic spectrum. For artificial introduction, selection of species that are well adapted to the local environment should be emphasized. Indigenous species are preferable to exotics because they are most likely to fit into fully functional ecosystem and are climatically adapted.

Considering the carbon sequestration potential in the mining zone of Odisha and Jharkhand, five native species are extremely important as highlighted in Table 9.4.

Recommendation 4

- Use of organic fertilizer (liquid form) is recommended (Table 9.5). This may be sourced/manufactured from several ingredients as stated in Table 9.5. Water hyacinth can also be cultivated in ponds/rain water harvested area and transformed in to organic fertilizer.

Table 9.3 Growth-promoting substances released by PGPR (Ahemad and Kibret 2014)

PGPR	Plant growth-promoting traits	References
Pseudomonas putida	IAA, siderophores, HCN, ammonia, exo-polysaccharides, phosphate solubilization	Ahemad and Khan (2011c, 2012a, c)
Pseudomonas aeruginosa	IAA, siderophores, HCN, ammonia, exo-polysaccharides, phosphate solubilization	Ahemad and Khan (2010d, 2011a, 2012e)
Klebsiella sp.	IAA, siderophores, HCN, ammonia, exo-polysaccharides, phosphate solubilization	Ahemad and Khan (2011b, f, g)
Enterobacter asburiae	IAA, siderophores, HCN, ammonia, exo-polysaccharides, phosphate solubilization	Ahemad and Khan (2010a, b)
Rhizobium sp. *(pea)*	IAA, siderophores, HCN, ammonia, exo-polysaccharides	Ahemad and Khan (2009b, 2010c, 2011i, 2012b)
Mesorhizobium sp.	IAA, siderophores, HCN, ammonia, exo-polysaccharides	Ahemad and Khan (2009a, 2010e, h, 2012d, e)
Acinetobacter spp.	IAA, phosphate solubilization, siderophores	Rokhbakhsh-Zamin et al. (2011)
Rhizobium sp.*(lentil)*	IAA, siderophores, HCN, ammonia, exo-polysaccharides	Ahemad and Khan (2010f, g, 2011e)
Pseudomonas sp. *A3R3*	IAA, siderophores	Ma et al. (2011b)
Psychrobacter sp. *SRS8*	Heavy metal mobilization	Ma et al. (2011a)
Bradyrhizobium sp.	IAA, siderophores, HCN, ammonia, exo-polysaccharides	Ahemad and Khan (2011d, h, 2012f)
Pseudomonas aeruginosa 4EA	Siderophores	Naik and Dubey (2011)
Bradyrhizobium sp. *750*, *Pseudomonas* sp., *Ochrobactrum cytisi*	Heavy metal mobilization	Dary et al. (2010)
Bacillus sp. *PSB10*	IAA, siderophores, HCN, ammonia	Wani and Khan (2010)
Paenibacillus polymyxa	IAA, siderophores	Phi et al. (2010)
Rhizobium phaseoli	IAA	Zahir et al. (2010)
Stenotrophomonas maltophilia	Nitrogenase activity, phosphate solubilization, IAA, ACC deaminase	Mehnaz et al. (2010)
Rahnella aquatilis	Phosphate solubilization, IAA, ACC deaminase	Mehnaz et al. (2010)
Pseudomonas aeruginosa, *Pseudomonas fluorescens*, *Ralstonia metallidurans*	Siderophores	Braud et al. (2009)
Proteus vulgaris	Siderophores	Rani et al. (2009)
Pseudomonas sp.	Phosphate solubilization, IAA, siderophore, HCN, biocontrol potential	Tank and Saraf (2009)

(continued)

Table 9.3 (continued)

PGPR	Plant growth-promoting traits	References
Azospirillum amazonense	IAA, nitrogenase activity	Rodrigues et al. (2008)
Mesorhizobium sp.	IAA, siderophores, HCN, ammonia	Wani et al. (2008)
Pseudomonas sp.	ACC deaminase, IAA, siderophore	Poonguzhali et al. (2008)
Serratia marcescens	IAA, siderophore, HCN	Selvakumar et al. 2008
Pseudomonas fluorescens	ACC deaminase, phosphate solubilization	Shaharoona et al. (2008)
Acinetobacter sp., *Pseudomonas* sp.	ACC deaminase, IAA, antifungal activity, N_2 fixation, phosphate solubilization	Indira Gandhi et al. (2008)
Enterobacter sp.	ACC deaminase, IAA, siderophore, phosphate solubilization	Kumar et al. (2008)
Burkholderia sp.	ACC deaminase, IAA, siderophore, heavy metal solubilization, phosphate solubilization	Jiang et al. (2008)
Pseudomonas jessenii	ACC deaminase, IAA, siderophore, heavy metal solubilization, phosphate solubilization	Rajkumar et al. (2008)
Pseudomonas aeruginosa	ACC deaminase, IAA, siderophore, phosphate solubilization	Ganesan (2008)
Pseudomonas sp.	ACC deaminase, IAA, siderophore, heavy metal solubilization, phosphate solubilization	Rajkumar et al. (2008)
Azotobacter sp., *Mesorhizobium* sp., *Pseudomonas* sp., *Bacillus* sp.	IAA, siderophore, antifungal activity, ammonia production, HCN	Ahemad and Khan (2009a)
Bradyrhizobium sp.	IAA, siderophores, HCN, ammonia	Wani et al. (2007a)
Rhizobium sp.	IAA, siderophores, HCN, ammonia	Wani et al. (2007b)
Mesorhizobium ciceri, *Azotobacter chroococcum*	IAA, siderophores	Wani et al. (2007c)
Pseudomonas, *Bacillus*	Phosphate solubilization, IAA and siderophores	Wani et al. (2007c)
Klebsiella oxytoca	IAA, phosphate solubilization, nitrogenase activity	Jha and Kumar (2007)
Bacillus spp., *Pseudomonas* spp., *Azotobacter* spp., *Rhizobium* spp.	IAA, ammonia production	Joseph et al. (2007)
Pseudomonas fluorescens	Induced systemic resistance, antifungal activity	Saravanakumar et al. (2007)

(continued)

Table 9.3 (continued)

PGPR	Plant growth-promoting traits	References
Pseudomonas chlororaphis	Antifungal activity	Liu et al. (2007)
Bacillus subtilis	Antifungal activity	Cazorla et al. (2007)
Gluconacetobacter diazotrophicus	Zinc solubilization	Saravanan et al. (2007)
Brevibacillus spp.	Zn resistance, IAA	Vivas et al. (2006)
Bacillus subtilis	IAA, phosphate solubilization	Zaidi et al. (2006)
Pseudomonas sp., *Bacillus* sp.	IAA, siderophore, phosphate solubilization	Rajkumar et al. (2006)
Pseudomonas putida	Antifungal activity, siderophore, HCN, phosphate solubilization	Pandey et al. (2006)
Brevibacterium sp.	Siderophore	Noordman et al. (2006)
Xanthomonas sp. RJ3, *Azomonas* sp. RJ4,	IAA	Sheng and Xia (2006)
Pseudomonas sp. RJ10, *Bacillus* sp. RJ31	P-solubilization	Canbolat et al. (2006)
Bradyrhizobium japonicum	IAA	Shaharoona et al. (2006)
Pseudomonas fluorescens PRS$_9$	IAA, siderophores, phosphate solubilization	Gupta et al. (2005)
Variovorax paradoxus, *Rhodococcus* sp.	IAA and siderophores	Belimov et al. (2005)
Sphingomonas sp., *Mycobacterium* sp.,	IAA	Tsavkelova et al. (2005)
Pseudomonas fluorescens	IAA, siderophores, antifungal activity	Dey et al. (2004)
Bacillus sp., *Azospirillum* sp.	IAA, P-solubilization	Yasmin et al. (2004)
Azospirillum brasilense, *Azospirillum amazonense*	IAA, P-solubilization, nitrogenase activity, antibiotic resistance	Thakuria et al. (2004)
Pseudomonas fluorescens	IAA, phosphate solubilization	Jeon et al. (2003)
Rhizobium sp., *Bradyrhizobium* sp.	HCN, siderophore, Siderophore, IAA, P-solubilization	Deshwal et al. (2003)
Bacillus sp., *Pseudomonas* sp., *Azotobacter* sp., *Azospirillum* sp., *Rhizobium* sp.	P-solubilization and IAA	Tank and Saraf (2003)
Mesorhizobium sp., *Bradyrhizobium* sp.	Siderophore	Khan et al. (2002)
Azotobacter chroococcum	Gibberellin, kinetin, IAA	Verma et al. (2001)
Rhizobium meliloti	Siderophore	Arora et al. (2001)
Kluyvera ascorbata	Siderophore	Burd et al. (2000)
Kluyvera ascorbata	ACC deaminase, siderophores, metal resistance	Genrich et al. (1998)
Bradyrhizobium sp., *Rhizobium* sp.	Siderophore	Duhan et al. (1998)
Rhizobium ciceri	Siderophore	Berraho et al. (1997)
Bradyrhizobium japonicum	Siderophore	Wittenberg et al. (1996)

(continued)

Table 9.3 (continued)

PGPR	Plant growth-promoting traits	References
Rhizobium leguminosarum	Cytokinin	Noel et al. (1996)
Rhizobium sp., *Bradyrhizobium* sp.	P-solubilization	Abd-Alla (1994)

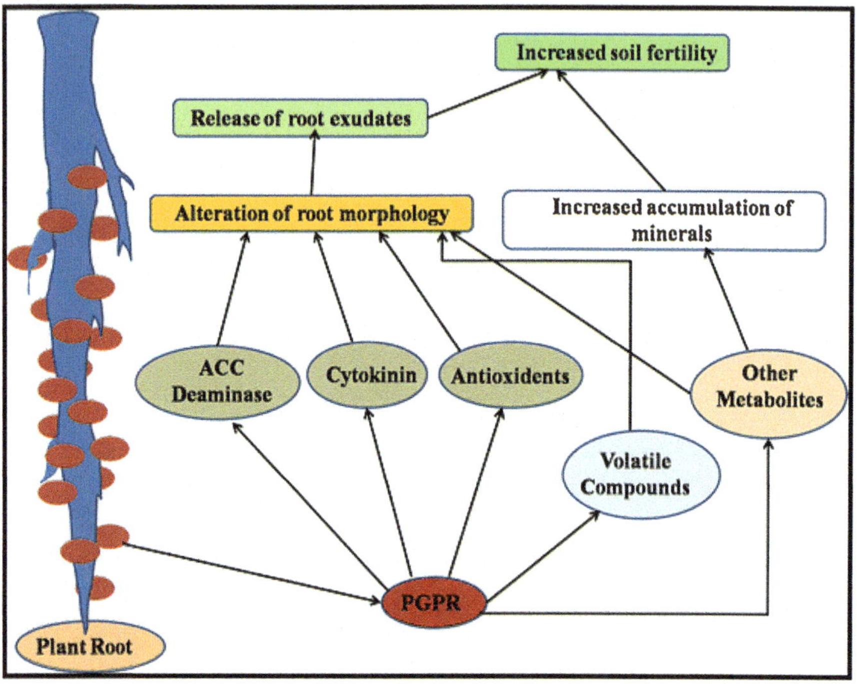

Fig. 9.16 Role of PGPR in plant growth through increasing soil nutrient availability

9.4 Take Home Messages

(A) Soil is an important component of the environment and vital for all living beings, as it produces food on which various life forms depend. Human activities have led to modification in soil characteristics through use of fertilizers and pesticides to increase the agricultural production. The other activities that have changed natural soil characteristics include deforestation and improper disposal of wastes, especially the garbage refuse and fly ash.

(B) Soil consists of solid materials, organic matter and air spaces. The four important components of soil are mineral matter (40%), soil water (25%), organic matter (10%) and soil air (25%). The topmost layer of soil is called A-horizon which is several inches in thickness. The layer has maximum biological activity

Table 9.4 Potential native species with considerable AGC values

AGC (t/ha) of five dominant species in and around the Mine					
PARTS	MAHUA	KARANJ	SAL	KUSUM	IMLI
STEM	19.79	5.83	20.22	21.14	6.70
BRANCH	7.92	2.56	9.91	8.67	3.15
LEAF	3.96	1.34	5.46	5.07	1.41
AGC	31.67	9.73	35.59	34.88	11.26

and has the organic matter humus. This top layer is important for the agricultural crops and vegetation. There is continuous leaching of metal ions. Due to deforestation this topmost layer is prone to soil erosion which in turn reduces the soil fertility and also the production. The next layer beneath to A-horizon is known as B-horizon. The organic matter, minerals and salts leach from A-horizon and reach B-horizon. The next layer is called as C-horizon which consists of weathered parent rocks from which soil originates.

(C) The major causes of soil pollution are due to use of agrochemicals. The other factors causing soil pollution include deforestation, improper waste disposal, etc. Fly ash generated from thermal power plants also pollutes the soil to a great extent. Mining activities lowers the soil pH that results in the loss of biodiversity.

(D) Soil pollution decreases the soil fertility, soil erodibility and loss of soil nutrients, thereby reducing the crop yield and imbalance of flora and fauna.

(E) Soil pollution can be controlled through several ways. The impact of pesticide can be lowered by reducing its use in agriculture and replacing the same with biofertilizer or biological pest control. Methods such as crop rotation and mixed cropping need to be promoted so as to reduce the use of agrochemicals. To reduce the impact of fertilizers, practices such as growing of leguminous crops (crops having property of nitrogen fixation), fallowing of land for some time and crop rotation should be followed. Use of organic fertilizers and farm yard manures should be promoted.

Table 9.5 Composition of organic fertilizer ingredients

Ingredient	N	P_2O_5	K_2O	Ca	Mg	S	Comments
Blood (dried)	12–15	2–3	1	BDL	BDL	BDL	Good source of nutrients
Blood meal (steamed)	15	0.5–1	1	BDL	BDL	BDL	Good source of nutrients
Bone meal (steamed)	~ 1.8	10–20	0	18–30	0	0	Low nitrogen and moderate source of nutrients
Compost (garden)	V	V	V	V	V	V	Depending on the ingredients and technology, the composition varies
Cotton seed meal	6–7	2.5	1.5	BDL	BDL	BDL	Good source of nutrients
Cotton seed hull ash	0	0	27	BDL	BDL	BDL	Noted for high potassium level
Fish scrap (Acidulated)	8.5–9.5	1–2	0	BDL	BDL	2.0	Rich in nitrogen, but bioaccumulated heavy metals and pesticides are causes of concern
Dried fish meal	8.5–10.0	0	0	5.8	BDL	BDL	Rich in nitrogen, but bioaccumulated heavy metals and pesticides are causes of concern
Legume	2–3	2.4	2.4	1.2	0.2	0.3	Balanced nutrient level
Cattle manure	0.5–2	1.5	1.1–1.2	1.1	0.3	BDL	–
Broiler litter	2–3	3.0	2.0	1.6–1.9	0.4	0.3	Balanced nutrient level

[a]*V* variable, *BDL* below detectable level

9.5 Brain Churners

Soil pollution is a burning issue in underdeveloped and developing nations of the world. It is intricately related to industrialization and urbanization and has severe adverse effects on human health, biodiversity, crop production and climate of the planet Earth. The readers can consider this chapter from the health security point of view. This Brain Churners section is an approach to understand the pulse of the readers through a sort of *self-appraisal test* as correct answers to all the questions are marked as black in the box at the end.

1. The study of soil science is known as
   (a) Zoology
   (b) Geography
   (c) Edaphology
   (d) Botany

2. The topmost layer of the soil is called
   (a) B-horizon
   (b) A-horizon
   (c) C-horizon
   (d) None of the above
3. C-horizon consists of
   (a) Weathered parent rocks, from which soil originates
   (b) Sand
   (c) Organic matter, humus
   (d) None of the above
4. DDT gets stored in the
   (a) Fat of organisms
   (b) Protein of organisms
   (c) Bones of organisms
   (d) None of the above
5. Benzene hexachloride is
   (a) Biopesticide
   (b) Nontoxic pesticide
   (c) Useful pesticide
   (d) Harmful pesticide
6. Humans are exposed to aldrin through
   (a) Dairy products and animal meats
   (b) Air needed for breathing
   (c) Water used for drinking
   (d) None of the above
7. Mirex is an insecticide which is used to combat
   (a) Locusts
   (b) Cockroaches
   (c) Bugs
   (d) Fire ants
8. Toxaphene is an insecticide which is used to control
   (a) Ticks and mites
   (b) Ants
   (c) Beetle
   (d) None of the above
9. Mining action results in
   (a) Increase of soil pH
   (b) Decrease of soil pH

(c) No change of soil pH
(d) None of the above

10. Calcium carbonate Equivalent (CCE) is a measure of the

(a) Impurity present in lime
(b) Sand present in lime
(c) Oxygen present in lime
(d) Purity of the liming material

11. The CCE value of 'slag' ($CaSiO_3$) is

(a) 100
(b) 108
(c) 86
(d) 179

12. PGPR results in the

(a) Decrease of plant growth
(b) Death of the plant
(c) Increase of plant growth
(d) None of the above

13. *Pseudomonas putida* releases

(a) IAA
(b) HCN
(c) ACC-d aminase
(d) None of the above

14. Bone meal has

(a) High nitrogen content
(b) Moderate nitrogen content
(c) Low nitrogen content
(d) none of the above

15. On 1968 consumption of PCB contaminated rice oil in Japan caused

(a) Cancer and death
(b) Pigmentation of nails and swelling of eyelids
(c) Polio
(d) Anaemia

16. Hexachlorobenzene (HCB) is used to kill

(a) Fungi
(b) Algae
(c) Insects
(d) None of the above

17. *Porphyria turcica* is
    - (a) A gastric problem caused by DDT
    - (b) A metabolic disorder caused by HCB
    - (c) A vision problem caused by vitamin deficiency
    - (d) None of the above
18. The half-life of dieldrin in soil is
    - (a) Approximately 20 years
    - (b) Approximately 5 years
    - (c) Approximately 1 year
    - (d) None of the above
19. The health of the soil is evaluated by
    - (a) NPK level
    - (b) Heavy metal level
    - (c) Clay content
    - (d) Sand content
20. Use of agrochemicals can be reduced by
    - (a) Promoting chemical fertilizer
    - (b) Applying lime on top soil
    - (c) Crop rotation
    - (d) None of the above

Answer Sheet

S. No.	a	b	c	d
1.			■	
2.		■		
3.	■			
4.	■			
5.				■
6.	■			
7.				■
8.	■			
9.		■		
10.				■
11.				■
12.			■	
13.	■			
14.			■	
15.		■		
16.	■			
17.		■		
18.		■		
19.	■			
20.			■	

References

Abd-Alla MH (1994) Solubilization of rock phosphates by *Rhizobium* and *Bradyrhizobium*. Folia Microbiol 39:53–56

Ahemad M, Khan MS (2009a) Effect of insecticide-tolerant and plant growth promoting *Mesorhizobium* on the performance of chickpea grown in insecticide stressed alluvial soils. J Crop Sci Biotechnol 12:213–222

Ahemad M, Khan MS (2009b) Toxicity assessment of herbicides quizalafop-p-ethyl and clodinafop towards *Rhizobium* pea symbiosis. Bull Environ Contam Toxicol 82:761–766

Ahemad M, Khan MS (2010a) Influence of selective herbicides on plant growth promoting traits of phosphate solubilizing *Enterobacter asburiae* strain PS2. Res J Microbiol 5:849–857

Ahemad M, Khan MS (2010b) Plant growth promoting activities of phosphate-solubilizing *Enterobacter asburiae* as influenced by fungicides. EurAsia J Biosci 4:88–95

Ahemad M, Khan MS (2010c) Comparative toxicity of selected insecticides to pea plants and growth promotion in response to insecticide-tolerant and plant growth promoting *Rhizobium leguminosarum*. Crop Prot 29:325–329

Ahemad M, Khan MS (2010d) Phosphate-solubilizing and plant-growth-promoting *Pseudomonas aeruginosa* PS1 improves green-gram performance in quizalafop-p-ethyl and clodinafop amended soil. Arch Environ Contam Toxicol 58:361–372

Ahemad M, Khan MS (2010e) Ameliorative effects of *Mesorhizobium* sp. MRC4 on chickpea yield and yield components under different doses of herbicide stress. Pestic Biochem Physiol 98:183–190

Ahemad M, Khan MS (2010f) Insecticide-tolerant and plant- growth promoting *Rhizobium* improves the growth of lentil (*Lens esculentus*) in insecticide-stressed soils. Pest Manag Sci 67:423–429

Ahemad M, Khan MS (2010g) Growth promotion and protection of lentil (*Lens esculenta*) against herbicide stress by *Rhizobium* species. Ann Microbiol 60:735–745

Ahemad M, Khan MS (2010h) Improvement in the growth and symbiotic attributes of fungicide-stressed chickpea plants following plant growth promoting fungicide-tolerant *Mesorhizobium* inoculation. Afr J Basic Appl Sci 2:111–116

Ahemad M, Khan MS (2011a) Toxicological assessment of selective pesticides towards plant growth promoting activities of phosphate solubilizing *Pseudomonas aeruginosa*. Acta Microbiol Immunol Hung 58:169–187

Ahemad M, Khan MS (2011b) Effects of insecticides on plant- growth-promoting activities of phosphate solubilizing rhizobacterium *Klebsiella* sp. strain PS19. Pestic Biochem Physiol 100:51–56

Ahemad M, Khan MS (2011c) Assessment of plant growth promoting activities of rhizobacterium *Pseudomonas putida* under insecticide-stress. Microbiol J 1:54–64

Ahemad M, Khan MS (2011d) Effect of pesticides on plant growth promoting traits of greengram-symbiont, *Bradyrhizobium* sp. strain MRM6. Bull Environ Contam Toxicol 86:384–388

Ahemad M, Khan MS (2011e) Ecotoxicological assessment of pesticides towards the plant growth promoting activities of Lentil (*Lens esculentus*)-specific *Rhizobium* sp. strain MRL3. Ecotoxicology 20:661–669

Ahemad M, Khan MS (2011f) Biotoxic impact of fungicides on plant growth promoting activities of phosphate-solubilizing *Klebsiella* sp. isolated from mustard (*Brassica campestris*) rhizosphere. J Pest Sci. https://doi.org/10.1007/s10340-011-0402-1

Ahemad M, Khan MS (2011g) Toxicological effects of selective herbicides on plant growth promoting activities of phosphate solubilizing *Klebsiella* sp. strain PS19. Curr Microbiol 62:532–538

Ahemad M, Khan MS (2011h) Insecticide-tolerant and plant growth promoting *Bradyrhizobium* sp. (vigna) improves the growth and yield of green gram [*Vigna radiata* (L.) Wilczek] in insecticide- stressed soils. Symbiosis 54:17–27

Ahemad M, Khan MS (2011i) Effect of tebuconazole-tolerant and plant growth promoting *Rhizobium* isolate MRP1 on pea-*Rhizobium* symbiosis. Sci Hortic 129:266–272

Ahemad M, Khan MS (2012a) Effect of fungicides on plant growth promoting activities of phosphate solubilizing *Pseudomonas putida* isolated from mustard (*Brassica campestris*) rhizosphere. Chemosphere 86:945–950

Ahemad M, Khan MS (2012b) Ecological assessment of biotoxicity of pesticides towards plant growth promoting activities of pea (*Pisum sativum*)-specific *Rhizobium* sp. strain MRP1. Emirates J Food Agric 24:334–343

Ahemad M, Khan MS (2012c) Evaluation of plant growth promoting activities of rhizobacterium *Pseudomonas putida* under herbicide-stress. Ann Microbiol 62:1531–1540

Ahemad M, Khan MS (2012d) Effects of pesticides on plant growth promoting traits of *Mesorhizobium* strain MRC4. J Saudi Soc Agric Sci 11:63–71

Ahemad M, Khan MS (2012e) Alleviation of fungicide-induced phytotoxicity in greengram [*Vigna radiata* (L.) Wilczek] using fungicide-tolerant and plant growth promoting *Pseudomonas* strain. Saudi J Biol Sci 19:451–459

Ahemad M, Khan MS (2012f) Productivity of greengram in tebuconazole-stressed soil, by using a tolerant and plant growth- promoting *Bradyrhizobium* sp. MRM6 strain. Acta Physiol Plant 34:245–254

Ahemad M, Kibret M (2014) Mechanisms and applications of plant growth promoting rhizobacteria: current perspective. J K S Univ-Sci 26:1–20

Arora NK, Kang SC, Maheshwari DK (2001) Isolation of siderophore producing strains of *Rhizobium meliloti* and their biocontrol potential against *Macrophomina phaseolina* that causes charcoal rot of groundnut. Curr Sci 81:673–677

Belimov AA, Hontzeas N, Safronova VI, Demchinskaya SV, Piluzza G, Bullitta S, Glick BR (2005) Cadmium-tolerant plant growth promoting rhizobacteria associated with the roots of Indian mustard (*Brassica juncea* L. Czern.). Soil Biol Biochem 37:241–250

Berraho EL, Lesueur D, Diem HG, Sasson A (1997) Iron requirement and siderophore production in *Rhizobium ciceri* during growth on an iron-deficient medium. World J Microbiol Biotechnol 13:501–510

Braud A, Je'ze'quel K, Bazot S, Lebeau T (2009) Enhanced phytoextraction of an agricultural Cr-, Hg- and Pb-contaminated soil by bioaugmentation with siderophore producing bacteria. Chemosphere 74:280–286

Burd GI, Dixon DG, Glick BR (2000) Plant growth promoting bacteria that decrease heavy metal toxicity in plants. Can J Microbiol 46:237–245

Canbolat MY, Bilen S, Cakmakc R, Sahin F, Aydın A (2006) Effect of plant growth-promoting bacteria and soil compaction on barley seedling growth, nutrient uptake, soil properties and rhizosphere microflora. Biol Fertil Soils 42:350–357

Cazorla FM, Romero D, Perez-Garcıa A, Lugtenberg BJJ, de Vicente A, Bloemberg G (2007) Isolation and characterization of antagonistic *Bacillus subtilis* strains from the avocado rhizoplane displaying biocontrol activity. J Appl Microbiol 103:1950–1959

Dary M, Chamber-Pe'rez MA, Palomares AJ, Pajuelo E (2010) "*In situ*" phytostabilisation of heavy metal polluted soils using *Lupinus luteus* inoculated with metal resistant plant-growth promoting rhizobacteria. J Hazard Mater 177:323–330

Deshwal VK, Pandey P, Kang SC, Maheshwari DK (2003) *Rhizobia* as a biological control agent against soil borne plant pathogenic fungi. Indian J Exp Biol 41:1160–1164

Dey R, Pal K, Bhatt DM, Chauhan SM (2004) Growth promotion and yield enhancement of peanut (*Arachis hypogaea* L.) by application of plant growth-promoting rhizobacteria. Microbiol Res 159:371–394

Duhan JS, Dudeja SS, Khurana AL (1998) Siderophore production in relation to N_2 fixation and iron uptake in pigeon pea-*Rhizobium* symbiosis. Folia Microbiol 43:421–426

Ganesan V (2008) Rhizoremediation of cadmium soil using a cadmium-resistant plant growth-promoting rhizopseudomonad. Curr Microbiol 56:403–407

Genrich IB, Dixon DG, Glick BR (1998) A plant growth- promoting bacterium that decreases nickel toxicity in seedlings. Appl Environ Microbiol 64:3663–3668
Glick BR, Patten CL, Holguin G, Penrose GM (1999) Biochemical and genetic mechanisms used by plant growth promoting bacteria. Imperial College Press, London
Gupta A, Rai V, Bagdwal N, Goel R (2005) *In situ* characterization of mercury resistant growth promoting fluorescent *pseudomonads*. Microbiol Res 160:385–388
Hynes RK, Leung GC, Hirkala DL, Nelson LM (2008) Isolation, selection, and characterization of beneficial rhizobacteria from pea, lentil and chickpea grown in Western Canada. Can J Microbiol 54:248–258
Indiragandhi P, Anandham R, Madhaiyan M, Sa TM (2008) Characterization of plant growth-promoting traits of bacteria isolated from larval guts of diamondback moth *Plutella xylostella* (Lepidoptera: Plutellidae). Curr Microbiol 56:327–333
Jahanian A, Chaichi MR, Rezaei K, Rezayazdi K, Khavazi K (2012) The effect of Plant Growth Promoting Rhizobacteria (PGPR) on germination and primary growth of artichoke (*Cynara scolymus*). Int J Agric Crop Sci 4:923–929
Jeon J, Lee S, Kim H, Ahn T, Song H (2003) Plant growth promotion in soil by some inoculated microorganisms. J Microbiol 41:271–276
Jha PN, Kumar A (2007) Endophytic colonization of *Typha australis* by a plant growth-promoting bacterium *Klebsiella oxytoca* strain GR-3. J Appl Microbiol 103:1311–1320
Jiang C, Sheng X, Qian M, Wang Q (2008) Isolation and characterization of a heavy metal-resistant *Burkholderia* sp. from heavy metal-contaminated paddy field soil and its potential in promoting plant growth and heavy metal accumulation in metal- polluted soil. Chemosphere 72:157–164
Joo GJ, Kin YM, Kim JT, Rhee IK, Kim JH, Lee IJ (2005) Gibberellins-producing rhizobacteria increase endogenous gibberellins content and promote growth of red peppers. J Microbiol 43:510–515
Joseph B, Patra RR, Lawrence R (2007) Characterization of plant growth promoting rhizobacteria associated with chickpea (*Cicer arietinum* L.). Int J Plant Prod 2:141–152
Khan AG (2005) Role of soil microbes in the rhizospheres of plants growing on trace metal contaminated soils in phytoremediation. J Trace Elem Med Biol 18:355–336
Khan MS, Zaidi A, Aamil M (2002) Biocontrol of fungal pathogens by the use of plant growth promoting rhizobacteria and nitrogen fixing microorganisms. Ind J Bot Soc 81:255–263
Kumar KV, Singh N, Behl HM, Srivastava S (2008) Influence of plant growth promoting bacteria and its mutant on heavy metal toxicity in *Brassica juncea* grown in fly ash amended soil. Chemosphere 72:678–683
Liu H, He Y, Jiang H, Peng H, Huang X, Zhang X, Thomashow LS, Xu Y (2007) Characterization of a phenazine- producing strain *Pseudomonas chlororaphis* GP72 with broad- spectrum antifungal activity from green pepper rhizosphere. Curr Microbiol 54:302–306
Ma Y, Rajkumar M, Vicente JA, Freitas H (2011a) Inoculation of Ni-resistant plant growth promoting bacterium *Psychrobacter* sp. strain SRS8 for the improvement of nickel phytoextraction by energy crops. Int J Phytoremediation 13:126–139
Ma Y, Rajkumar M, Luo Y, Freitas H (2011b) Inoculation of endophytic bacteria on host and non-host plants-effects on plant growth and Ni uptake. J Hazard Mater 195:230–237
Mayak S, Tirosh T, Glick BR (2004) Plant growth-promoting bacteria confer resistance in tomato plants to salt stress. Plant Physiol Biochem 42:565–572
Mehnaz S, Baig DN, Lazarovits G (2010) Genetic and phenotypic diversity of plant growth promoting rhizobacteria isolated from sugarcane plants growing in Pakistan. J Microbiol Biotechnol 20:1614–1623
Murphy JF, Zehnder GW, Schuster DJ, Sikora EJ, Polston JE, Kloepper JW (2000) Plant growth-promoting rhizobacterial mediated protection in tomato against tomato mottle virus. Plant Dis 84:779–784
Naik MM, Dubey SK (2011) Lead-enhanced siderophore production and alteration in cell morphology in a Pb-resistant *Pseudomonas aeruginosa* strain 4EA. Curr Microbiol 62:409–414

Noel TC, Sheng C, Yost CK, Pharis RP, Hynes MF (1996) *Rhizobium leguminosarum* as a plant growth promoting rhizobacterium: direct growth promotion of canola and lettuce. Can J Microbiol 42:279–283

Noordman WH, Reissbrodt R, Bongers RS, Rademaker ILW, Bockelmann W, Smit G (2006) Growth stimulation of *Brevibacterium* sp. by siderophores. J Appl Microbiol 101:637–646

Pandey A, Trivedi P, Kumar B, Palni LMS (2006) Characterization of a phosphate solubilizing and antagonistic strain of *Pseudomonas putida* (B0) isolated from a sub-alpine location in the Indian Central Himalaya. Curr Microbiol 53:102–107

Phi Q-T, Yu-Mi P, Keyung-Jo S, Choong-Min R, Seung-Hwan P, Jong-Guk K, Sa-Youl G (2010) Assessment of root-associated *Paenibacillus polymyxa* groups on growth promotion and induced systemic resistance in pepper. J Microbiol Biotechnol 20:1605–1613

Poonguzhali S, Madhaiyan M, Sa T (2008) Isolation and identification of phosphate solubilizing bacteria from Chinese cabbage and their effect on growth and phosphorus utilization of plants. J Microbiol Biotechnol 18:773–777

Rajkumar M, Nagendran R, Kui JL, Wang HL, Sung ZK (2006) Influence of plant growth promoting bacteria and Cr (VI) on the growth of Indian mustard. Chemosphere 62:741–748

Rajkumar M, Ma Y, Freitas H (2008) Characterization of metal- resistant plant-growth promoting *Bacillus weihenstephanensis* isolated from serpentine soil in Portugal. J Basic Microbiol 48:500–508

Rani A, Souche YS, Goel R (2009) Comparative assessment of in situ bioremediation potential of cadmium resistant acidophilic *Pseudomonas putida* 62BN and alkalophilic *Pseudomonas monteilii* 97AN strains on soybean. Int Biodeterior Biodegrad 63:62–66

Rodrigues EP, Rodrigues LS, de Oliveira ALM, Baldani VLD, Teixeira KRS, Urquiaga S, Reis VM (2008) *Azospirillum amazonense* inoculation: effects on growth, yield and N_2 fixation of rice (*Oryza sativa* L.). Plant Soil 302:249–261

Rokhbakhsh-Zamin F, Sachdev D, Kazemi-Pour N, Engineer A, Pardesi KR, Zinjarde S, Dhakephalkar PK, Chopade BA (2011) Characterization of plant-growth-promoting traits of *Acinetobacter* species isolated from rhizosphere of *Pennisetum glaucum*. J Microbiol Biotechnol 21:556–566

Russo A, Vettori L, Felici C, Fiaschi G, Morini S, Toffanin A (2008) Enhanced micropropagation response and biocontrol effect of *Azospirillum brasilense* Sp245 on *Prunus cerasifera* L. clone Mr.S 2/5 plants. J Biotechnol 134:312–319

Saravanakumar D, Vijayakumarc C, Kumarb N, Samiyappan R (2007) PGPR-induced defense responses in the tea plant against blister blight disease. Crop Prot 26:556–565

Saravanan VS, Madhaiyan M, Thangaraju M (2007) Solubilization of zinc compounds by the diazotrophic, plant growth promoting bacterium *Gluconacetobacter diazotrophicus*. Chemosphere 66:1794–1798

Selvakumar G, Mohan M, Kundu S, Gupta AD, Joshi P, Nazim S, Gupta HS (2008) Cold tolerance and plant growth promotion potential of *Serratia marcescens* strain SRM (MTCC 8708) isolated from flowers of summer squash (*Cucurbita pepo*). Lett Appl Microbiol 46:171–175

Shaharoona B, Arshad M, Zahir ZA (2006) Effect of plant growth promoting rhizobacteria containing ACC-deaminase on maize (*Zea mays* L.) growth under axenic conditions and on nodulation in mung bean. Lett Appl Microbiol 42:155–159

Shaharoona B, Naveed M, Arshad M, Zahir ZA (2008) Fertilizer-dependent efficiency of Pseudomonads for improving growth, yield, and nutrient use efficiency of wheat (*Triticum aestivum* L.). Appl Microbiol Biotechnol 79:147–155

Sheng XF, Xia JJ (2006) Improvement of ripe (*Brassica napus*) plant growth and cadmium uptake by cadmium-resistant bacteria. Chemosphere 64:1036–1042

Tank N, Saraf M (2003) Phosphate solubilization, exopolysaccharide production and indole acetic acid secretion by rhizobacteria isolated from *Trigonella graecum*. Indian J Microbiol 43:37–40

Tank N, Saraf M (2009) Enhancement of plant growth and decontamination of nickel-spiked soil using PGPR. J Basic Microbiol 49:195–204

Tank N, Saraf M (2010) Salinity-resistant plant growth promoting rhizobacteria ameliorates sodium chloride stress on tomato plants. J Plant Interact 5:51–58

Thakuria D, Talukdar NC, Goswami C, Hazarika S, Boro RC, Khan MR (2004) Characterization and screening of bacteria from rhizosphere of rice grown in acidic soils of Assam. Curr Sci 86:978–985

Tian F, Ding Y, Zhu H, Yao L, Du B (2009) Genetic diversity of siderophore-producing bacteria of tobacco rhizosphere. Braz J Microbiol 40:276–284

Tsavkelova EA, Cherdyntseva TA, Netrusov AI (2005) Auxin production by bacteria associated with orchid roots. Microbiology 74:46–53

Verma A, Kukreja K, Pathak DV, Suneja S, Narula N (2001) *In vitro* production of plant growth regulators (PGRs) by *Azotobacter chroococcum*. Indian J Microbiol 41:305–307

Vivas A, Biro B, Ruiz-Lozano JM, Barea JM, Azcon R (2006) Two bacterial strains isolated from a Zn-polluted soil enhance plant growth and mycorrhizal efficiency under Zn toxicity. Chemosphere 52:1523–1533

Wani PA, Khan MS (2010) Bacillus species enhance growth parameters of chickpea (*Cicer arietinum* L.) in chromium stressed soils. Food Chem Toxicol 48:3262–3267

Wani PA, Khan MS, Zaidi A (2007a) Effect of metal tolerant plant growth promoting *Bradyrhizobium* sp. (vigna) on growth, symbiosis, seed yield and metal uptake by greengram plants. Chemosphere 70:36–45

Wani PA, Khan MS, Zaidi A (2007b) Co inoculation of nitrogen fixing and phosphate solubilizing bacteria to promote growth, yield and nutrient uptake in chickpea. Acta Agron Hung 55:315–323

Wani PA, Khan MS, Zaidi A (2007c) Synergistic effects of the inoculation with nitrogen fixing and phosphate solubilizing rhizobacteria on the performance of field grown chickpea. J Plant Nutr Soil Sci 170:283–287

Wani PA, Khan MS, Zaidi A (2008) Chromium-reducing and plant growth-promoting *Mesorhizobium* improves chickpea growth in chromium-amended soil. Biotechnol Lett 30:159–163

Wittenberg JB, Wittenberg BA, Day DA, Udvardi MK, Appleby CA (1996) Siderophore bound iron in the peribacteroid space of soybean root nodules. Plant Soil 178:161–169

Yasmin S, Rahman M, Hafeez FY (2004) Isolation, characterization and beneficial effects of rice associated plant growth promoting bacteria from Zanzibar soils. J Basic Microbiol 44:241–252

Zahir ZA, Shah MK, Naveed M, Akhter MJ (2010) Substrate- dependent auxin production by *Rhizobium phaseoli* improves the growth and yield of *Vigna radiata* L. under salt stress conditions. J Microbiol Biotechnol 20:1288–1294

Zaidi S, Usmani S, Singh BR, Musarrat J (2006) Significance of *Bacillus subtilis* strain SJ 101 as a bioinnoculant for concurrent plant growth promotion and nickel accumulation in *Brassica juncea*. Chemosphere 64:991–997

Chapter 10
Oil Pollution

Abstract Oil is one of the main sources of energy and is transported by ships and tankers across the oceans and seas and through pipelines across the land. Pollution caused by oil has severe impacts on the biodiversity that have been detailed in the chapter with relevant case studies. The various technologies to control oil pollution have been discussed in a comprehensive style with special thrust on the mechanism of bioremediation, where microbes act as the game changer.

Keywords Oil spill · Boom · Skimmer · Bioremediation · Microbes

10.1 Oil Pollution: An Overview

il is one of the world's main sources of energy, and it is transported by ship across oceans and by pipelines across land. This results in accidents while transferring oil to vessels, when transporting oil and when pipelines break, as well as when drilling for oil. While massive and catastrophic oil spills receive most of the attention, smaller and chronic oil spills and seeps occur almost at regular intervals. These spills contaminate coasts and estuaries, and they can cause harm to human. Oil can refer to many different materials, including crude oil, refined petroleum products (such as gasoline or diesel fuel) or by-products, ships bunkers, oily refuse or oil mixed in waste. The contribution of oil to the marine environment arises from the disposal of waste oil from domestic and industrial automotive sources and from industrial operations (e.g. lubrication, hydraulic, cutting oils, etc.). Most of the oil comes from unauthorized sources such as illegal cleaning of oil tankers and residue dumping. For example, only 5 out of 14 harbours in Mediterranean have the facilities for receiving oil residues from tankers. Thus, in most cases, the residues are dumped into the sea. It is estimated that in this particular dumping area, 300,000 tonnes of oil are released annually. Oil is also released into the environment from natural geologic seeps on the seafloor, as along the California coastline. Overall production of petroleum products rose from about 500 million tonnes in 1950 to over 2500 million tonnes by the mid-1990s, resulting in massive transport and associated oil spills.

A. Mitra, S. Zaman, *Environmental Science - A Ground Zero Observation on the Indian Subcontinent*, https://doi.org/10.1007/978-3-030-49131-4_10

Oil is a mixture of hydrocarbon compounds. Crude oil is a mixture of gas, naphtha, kerosene, light gas and residuals, which have different adverse effects on the health of marine biota. The fate, behaviour and environmental effects of spilled oil can vary, depending upon the type and amount of material spilled. In general, lighter refined petroleum products such as diesel and gasoline are more likely to mix in the water column and are more toxic to marine life but tend to evaporate more quickly and do not persist long in the environment. Heavier crude or fuel oil, while of less immediate toxicity, can remain on the water surface or stranded on the shoreline for much longer period. A heavy oil spill across the shore blankets rock pools, etc., preventing gas exchange and eliminating light as well as directly leaching toxins into the water and can also become mixed deeply into pebble, shingle or sandy beaches, where it may remain for months or even years. Fishes often escape from toxicity by swimming away from the polluted area, but the benthic organisms suffer large numbers of mortalities as the oil settles with the sediments. Coral reefs often get damaged both by oil deposition and by the water-soluble chemicals contained in oil. Seabirds are adversely affected as the oil penetrates and opens up the structure of their plumage, so they become chilled, lose the ability to fly and lose their buoyancy in water. They are thus unable to feed normally but ingest the oil as they attempt to preen. This condition finally draws these avian faunas towards death.

Oil in the seas also affects human health. Short-term public health impacts from oil spills include accidents suffered by those on damaged tankers or those involved in the cleanup and illnesses caused by toxic fumes or by eating contaminated fish or shellfish. However, there are other less obvious public health impacts, including losses and disruptions of commercial and recreational fisheries, seaweed culture units, boating and a variety of other uses of affected water. The concern involves the ingestion of carcinogenic compounds, especially the polycyclic aromatic hydrocarbons present in petroleum.

The lower Gangetic region at the apex of Bay of Bengal is very sensitive in terms of the existence of mangrove chunks. The pneumatophores, if choked by oil, may cause death of the tree. Seedlings are much more vulnerable to oil pollution because of stomatal blocking by oil. However, tidal flushing is a boon to the system as it helps in breaking the oil mat into fragments. Four oil spills have been reported in the vicinity of Sundarbans. In February 1992 by MV Wash near New Moore Island, in June 1997 by MV Green Opal in Hooghly River, in July 2000 by MV Prime Value off Sagar island and in May 2003 by MV Segitega Biru off Haldia (145 T of FFO). In first three cases, the quantity of oil spilt could not be assessed. The effect of this spillage in Sundarbans area was minimal as they were minor spills and oil did not enter in the Reserve Forest area or core zone.

10.2 Effects of Oil Pollution

The various components of oil have varied effects on the biotic community. Insoluble petroleum fractions can be very damaging as they may coat the organisms and thereby cause suffocation, or they may cause tainting of edible species. The toxicity

Table 10.1 Effects of oil on marine populations and communities

Community or population type	Expected degree of initial impact	Expected recovery rate
Plankton	Light to moderate	Fast to moderate
Benthic communities:		
1. Rocky intertidal 2. Sandy or muddy intertidal 3. Subtidal, offshore	Light Moderate Heavy	Fast Moderate Slow
Fish	Light to moderate	Fast to moderate
Birds	Heavy	Slow
Marine mammals	Light	Slow, if population seriously affected

due to oil is caused by the soluble fractions. Aliphatic hydrocarbons having low molecular weight, which are readily soluble in seawater, produce narcosis and anaesthesia at low concentration. At higher concentration they can cause cell damage and death, particularly to the larval forms of marine life. Low boiling point aromatic hydrocarbons have been found to be the most toxic fractions and to be the primary cause of organism mortality. Certain heterocyclic compounds are also toxic at very high concentrations such as 3, 4-benzopyrene, and other polycyclic aromatic compounds are known to be carcinogenic.

Susceptibility to oil toxicity has been best correlated with habitat. It has been found that pelagic fish and shrimp are most sensitive to crude oil or No. 2 fuel oil, and the intertidal animals (fish, crabs, starfish and molluscs) are the most tolerant.

The degree of initial impact by oil on marine and estuarine communities and the expected recovery rate are summarized in Table 10.1.

10.2.1 Effects of Oil Spills on Marine Biota

Oil spills have considerable adverse impact on marine organisms. Birds die from oil spills if their feathers are covered with oil. Animals may die because they get hypothermia, causing their body temperature to be really low. Oil may also cause the death of an animal by entering the animal's lungs or liver. Oil also can kill an animal by blinding it. The animal will not be able to see and be aware of their predators. Oil spills sometimes are the reason for animals becoming endangered.

Seabirds

Seabirds are strongly affected by oil spills. A seabird may get covered with the oil. The thick black oil is too heavy for the birds to fly, and so they attempt to clean themselves. The bird then eats the oil to clean its feathers and poisons itself.

Marine Mammals

Seals, sea lions and walruses are particularly vulnerable to oiling because they haul out, often in very large numbers on the shorelines of small islands, rocks or remote coasts with few options for new territory. For example, in 2011, 30,000 walruses were hauled out on a 1.6 km of beach in the Chukchi Sea, Alaska. Factors that influence the sensitivity of marine mammals to oil exposure include whether they rely on fur or blubber for thermoregulation; whether they form dense groups on land or in water; how and on what they feed; and inhalation risks.

External oiling of young seals or sea lions generally causes death because their coats are not developed enough to provide insulation in an oiled state. Oil is often absorbed or ingested and mothers may not feed their young when they are oiled. After a large oil spill in South America, about 10,000 baby seals perished when the beaches of their island were contaminated by oil. Not many adult seals perished at the same site, and those who did probably drowned. Most of the oiled young seals died by starvation after their mothers refused to feed them. Older seals, sea lions and walruses can take a large amount of oiling without causing death. If lightly oiled, adult seals survive, and the oil is slowly lost. Oiling of both adult and young causes the fur to lose the water proofing and buoyancy. It is not known if seals or their relatives would avoid oil if they could, as this has not been observed at spill sites.

Oil spills are one of the many ways killer whales have become endangered. The oil may be eaten or enter the whale's blowhole. A blowhole is a hole to help them breath. Plugging of the blowhole by oil will inhibit the entry of atmospheric air inside the whale's system. This may result in the death of the animal.

Plankton

Plankton are drifting community and are fully dependent on exchange of gases for their survival. Oil, being lighter than the seawater, floats on the surface and cuts off the diffusion of atmospheric gases into the water. This results in the shortage of oxygen resulting in the death of the planktonic forms. The layer of oil also prevents the solar radiation to reach the seawater due to which the photosynthetic process in phytoplankton is also affected.

Mangroves

Mangroves are unique vegetation that are found at the land–sea interface and are characterized with some peculiar features (unlike common terrestrial vegetation) like presences of pneumatophores, prop root, stilt root, viviparous germination, etc. Mangroves are highly adapted to cope with the saline habitats through salt exclusion/excretion mechanism (Tomlinson 1986; Hogarth 2007). Oil pollution from oil and gas exploration, petroleum production and accidental spills severely damages mangrove ecosystems. The degree of damages of oil pollution on mangroves is a function of beach types, and on this basis a vulnerability index has been constructed considering the nature of substratum on which the mangroves survive (Table 10.2).

The toxicity of oil may also differ among mangrove species, particularly with respect to magnitude of adverse impact. For example, along the coast of Sao Paulo, Brazil, an oil spill caused 25.9% defoliation of *Rhizophora mangle*, 43.9% defoliation of *Laguncularia racemosa* and 64.5% defoliation of *Avicennia schaueriana* (Lamparelli et al. 1997).

The extent of mangrove damage from oil pollution usually depends on the kind of oil and the magnitude and frequency of spilling. For example, fresh oil causes more leaf loss in *Avicennia* sp. seedlings than does aged oil (Martin et al. 1990; Grant et al. 1993). Boeer (1996) measured the effects from a mineral oil spill in the Arabian Sea off Fujairah. He observed that the mangroves were relatively unaffected and all signs of the spill were nearly gone only 7 months after the spill.

Table 10.2 Vulnerability index in order of increasing vulnerability to damage from oil spills (After Gundlach and Hayes 1978)

Vulnerability index	Shoreline type	Comments
1	Exposed rocky headlands	Exposed to high wave energy. Wave reflection keeps most of the oil offshore. Vigorous mixing aids in natural breakdown of the oil. Oil spill control and cleanup are not necessary
2	Eroding wave-cut platforms	High wave action. Natural cleansing of the beach occurs rapidly, usually within weeks. Oil spill control and cleanup are not necessary
3	Flat, fine-grained sandy beaches	Beaches are flat and hard packed with grain size ranging between 0.0625 and 0.25 mm. Oil does not penetrate facilitating mechanical removal if necessary. Initial biological damage may be great, but revival of the biological community will occur within a year. Oil may persist for several months if not removed by scraping off the top few centimetres of the sand
4	Steeper, medium to coarse-grained sand beaches	The grain size ranges within 0.25 to 2.0 mm. Biological activity is relatively low and generally not a major consideration. Oil may sink 15 to 25 cm into the sand or be buried by natural processes making oil spill cleanup very difficult. Attempts at mechanical removal could be very damaging because of the depth of the oil
5	Exposed, compacted tidal flats	Usually fine-grained (mud or sand) tidal flats exposed to winds, waves and currents. Most oil will not adhere to or penetrate into the tidal flat. Cleanup is not usually necessary. If necessary to protect the biological community, activities should concentrate on manual removal of small oil pools left after each tidal cycle
6	Mixed sand and gravel beaches	Oil readily penetrates 10 to 20 cm into the sediment, and burial may be rapid. Biological community is relatively limited because of instability of the environment. Oil may persist for many years. Removal of the oil is extremely difficult without damaging the beach.
7	Gravel beaches	Grain size > 2 mm. Oil penetrates rapidly and deeply. A solid asphalt pavement may from under heavy oil accumulation. Cleanup is practically impossible without severe damage to the beach. Cleanup should concentrate on the high-tide swash area
8	Sheltered rocky coasts	Areas of reduced wave action. Oil will coat the rough surfaces and tidal pools. Oil may persist for many years. The resident biological community is extensive and vulnerable to oil spill damage. Cleanup is difficult, very expensive and not recommended unless oil concentration is very heavy
9	Sheltered estuarine tidal flats	Areas of great biological productivity and low wave energy. Oil spills may have long-term deleterious effects. Oil may persist for many years. Cleanup is not recommended unless oil accumulation is very heavy, as oil removal activities will destroy much of the resident biological community. These areas should receive priority protection by using booms or oil sorbent materials to prevent oil from entering the environment

(continued)

Table 10.2 (continued)

Vulnerability index	Shoreline type	Comments
10	Sheltered estuarine salt marshes and mangrove coasts	Most productive of all aquatic environments. Oil pollution may persist with detrimental effects for many years. Cleaning of salt marshes by burning or cutting should be undertaken only if heavily oiled. Mangroves should not be altered. Physical marine processes should normally be allowed to naturally cleanse the marsh. Protection of these environments by booms or sorbents should receive first priority

Proffitt and Devlin (1998) tested the effects of sequential oilings of potted *Rhizophora mangle* seedlings, first by treating the seedlings with No.6 fuel oil, followed by 34 months later, with crude oil. They found no evidence of cumulative and synergistic effects, but this conclusion has been challenged because of unnatural laboratory condition (which has been far from real practical environment) and low statistical power (Ellison et al. 1999). With a clear idea on the sensitivity of mangroves to soil conditions, it is estimated to study the effects of oil under conditions that reflect the natural environment as closely as possible. For example, salinity should be held at field levels, and realistic oil concentrations should be used to model chronic exposure of plants in oil-contaminated soil (Ellison et al. 1999).

The most realistic measure of repeated oil exposure comes from field habitats exposed to natural oiliness. Two large oil spills (the first in 1968 and the second in 1986) have caused large-scale damage to mangrove forests in Panama. In addition to killing trees outright, the oil retained in the sediments caused apparent sublethal effects (Duke et al. 1998). The residual effects of oiling may make the mangroves more vulnerable to future damage. More careful, long-term laboratory experiments under natural conditions are necessary to understand the responses of mangroves to oil and the consequences of oiling.

It has been reported that animals living in mangal sediments or submerged mangrove roots may be contaminated by oil (Mackey and Hodgkinson 1996). Five years after the Galeta oil spill in Panama, there was a 60% decrease in the number of isopods on submerged *Rhizophora mangle* prop roots and a 40–50% drop in the number of spiny lobsters (Levings and Garrity 1995).

The entire spectrum of biological effects that might occur due to oil pollution may be of the following types:

- Lethal toxic effects
- Sublethal effects that disrupt physiological or behavioural activities, but do not cause immediate mortality
- Uptake of the oil or certain fractions of it by organisms causing tainting or in some cases carcinomas
- Ingestion by organisms which leads oil to enter the food web

- Direct smothering and suffocation or interference with movements to obtain food or escape from predators as a result of being coated by oil
- Alteration to the chemical and physical properties of marine habitat
- Mortalities caused by indiscriminate use of detergents to dispose the oil

10.3 Control of Oil Pollution

The entire domain of combating the oil spill may be discussed under three broad headings, namely, [A] oil spill prevention, [B] counter measures and [C] technological processes

[A] Oil Spill Prevention

Field-based equipment, procedures and capacity building are the three key components of oil spill prevention systems. Regulated facilities are required by the Environmental Protection Agencies of each nation to combat oil spills, control and counter plans to prevent oil discharges to ambient aquatic phase and adjoining shorelines. Such plans require operators to identify and analyse potential spill hazards and previous spills, identify and ensure availability of resources to remove spilled oil, to the maximum extent practicable. The Coast Guard of almost all nations of the world has similar requirements for marine transportation-related facilities. Double hull tankers are thought to offer a higher level of oil spill prevention because both hulls must be penetrated to cause a release of cargo. A double hull design on average reduces the size of oil spills from tanker vessels by 62% and from tank barges by 20% for spills as investigated by the US Coast Guard from 2001 to 2008 (Yip et al. 2011). Use of single-hull tankers for the transport of oil was phased out worldwide as of 2010 under regulations of the International Maritime Organization, though such tankers may continue to operate until 2015 if they satisfy certain condition assessments. Flag states may grant life extensions to certain types of single-hull tankers; however, port states are permitted to deny entry to their ports and offshore terminals to single-hull tankers operating under such life extensions. Single-hull tankers are prevented from carrying heavy grade oils.

Industry, agencies and organizations are dealing with three emerging oil spill risks: oil and gas development and maritime transportation in the Arctic; deepwater offshore oil and gas development; and the rapid expansion of rail transportation of crude oil. Prevention actions for each of these are summarized here.

10.4 Oil and Gas Development and Maritime Transportation in the Arctic

The ice-covered Arctic region is becoming increasingly accessible as a result of climate change, offshore oil and gas operations and maritime shipping, and hence tourist pressure is expected to increase in the near future. The Arctic is estimated to contain 13% of the world's undiscovered oil and 30% of the world's undiscovered gas reserves. There has been a 118% increase in maritime transit through the Bering Strait between 2008 and 2012 (US Coast Guard 2013). The Emergency Prevention, Preparedness and Response (EPPR) Working Group is one of six working groups of the Arctic Council, which was established in 1996 to foster international cooperation on environmental protection and sustainable development in the Arctic. The EPPR Working Group identified the few major recommendations and best practices for the prevention of marine oil pollution in the Arctic (EPR 2013) as highlighted here.

- **Health, Safety and Environment (HSE) Systems**: A well-integrated HSE system adapted to Arctic conditions will lead to safer operations. A strong safety culture, characterized by healthy attitudes towards reducing risk and preventing accidents at all levels in the organization is needed.
- **Common International Standards**: There is a lack of consistency between standards and national requirements, as well as inconsistent enforcement of rules and regulations among Arctic nations. A combination of prescriptive and goal-based requirements is needed.
- **Monitoring**: An improvement in monitoring methods and technology (i.e. satellites, unmanned aerial vehicles, subsurface vehicles) would provide more accurate data on weather and ice conditions detect accidents and spill incidents and deter illegal actions.
- **Risk Analysis**: Identifying hazards and conducting a risk analysis of the specific operation and implementing risk-mitigating measures are key actions in preventing oil spills.
- **Technology Development and Best Practice Sharing**: Easy access to information regarding previous experience, best available technologies, best practices and sharing results of Research and Development projects is of high priority.
- **Oil and Gas Facilities**: Oil and gas facilities and associated equipment should be adequately winterized, certified for operations in sub-freezing temperatures and potential interaction with sea ice and should incorporate leak detection and in-line inspection systems for pipelines.
- **Qualified Personnel and Continuous Improvement of Skills**: Lack of competence and a need for robust Arctic-specific training requirements was identified as needs to ensure the ability to work in remote areas as well as the continuous improvement of skills.

1. **Deepwater Offshore Oil and Gas Development**

The risks of deepwater oil and gas exploration and development were demonstrated by the *Deepwater Horizon* oil spill in 2010, where an estimated 800 million L of oil were released into the Gulf of Mexico during the well blow out. Thus, industry has developed new technologies in blow out prevention systems as well as a focus on safe operating processes, training and drills. In 2012, BSEE issued rules for Increased Safety Measures for Energy Development on the Outer Continental Shelf as highlighted here.

- Establishes new casing installation requirements
- Establishes new cementing requirements
- Requires independent third-party verification of blind-shear ram capability
- Requires independent third-party verification of subsea blow out prevention stack compatibility
- Requires new casing and cementing integrity tests
- Establishes new requirements for subsea secondary blow out prevention intervention
- Requires function testing for subsea secondary blow out prevention intervention
- Requires documentation for blow out prevention, inspections and maintenance
- Requires a Registered Professional Engineer to certify casing and cementing requirements
- Establishes new requirements for specific well control training to include deep-water operations

2. **Rapid Expansion of Rail Transportation of Crude Oil**

The use of hydraulic fracturing or 'fracking' has dramatically increase US energy production, with crude oil production approaching its highest level in decades. Much of this new oil production is in the mid-continent with the Bakken field in North Dakota that, as of 2014, was producing over 1,000,000 barrels per day (EIA, http://www.eia.gov/.) The production of oil from Canadian oils is expected to double by 2020 (Reuters 2013). These shifts in crude oil production areas have changed the patterns and means of oil transportation to refineries. Because the capacity of existing pipelines has been strained by this increased production, producers have turned to transport by rail. In 2014, transport of oil by railcar to refineries on the West, East and Gulf of Mexico coasts is expected to reach 650,000 carloads or roughly 450 million barrels (Congressional Research Service 2014).

Efforts to prevent oil spills from rail transportation are focused on tank car design, prevention of derailments and railroad operations. New tank car design standards are underdevelopment to include an outer steel jacket around the tank car and thermal protection, full-height head shields and high flow capacity pressure relief valves; there is a plan to aggressively phase out older model tank cars used to move flammable liquids that cannot be retrofitted to meet these standards. In February 2014, a rail operations safety initiative was agreed to between the railroads and the US government to institute new voluntary operating practices for moving crude oil by rail. These measures include:

- Addition of braking power to allow train crews to apply emergency brakes from both ends of the train in order to stop the train faster
- Increased track inspections, with at least two high-tech track geometry inspections each year on main line routes over which trains with 20 or more loaded cars of crude oil are moving
- Use of Rail Traffic Routing Technology to aid in the determination of the safest and most secure rail routes for trains with 20 or more cars of crude oil
- Reduced train speeds to 40 mph through urban areas
- Installation of additional wayside wheel bearing detectors

[B] Counter Measures

Containment of an oil spill refers to the process of confining the oil, to prevent it from either spreading or expanding to a particular area, to divert it to another area where it can be recovered or treated, or to concentrate the oil, so it can be recovered or burned (Fingas 2012). Containment booms are the basic and most frequently used piece of equipment for containing an oil spill on water. Booms are generally the first equipment mobilized at a spill and are often used as long as the oil persists on the water surface. While many pipeline spills occur on land, most of the oil ends up in water bodies, and, as such, booms are frequently used.

A boom is a floating mechanical barrier designed to stop or divert the movement of oil on water. Booms resemble a vertical curtain with portions extending above and below the water line. Booms are constructed in sections, usually 15 or 30 metres (m) long, with connectors installed on each end so that sections of the boom can be attached to each other, towed or anchored. The three basic types of booms are fence and curtain booms, which are common, and shoreline seal booms, which are seldom used. Booms are used to enclose floating oil and prevent it from spreading, to protect biologically sensitive areas, to divert oil to areas where it can be recovered or treated and to concentrate oil and maintain an adequate thickness so that skimmers can be used or other cleanup techniques, such as in situ burning, can be applied. Booms are used primarily to contain oil, although they are also used to deflect oil. When used for containment, booms are often arranged in a U configuration. The U-shape is created by the current pushing against the centre of the boom. The critical requirement is that the current in the apex of the 'U' does not exceed 0.35 m per second (m/s) or 0.7 knots, which is referred to as the critical velocity.1 This is the speed of the current flowing perpendicular to the boom, above which oil will be lost from the boom by entrainment into the water and under the boom. If used in areas where the currents are likely to exceed 0.35 m/s, such as in rivers and estuaries, booms must be deployed at an angle to the current. The oil can then be deflected out of the strong currents to areas where it can be collected or to less sensitive areas. If strong currents prevent the best positioning of the boom in relation to the current, several booms can be deployed in a cascading pattern to progressively move oil towards one side of the watercourse. This technique is effective in wide rivers or where strong currents may cause a single boom to fail. When booms are used for deflection, the forces of the current on the boom are usually so powerful that stronger booms are required, and they must be anchored along their entire length. A boom's performance and its ability to contain oil are affected by water currents, waves and winds. Either alone or in combination, these forces often lead to boom failure and loss of oil. The most critical factor is the current speed relative to the boom. Failures will occur when this exceeds 0.35 m/s (0.7 knots) (Fingas 2012).

Recovery is the next step after containment in an oil spill cleanup operation. Even in the case of land spills, the oil most often flows to a waterbody from where it is recovered. An important objective of containment is to concentrate oil into thick layers to facilitate recovery. In fact, the containment and recovery phases of an oil spill cleanup operation are often carried out at the same time. As soon as booms are deployed at the site of a spill, equipment and personnel are mobilized to take advantage of the increased oil thickness, favourable weather and less weathered oil. After oil spreads or becomes highly weathered, recovery becomes less viable and is sometimes impossible.

Skimmers are mechanical devices designed to remove floating oil from a water surface. They vary greatly in size, application and capacity, as well as in recovery efficiency. Skimmers are classified according to the area where they are used, for example, inshore, offshore, in shallow water or in rivers, and by the viscosity of the oil they are intended to recover.

Most skimmers function best when the oil slick is relatively thick. The oil must, therefore, be collected in booms or against a shoreline or floating ice before skimmers can be used effectively. The skimmer is placed wherever the oil is most concentrated in order to recover as much oil as possible. Weather conditions at a spill site have a major effect on the efficiency of skimmers. Depending on the skimmer type, most will not work effectively in waves greater than 1 m in height or in currents>0.5 m/s. Most skimmers do not operate effectively in water with ice or debris such as branches, seaweed and floating waste. Some skimmers have screens around the intake to prevent debris or ice from entering, conveyors or similar devices to remove or deflect debris and cutters to deal with sea weed. Very viscous oils, tar balls or oiled debris can clog the intake or entrance of skimmers and make it impossible to pump oil from the skimmer's recovery system.

Skimmers are also classified according to their basic operating principles: oleophilic surface skimmers; weir skimmers; suction skimmers or vacuum devices; elevating skimmers; and submersion skimmers. Over time, all skimming systems become less effective because of the oil's spread into thinner slicks and weathering into more a viscous liquid.

Sorbents are materials that recover oil through either absorption or adsorption. They play an important role in oil spill cleanup and are used in the following ways: to cleanup the final traces of oil spills on water or land; as a backup to other containment means, such as sorbent booms; as a primary recovery means for very small spills; and as a passive means of cleanup. An example of such passive cleanup is when sorbent booms are anchored off lightly oiled shorelines to absorb any remaining oil released from the shore and prevent further contamination or re-oiling of the shoreline.

Sorbents can be natural or synthetic materials. Natural sorbents are divided into organic materials, such as peat moss or wood products, and inorganic materials, such as vermiculite or clay. Sorbents are available in a loose form, which includes granules, powder, chunks and cubes, often contained in bags, nets or socks. Sorbents are also available formed into pads, rolls, blankets and pillows. The use of synthetic sorbents in oil spill recovery has increased in the last few years. These sorbents are often used to wipe other oil spill recovery equipment, such as skimmers and booms, after a spill cleanup operation. Sheets of sorbent are often used for this purpose. Sorbent booms are deployed on the water when the oil slick is relatively thin, i.e. for the final 'polishing' of an oil spill, to remove small traces of oil or sheen, or as a backup to other booms. Sorbent booms can be placed off a shoreline to recover oil that is mobilized by wave or tidal action; this strategy is often used for marshes where other response options are likely to cause additional harm. They are not absorbent enough to be used as a primary countermeasure technique for any significant amount of oil.

Treating the oil with specially formulated chemicals is another option for dealing with oil spills. An assortment of chemical spill-treating agents is available to assist in cleaning up oil. Approval must be obtained from appropriate authorities before these chemical agents can be used. In addition, these agents are not always effective, and the treated oil may still have toxic effects on aquatic and other wildlife. Dispersants are chemical spill-treating agents that promote the formation of small droplets of oil that are dispersed throughout the top layer of the water column by wave action and currents. These small oil droplets are less likely to resurface into slicks; instead, they are subject to dilution, transport by currents and natural weathering process that include dissolution, microbial degradation and sedimentation. The use of dispersants remains a controversial issue, and special permission is required in most jurisdictions. Generally, in freshwater or land applications, their use is banned.

Dispersants have a relatively narrow window of effectiveness because oil becomes less dispersible as it weathers, mostly due to an increase in viscosity. There must be sufficient mixing energy from waves or currents, and the wind must not be too strong to prevent the dispersant droplets from landing on the oil slick. There are many restrictions on when and where dispersants may be used as an oil spill countermeasure, including water depth (usually >10 m), distance from shore (often >5 km), season, no spray over birds, sea turtles or marine mammals and duration of application (Table 10.3).

In situ burning is an oil spill cleanup technique that involves controlled burning of the oil at or near the spill site. The major advantage of this technique is its potential for removing large amounts of oil over an extensive area in less or about the same time than other techniques but with the distinct advantage of being a final solution. The technique has been used at actual spill sites for some time, especially on land (including wetlands) and in ice-covered waters where the oil is contained by the ice. During the 2010 *Deepwater Horizon* oil spill in the Gulf of Mexico, it was used extensively and removed 6% of oil from the water surface (The Federal Interagency Solutions Group 2010).

Table 10.3 Advantages and disadvantages of using chemical dispersants on oil spill

Advantages	Disadvantages
Helps in removing surface oil and diluting the oil in the aquatic phase	Probability of increasing the availability to aquatic lives increases (bioavailability)
Plays a positive role in natural oil biodegradation	Increases the chance of mortality to entrained species preferably to plankton and benthic organisms like annelids, oysters, etc.
Retards the adverse impacts on shoreline habitats	Can be used considering the congenial weather conditions
Minimizes impacts on populations with long life spans (birds, mammals, turtles)	Cannot be used on all oil types and are less efficient on weathered oils

Shoreline Cleanup

Oil spilled on water is it at sea or inland, is seldom completely contained and recovered, and some of it eventually reaches the shoreline or margin of a waterbody. It is more difficult and time-consuming to cleanup shoreline areas than it is to carry out containment and recovery operations at sea. Physically removing oil from some types of shoreline can also result in more ecological and physical damage than if oil removal is left to natural processes. The decision to initiate cleanup and restoration activities on oil-contaminated shorelines should be based on careful evaluation of socio-economic, aesthetic and ecological factors. These include the behaviour of oil in shoreline regions, the types of shoreline and their sensitivity to oil spills, the assessment process, shoreline protection measures and recommended cleanup methods. Similarly some of the shoreline types are related to land types and the same cleanup applies. The type of shoreline or similar land surface is crucial in determining the fate and effects of an oil spill as well as the cleanup methods to be used. In fact, the shoreline's basic structure and the size of material present are the most important factors in terms of oil spill cleanup. There are many types of shorelines, all of which are classified in terms of sensitivity to oil spills and ease of cleanup. The shoreline types in order of increasing sensitivity are exposed bedrock and man-made solid structures, sand beaches, mixed sand and gravel beaches, gravel beaches, riprap, exposed tidal flats, sheltered rocky shores, peat, sheltered tidal flats, marshes, mangroves, swamps and low-lying tundra. These types occur on both seashore and freshwater shores.

Many methods are available for removing oil from shorelines or landforms. All of them are costly and take a long time to carry out. The selection of the appropriate cleanup technique is based on the type of substrate, the depth of oil penetration or burial in the sediments, the amount and type of oil and its present form/condition, the ability of the shoreline to support foot and vehicular traffic, the environmental, human and cultural sensitivity of the shoreline and the prevailing ocean and weather conditions. Some recommended shoreline cleanup methods are natural recovery, manual removal, flooding or washing, use of vacuums, mechanical removal, tilling and aeration, sediment reworking, sorbents and chemical cleaning agents. Surface-washing agents are intended to be applied to shorelines or surfaces to increase the amount of oil released from the surface, or to reduce the pressure and temperature of flushing. Only those agents that cause the released oil to float are used for use on the shoreline, so that the released oil can be recovered. Other agents are essentially industrial cleaners that may be appropriate for cleaning equipment when removed from the water, but not on shorelines. Numerous guides have been developed to assist responders in selecting the most appropriate cleanup method depending on the type and degree of oiling, the habitat type and environmental considerations. Table 10.4 shows a matrix for selection of cleanup methods for marshes. Sometimes the best response to an oil spill on a shoreline may be to leave the oil and monitor the natural recovery of the affected area. This would be the case if more damage would be caused by cleanup than by leaving the environment to recover on its own. This option is suitable for small spills in sensitive environments and on a beach that will recover quickly on its own such as on exposed shorelines and with non-persistent oils such

Table 10.4 Recommendations for response options in oiled marshes by oil group (Michel and Rutherford 2014)

Oil group description	Oil group				
	Response method	I	II	III	IV
I – Gasoline products	Natural recovery	A	A	B	B
II – Diesel-like products and light crudes	Barriers/berms	B	B	B	B
III – Medium-grade crudes and intermediate products	Manual oil removal/cleaning	D	C	B	B
IV – Heavy crudes and residual products	Mechanical oil removal	D	D	C	C
	Sorbents	–	A	A	B
	Vacuum	–	B	B	B
	Debris removal	–	B	B	B
	Sediment/reworking tilling	D	D	D	D
	Vegetation cutting removal	D	D	C	C
	Flooding/deluge	B	B	B	B
	Low pressure ambient water flushing	B	B	B	B
	Shoreline cleaning agents	–	–	B	B
	Nutrient enrichment	–	B	B	C
	Natural microbe seedling	–	I	I	I
	In situ burning	–	B	B	B

Note: *The following codes compare the relative environmental impact of each response method for each oil type:* ***A, the least adverse habitat impact; B, some adverse habitat impact; C, significant adverse habitat impact; D,*** *the most adverse habitat impact; I, insufficient information to evaluate impact or effectiveness*

as diesel fuel on impermeable beaches. This is not an appropriate response if important ecological or human resources are threatened by long-term persistence of the oil.

Cleanup of Oil Spills on Land

When dealing with oil spills on land, cleanup operations should begin as soon as possible. It is important to prevent the oil from spreading by containing it and to prevent further contamination by removing the source of the spill. It is also important to prevent the oil from penetrating the surface and possibly contaminating the groundwater. Berms or dikes can be built to contain oil spills and prevent oil from spreading horizontally. Caution must be exerted, however, that the oil does not backup behind the berm and permeate the soil. Berms can be built with soil from the area, sand bags or construction materials. Berms are removed after cleanup to restore the area's natural drainage patterns. Sorbents can also be used to recover some of the oil and to prevent further spreading. The contaminated area can sometimes be flooded with water to slow penetration and possibly float oil to the surface, although care must be taken not to increase spreading and to ensure that water-soluble components of the oil are not carried down into the soil with the water. Shallow trenches can be dug as a method of containment, which is particularly effective if the water table is high and oil will not permeate the soil. Oil can either be recovered directly from the trenches or burned in the trenches. After the cleanup, trenches are filled in to restore natural water levels and drainage patterns. Suction hoses, pumps, vacuum trucks and certain skimmers and sorbents are generally effective in removing excess oil from the surface, especially from ditches or low areas. The use of sorbents can complicate cleanup operations, however, as contaminated sorbents must be disposed of appropriately. Sorbents are best used to remove the final traces of oil from a water surface. Any removal of surface soil or vegetation also entails replanting. Mechanical recovery equipment, such as bulldozers, scrapers and front-end loaders, can cause severe and long-lasting damage to sensitive environments. These devices can be used in a limited capacity to clean oil from urban areas, roadsides and possibly on agricultural land. The unselective removal of a large amount of soil leads to the problem of disposing of the contaminated material. Contaminated soil must be treated, washed or contained before it can safely be disposed of in a landfill site, and this requires huge expenditure.

[C] Technological Processes

Bacterial degradation of petroleum has been reported and known for around last five decades, and the bacteria have been isolated from oil spills in soil and water (Rehm and Reiffl 1981). In recent years many ecologist and biologists have found many bacteria that can degrade these hydrocarbons efficiently. Many oil-degrading bacteria have been isolated, and their degradation potential is investigated. Major bioremediation studies have been conducted using pure cultures, while these bacteria's role remains silent in a natural environment (Harayama et al. 1999).

Numerous reports have been there on the bacteria as a proof and evidence that has the ability to degrade the oil spill water and soil that is contaminated due to petroleum hydrocarbon, which has happened to be there due to oil spills only (Chaerun et al. 2005).

Biological Method of Cleanup

Biological methods are cheaper and better methods to degrade the oil spills and petroleum containing soil (Leahy and Colwell 1990). Biological methods including the use of microbes or by using the enzymes of the microbes to detoxify and remove the pollutants from soil and water have diverse metabolic capabilities (Bhasheer et al., 2014; Medina et al. 2005).

Bioremediation

Bioremediation is not an out of the blue-type technology. It is known that bacteria have got the potential to degrade almost every type of compound. The time of degradation varies basically. As plastic takes around 400 years to degrade, some compounds take much lesser type depending on the physiological condition and type of bacterial strain degrading the compound. Biodegradation of hydrocarbons is possible through many strains of bacteria (Bonaventura and Johnson 1997).

Environment contamination via these hydrocarbons was reported around three decades ago, and still, there are numerous reports on how these hydrocarbons are aiding in environmental pollution, degrading the quality of soil, affecting microbes that are living in the deep Earth bed and affecting the lives of animals and plants. With the help of microbes, the waste material could be converted to organic material like carbon dioxide, methane and water (Martello 1991).

Contamination of environment with these hydrocarbons has caused critical health defects and due to this, there is increased consideration been centered on creating and executing imaginative innovative technology and techniques for cleaning the environment (Rahman et al. 2002). Bioremediation strategies over the years have proven to be useful and have effectively got great attention as a promising mother Earth-friendly method for the remediation of hydrocarbon by utilizing these microbes (Desai and Banat 1997). This approach could be set because of the microbial availability and the ability of microbes to degrade such types of dangerous PHCs. These microbes utilize them as their energy source and degrade them using their enzymes (Foght and Westlake 1987).

Biological Method Using Microbes: An Efficient Approach

Bacterial isolates produce their source of carbon and energy by degrading the hydrocarbons (Venosa and Zhu 2003). Biodegradation by the microbial load is the basic mechanisms through which all the diverse petroleum hydrocarbons along with minor constituents could be degraded and removed successfully by converting the complex and hazardous substances to into simpler and nontoxic forms (Leahy and Colwell 1990; Ulrici 2000; Lidderdale 1993). Microbes should be provided with adequate amount of nutrients, oxygen and pH between 6 and 9 so that they can exploit all their capabilities to degrade hydrocarbons efficiently (Das and Chandran 2011). Bacterial cell wall has some oil-containing molecules called hopanoids indicating the biodegradation of oil products (Panda et al. 2013; Gold 1985). Use of surfactants has been recorded to enhance degradation of crude oil (Urum et al. 2003; Yateem et al. 2002). Microbes degrade the petroleum hydrocarbons like benzene aerobically much faster than anaerobically as the rate of degradation is slow and poor in an anaerobic environment (Chaillan et al. 2006). There are around 22 species of bacteria that degrade the petroleum hydrocarbon naturally like *Pseudomonas*, *Aeromonas*, *Bacillus*, *Flavobacterium*, *Corynebacterium*, *Micrococcus*, etc. Based on the oil-degrading capacity, *Pseudomonas* species is the most efficient petroleum degrader (Saadoun 2002; Banat et al. 2000).

There are four approaches to biological oil spill bioremediation:

1. **Bioaugmentation**: Oil-degrading bacteria are added as a supplement to help the already-existing microbial population to degrade the hydrocarbon. Addition of specific microorganisms, native or exogenous to the contaminated sites, is an effective bioremediation process. Bioaugmentation can be carried out by inoculating whole cells or encapsulating the cells in a carrier material.
2. **Biostimulation**: Addition of nutrients and growth limiting co-substrates to support the growth of indigenous oil degraders. Here there is the utilization of an indigenous variety of microbes. No additional supplementation of nutrients required.
3. **Bioventing** uses microbes to degrade those organic constituents that are present on soil or either adsorbed on the soil. This bacterial activity is even increased by providing aeration condition to bacteria. So aerobic degradation is more successful in the case of venting. Lee and Swindoll showed the accuracy of bioventing for the in situ degradation of light hydrocarbons like petrol and diesel and other forms of aromatic compounds.

4. **Biosparging** is similar to the bioventing concept. It involves the supply of air under pressure in a zone that will be saturated so as to vaporize the volatile contaminants. These vaporized contaminants get carried away to the unsaturated zone where the further work is done by microbes as microbes degrade these volatile contaminants in that unsaturated zone. It is used for the degradation of aromatic and cyclic hydrocarbons. Compounds like gasoline, diesel and kerosene could be degraded using this biosparging technique. Natural microbes like *Candidatus magneto bacterium* and *Flavobacteriales bacterium* have the potential to break these volatile harmful compounds.

10.5 Take Home Messages

[A] Oil can refer to many different materials, including crude oil, refined petroleum products (such as gasoline or diesel fuel) or by-products, ships bunkers, oily refuse or oil mixed in waste. The contribution of oil to the marine environment arises from the disposal of waste oil from domestic and industrial automotive sources and from industrial operations (e.g. lubrication, hydraulic and cutting oils, etc.). Most of the oil comes from unauthorized sources such as illegal cleaning of oil tankers and residue dumping.

[B] Oil is a mixture of hydrocarbon compounds. Crude oil is a mixture of gas, naphtha, kerosene, light gas and residuals, which have different adverse effects on the health of marine biota. The fate, behaviour and environmental effects of spilled oil can vary, depending upon the type and amount of material spilled. In general, lighter refined petroleum products such as diesel and gasoline are more likely to mix in the water column and are more toxic to marine life but tend to evaporate more quickly and do not persist long in the environment. Heavier crude or fuel oil, while of less immediate toxicity, can remain on the water surface or stranded on the shoreline for much longer period. A heavy oil spill across the shore blankets rock pools, etc., preventing gas exchange and eliminating light as well as directly leaching toxins into the water and can also become mixed deeply into pebble, shingle or sandy beaches, where it may remain for months or even years. Fishes often escape from toxicity by swimming away from the polluted area, but the benthic organisms suffer large numbers of mortalities as the oil settles with the sediments. Coral reefs often get damaged both by oil deposition and by the water-soluble chemicals contained in oil. Seabirds are adversely affected as the oil penetrates and opens up the structure of their plumage, so they become chilled, lose the ability to fly and lose their buoyancy in water. They are thus unable to feed normally but ingest the oil as they attempt to preen. This condition finally draws these avian faunas towards death.

Oil in the seas also affects human health. Short-term public health impacts from oil spills include accidents suffered by those on damaged tankers or those involved in the cleanup and illnesses caused by toxic fumes or by eating contaminated fish or shellfish. However, there are other less obvious public health impacts, including losses and disruptions of commercial and recreational fisheries, seaweed culture units, boating and a variety of other uses of affected water. The concern involves the ingestion of carcinogenic compounds, especially the polycyclic aromatic hydrocarbons present in petroleum.

[C] The entire spectrum of biological effects that might occur due to oil pollution may be categorized as lethal toxic effects, sublethal effects that disrupt physiological or behavioural activities, but do not cause immediate mortality, uptake of the oil or certain fractions of it by organisms causing tainting or in some cases carcinomas, ingestion by organisms which leads oil to enter the food web, direct smothering and suffocation or interference with movements to obtain food or escape from predators as a result of being coated by oil, alteration to the chemical and physical properties of marine habitat and mortalities caused by indiscriminate use of detergents to dispose the oil.

[D] Biological methods are cheaper and better methods to degrade the oil spills and petroleum containing soil, and these include the use of microbes or by using the enzymes of the microbes to detoxify and remove the pollutants from soil and water which have diverse metabolic capabilities. There are around 22 species of bacteria that degrade the petroleum hydrocarbon naturally like *Pseudomonas*, *Aeromonas*, *Bacillus*, *Flavobacterium*, *Corynebacterium*, *Micrococcus*, etc. Based on the oil-degrading capacity, *Pseudomonas* species is the most efficient petroleum degrader.

10.6 Brain Churners

Oil pollution is a burning issue in underdeveloped and developing nations of the world. It is intricately related with industrialization and urbanization and has severe adverse effects on human health, biodiversity and climate of the planet Earth. This section is a sort of 'self-appraisal approach' as correct answers to all the questions are marked as black in the box at the end of this section.

1. Oil is a mixture of
   (a) Different heavy metals
   (b) Acid and hydrocarbons
   (c) Hydrocarbon compounds
   (d) None of the above
2. Crude oil is a mixture of
   (a) Heavy metal naphtha
   (b) Metal and non-metal
   (c) Gas, naphtha, kerosene, light gas and residuals
   (d) Zn, Cu, Pb and Fe

3. Diesel and gasoline are
   (a) Heavy oil
   (b) Lighter refined petroleum products
   (c) Nontoxic petroleum products
   (d) None of the above
4. 3,4-benzene pyrene is
   (a) Nontoxic and volatile
   (b) Toxic and solid in nature
   (c) Carcinogenic
   (d) None of the above
5. The vulnerability of oil pollution is maximum in
   (a) Rocky shore
   (b) Fine-grained sandy shore
   (c) Coarse-grained sand beaches
   (d) Mangrove dominated beaches
6. Booms which are used to control oil pollution resemble
   (a) Big box
   (b) A vertical curtain
   (c) Broom
   (d) None of the above
7. 'Booms' are used to control oil pollution by
   (a) Enclosing floating oil
   (b) Dispersing floating oil
   (c) Sinking floating oil
   (d) None of the above
8. Skimmers used for controlling oil pollution will not work effectively in
   (a) Waves less than 0.5 m in height
   (b) Waves equal to 0.5 m in height
   (c) Waves less than 1 m in height
   (d) Waves greater than 1 m in height
9. Sorbents are materials that recover oil from polluted aquatic phase through
   (a) Dispersion
   (b) Absorption or adsorption
   (c) Sinking
   (d) None of the above
10. Peat moss is a
    (a) Synthetic sorbent
    (b) Natural sorbent
    (c) Inorganic sorbent
    (d) None of the above

11. Use of chemical dispersions on oil spill helps in

    (a) Natural oil biodegradation
    (b) Sinking the oil
    (c) Absorbing the oil
    (d) None of the above

12. Berms or dykes are built to

    (a) Prevent oil from spreading horizontally
    (b) To allow oil to sink
    (c) Degrade the oil into small particles
    (d) None of the above

13. Suction hoses are used to remove excess oil from

    (a) Sea shore
    (b) Surface water
    (c) Rocky shore
    (d) Ditches or low areas

14. Biological method of controlling oil pollution involves the use of

    (a) Insects
    (b) Enzymes of microbes
    (c) Fungi
    (d) Algae

15. Microbes used for controlling oil pollution work best between pH

    (a) 8–9
    (b) 5–6
    (c) 6–9
    (d) 9–10

16. Hopanoids are

    (a) Oil-containing molecules present in fungi
    (b) Oil-containing molecules present in algae
    (c) Oil-containing molecules present in bacterial cell wall
    (d) None of the above

17. Petroleum hydrocarbon degrading microbes are

    (a) *Azotobacter* sp. and *Clostridium* sp.
    (b) *Penicillium* sp. and *Aspergillus* sp.
    (c) *Oscillatoria* sp. and *Coscinodiscus* sp.
    (d) *Pseudomonas* sp. and *Aeromonas* sp.

18. The process of bioventing is highly successful in.

    (a) Aerobic condition
    (b) Anaerobic condition
    (c) Both a and b
    (d) None of the above

19. Biosparging involves
    (a) Supply of air under low pressure
    (b) Supply of air under medium pressure
    (c) Supply of air under high pressure
    (d) None of the above

20. Oil pollution causes seals
    (a) To lose vision
    (b) To lose fur
    (c) To lose nails
    (d) To lose eyelids

Answer Sheet

S. No.	a	b	c	d
1.			■	
2.			■	
3.		■		
4.			■	
5.				■
6.		■		
7.	■			
8.				■
9.		■		
10.		■		
11.	■			
12.	■			
13.				■
14.		■		
15.			■	
16.			■	
17.				■
18.	■			
19.			■	
20.		■		

References

Banat IM, Makkar RS, Cameotra SS (2000) Potential commercial applications of microbial surfactants. Appl Microbiol Biotech 53:495–508

Bhasheer SK, Umavathi S, Banupriya D, Thangavel M, Thangam Y (2014) Diversity of diesel degrading bacteria from a hydrocarbon contaminated soil. Int J Curr Microbiol App Sci 3:363–369

Boeer B (1996) Impact of a major oil spill off Fujairah. Fresenius Environ Bull 5(1–2):7–12

Bonaventura C, Johnson FM (1997) Healthy environments for healthy people: bioremediation today and tomorrow. Environ Health Perspect 105:5–20

Chaerun SK, Tazaki K, Asada R, Kogure (2005). Interaction between clay minerals and hydrocarbon-utilizing indigenous microorganisms in high concentrations of heavy oil: implications for bioremediation. Clay Miner 40:105–114

Chaillan F, Chaineau CH, Point V, Saliot A, Oudot J (2006) Factors inhibiting bioremediation of soil contaminated with weathered oils and drill cuttings. Environ Pollut 144:255–265

Congressional Research Service (2014) US rail transportation of crude oil: Background and issues for Congress. Available at: http://www.fas.org/sgp/crs/10misc/R43390.pdf

Das N, Chandran P (2011) Microbial degradation of petroleum hydrocarbon contaminants: an overview. Biotechnol Res Int:941–810

Desai JD, Banat IM (1997) Microbial production of surfactants and their commercial potential. Microbiol Mol Biol Rev 61:47–64

Duke NC, Ball MC, Elisson JC (1998) Factors influencing biodiversity and distributional gradients in mangroves. Globl Ecol Biogeogr Lettr 7:27–47

Ellison AM, Farnsworth EJ, Merkt RE (1999) Origins of mangrove ecosystems and the mangrove biodiversity anomaly. Glob Ecol Biogeogr 8:95–115

EPPR (2013) Summary report and recommendations on the prevention of marine oil pollution in the Arctic

Fingas MF (2012) The basics of oil spill cleanup. Taylor and Francis, New York

Foght JM, Westlake DWS (1987) Biodegradation of hydrocarbons in fresh water. In: Oil in freshwater: chemistry. Biology, Countermeasure Technology, pp 217–230

Gold T (1985) The origin of natural gas and petroleum, and the prognosis for future supplies. Annu Rev Energy 10:53–77

Grant DL, Clarke PJ, Allaway WG (1993) The response of grey mangrove (*Avicennia marina* (Forsk.) Vierh.) seedlings to spills of crude oil (J). J Exp Mar Biol Ecol 171:273–295

Gundlach ER, Hayes MO (1978) Vulnerability of coastal environments to oil spill impacts. Mar Technol Soc J 12:18–27

Harayama S, Kishira H, Kasai Y, Shutsubo K (1999) Petroleum biodegradation in marine environments. J Mol Microbiol Biotechnol 1:63–70

Hogarth PJ (2007) The biology of mangroves and seagrasses, 2nd edn. Oxford University Press, Oxford

Lamparelli CC, Rodrigues FO, Moura DO (1997) A long-term assessment ofan oil spill in a mangrove forest in Sao Paulo, Brazil. In: Kjerfve B, Lacerda LD, Diop EHS (eds) Mangrove ecosystem studies in Latin America and Africa. United Nations Educational, Scientific and Cultural Organization, Paris, pp 191–203

Leahy JG, Colwell RR (1990) Microbial degradation of hydrocarbons in the environment. Microbiol Mol Biol Rev 54:305–315

Levings SC, Garrity SD (1995) Oiling of mangrove keys in the 1993 Tampa Bay of spill. In: Proceedings of the 1995 International Oil spill Conference 421–428

Lidderdale T (1993) Demand, supply, and price outlook for low-sulfur diesel fuel. Energy Information Administration Short Term Energy Outlook Annual Supplement DOE-E & 0202

Mackey AP, Hodgkinson (1996) Assessment of the impact of naphthalene contamination on mangrove fauna using behavorial bioassays. Bull Environ Contam Toxicol 56:279–286

Martello A (1991) Bioremediation-cleaning up with biology and technology. Scientist:18–19

Martin F, Dutrieux E, Derby A (1990) Natural recolonization of a chronically oil-polluted mangrove soil after a de-pollution process. Oce Shor Manage 14(3):173–190
Medina-Bellver JI, Marín P, Delgado A, Rodríguez-Sánchez A, Reyes E et al (2005) Evidence for in situ crude oil biodegradation after the prestige oil spill. Environ Microbiol 7:773–779
Michel J, Rutherford N (2014) Impacts, recovery rates, and treatment options for spilled oil in marshes. Marine Pollution Bull 82:19–25
Panda SK, Kar RN, Panda CR (2013) Isolation and identification of petroleum hydrocarbon degrading microorganisms from oil contaminated environment. Int J Environ Sci 3:1314
Proffitt CE, Devlin DJ (1998) Are there cumulative effects in red mangroves from oil spills during seedling and sapling stages? (J). Ecol Appl 8:121–127
Rahman KSM, Banat IM, Thahira J, Thayumanavan T, Lakshmanaperumalsamy P (2002) Bioremediation of gasoline contaminated soil by a bacterial consortium amended with poultry litter, coir pith and rhamnolipid biosurfactant. Bioresour Technol 81:25–32
Rehm HJ, Reiff I (1981) Mechanisms and occurrence of microbial oxidation of long-chain alkanes. In: Reactors and reactions. Springer, Berlin/Heidelberg, pp 175–215
Reuters (2013) Alberta oil sands production likely to double by 2022 regulator Available at: http://www.reuters.com/article/2013/05/08
Saadoun I (2002) Isolation and characterization of bacteria from crude petroleum oil contaminated soil and their potential to degrade diesel fuel. J Basic Microbiol 42:420–428
Tomlinson PB (1986) The botany of mangroves. Cambridge University Press, Cambridge, pp l–413
Ulrici W (2000) Contaminated soil areas, different countries and contaminants, monitoring of contaminants. Biotechnol Environ Proces II 11:5–41
Urum K, Pekdemir T, Gopur M (2003) Optimum conditions for washing of crude oil-contaminated soil with biosurfactant solutions. Process Saf Environ Prot 81:203–209
US Coast Guard, *Arctic Strategy* (2013) US Coast Guard Headquarters, Washington, D
Venosa AD, Zhu X (2003) Biodegradation of crude oil contaminating marine shorelines and freshwater wetlands. Spill Sci Technol Bull 8:163–178
Yateem A, Balba MT, Al-Shayji Y, Al-Awadhi N (2002) Isolation and characterization of biosurfactant-producing bacteria from oil-contaminated soil. Soil Sediment Contam 11:41–55
Yip TL, Talley WK, Jin D (2011) The effectiveness of double hulls in reducing vessel-accident oil spillage. Marine Pollution Bull 62:2427–2432

Internet References

EIA, US Energy Information Administration, Available at: http://www.eia.30gov/
The Federal Interagency Solutions Group, Oil Budget Calculator Science and Engineering Team, Oil Budget Calculator, A Report to the National Incident Command, November 2010. Available at: http://www.restorethegulf.gov/sites/default/files/documents/pdf/OilBudgetCalc_Full_HQ-Print/111110.pdf

Chapter 11
Human Population and the Environment

Abstract One of the burning issues of the present world is the explosion of human population due to which shortage of resources in terms of food, space, employment and other basic needs has cropped up. The chapter discusses a comprehensive population policy, which is based not on the traditional foundation of family planning, but on the pillars of women's access to education, health care, economic and political decisions. To solve the acute unemployment problem, a non-conventional alternative livelihood has also been presented as annexure to the chapter.

Keywords COVID-19 · Family planning · J-shaped growth curve · Population explosion · S-shaped growth curve

11.1 Population Growth Forms

Population is a more or less a permanent aggregation of individuals of the same species inhabiting a specific geographic area at a given time. Population ecology is the study of various factors affecting growth, distribution, natality, dispersal and mortality of individuals constituting the population. Population of a species at a specific place is never static. It is extremely dynamic and depicts variation of its size and density with time due to influence of various abiotic and biotic factors. Favourable environmental condition increases population growth whereas unfavourable conditions decrease population growth. Populations of different species inhabiting a specific common area constitute biotic community.

The accelerating pace of population growth in the last century was not due to any undue rise in birth rate of world population, but, because in the last century, there was a sharp fall in death rate, advancements in healthcare control over fatal diseases such as small pox, plague and cholera and improved food distribution system (result of food security). The average number of children born to a mother has declined from 5 to 3.5 since 1950, though the size of world population has more than doubled during the same period. However, when we are in the process of writing down this chapter the appalling shadow of Corona virus attack has

A. Mitra, S. Zaman, *Environmental Science - A Ground Zero Observation on the Indian Subcontinent*, https://doi.org/10.1007/978-3-030-49131-4_11

already taken the lives of 37,820 individuals throughout the world (https://www.worldometers.info/coronavirus/ Last updated: March 31, 2020, 04:32 GMT), which is a phase of retardation in the growing population scenario of the planet. Country-wise report of this pandemic event is highlighted in Annexure 11B.

11.1.1 Characteristics of Population

11.1.1.1 Natality (Birth Rate)

Natality is average rate of reproduction or birth per unit time. It increases the size of population. The offspring may be produced by birth, hatching or germination. The maximum natality rate achieved under ideal conditions is termed fecundity or potential natality or biotic natality.

The factors (shortage of food and living space, predation, competition, emigration, natural death, natural calamities and carrying capacity) preventing a species from achieving a potential natality are called environmental resistance or population regulation.

11.1.1.2 Mortality (Death Rate)

Mortality is the average number of individuals that die or get killed naturally per unit time. It is basically the ratio of deaths in an area to the population of that area and is expressed per 1000 per year.

11.1.1.3 Density

Density is the number of individuals of a particular species per unit area at a given time. Population density (D) can be calculated by counting all individuals present at a given time in a specific space and dividing it by the number of units of area or space (S). The units of space may be cm^2, m^2 or km^2. For example, earthworms can be in thousands in an acre of land.

11.1.1.4 Dispersal

Population of a place is never static. Its size goes on changing due to movement of organisms into or out of a population under the influence of various abiotic and biotic factors. This process is termed population dispersal. Dispersal is of three types:

- Emigration, i.e. permanent exit of some individuals from local population. The size of local population decreases.
- Immigration, i.e. permanent entry of addition individuals from outside into a given population. It increases the size of local population.
- Migration: migration involves two-way moment of the entire population. It is very common in birds and fishes.

11.1.1.5 Sex Ratio

The ratio of females to males in a population is termed as sex ratio. In 1981 in India, it was 934 females per 1000 males, while in 1991 it came down to 929 females per 1000 males. But after two decades in 2012, it rose to 940 females per 1000 males.

11.1.1.6 Age Ratio

The age structure if a given population refers to the proportion of individuals of different ages within that population. It indicates the ratio of different age groups recognized on the basis of the ability to produce. They are as follows:

11.1.1.7 Pre-reproductive Age

This is the juvenile stage of population. This comprises infants/adolescents who have not attained puberty and thus they are not capable of reproducing.

11.1.1.8 Reproductive Age

This is the age group comprising individuals capable of producing young ones.

11.1.1.9 Post-reproductive Age

This age group includes individuals who have lost the capacity to produce young ones.

Age distribution is an important characteristic of population. It affects both natality and mortality rate. Age ratio determines the reproductive capacity of the population.

11.1.1.10 Age Pyramids

The graphic representation of percentage of different age groups mentioned above is termed age pyramid.

Age pyramids are of three types:

- Triangular age pyramid (population growth is positive, e.g. India)
- Bell-shaped age pyramid (population growth is zero)
- Urn-shaped age pyramid (population growth is negative)

11.1.1.11 Population Growth

Population growth is the increase in number of individuals. The rate of growth is measured by an increase in the number of individuals in a population per unit time. For controlling the growth of population, natality and mortality are important factors. The percentage ratio of natality and mortality is termed vital index.

$$\text{Vital index} = \text{Natality} / \text{Mortality} \times 100$$

- If natality + immigration is greater (>) than mortality + emigration, the growth is positive.
- If natality + immigration is lesser (<) than mortality + emigration, the growth is negative.
- If natality + immigration is equal (=) to mortality + emigration, the growth is zero.

Average annual growth rate of human population can be calculated as follows:

$$\text{Average annual growth rate}(\%) = (\text{P1} - \text{P2}) / (\text{P1} \times \text{N}) \times 100$$

P1 is population size in the previous census.
P2 is population size in the present census.
N is the number of years between the two censuses.

The size of local population is affected not only by immigration, emigration, natality, mortality and biotic factors but also by environmental resistance and carrying capacity. Population growth is the result of interaction of biotic potential and environmental resistance. Two distinct patterns of population growth are identified.

11.1.1.12 S-Shaped Growth (Sigmoid)

This is the most common growth pattern (Fig. 11.1). Sigmoid growth has been divided into four phases.

- Early or lag phase (population increases slowly and population size small)
- Exponential or log phase (population growth increases)
- Diminishing growth phase (population growth rate decrease)
- Equilibrium/stationary phase (natality and mortality rates are at equilibrium, population growth zero)

11.1.1.13 J-Shaped Growth

J-shaped growth pattern has been distinct phases.

- Early lag phase
- Exponential (log) phase
- Crash phase

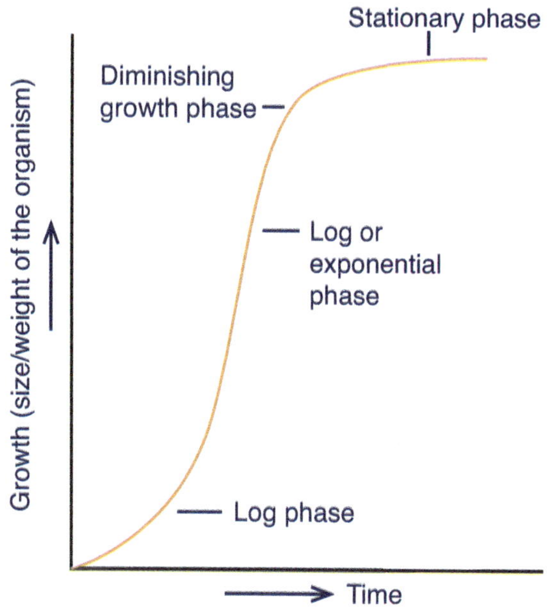

Fig. 11.1 S-Shaped Growth (Sigmoid)

This type of growth (Fig. 11.2) is found in algal bloom, seasonal annual plants and insect, etc.

11.2 Scenario of Global Population

The distribution of population in the world is highly non-uniform. Some areas are very thickly populated while some are sparsely populated. There are various reasons behind uneven distribution of population that can be can be categorized in to geographical factors, social and cultural factors and economic factors. The geographical factors include favourable topography, availability of mineral and freshwater resources, favourable climate soil fertility and food security that invite population and help to flourish. Indo-Gangetic Plains, Diamond Mines of South Africa, etc. are densely populated regions. The social and cultural factors encompass areas of better housing, education and health facilities and religious harmony. Places of religious and cultural significance also attract people, e.g. Varanasi, Jerusalem, etc. The economic factors also invite population due to presence of more industries, transportation and communication facilities, employment opportunities, etc.

The scenario of the present day world has changed a lot since the last few centuries. The population has increased exponentially (Table 11.1), while the area (space) remained constant (Table 11.2).

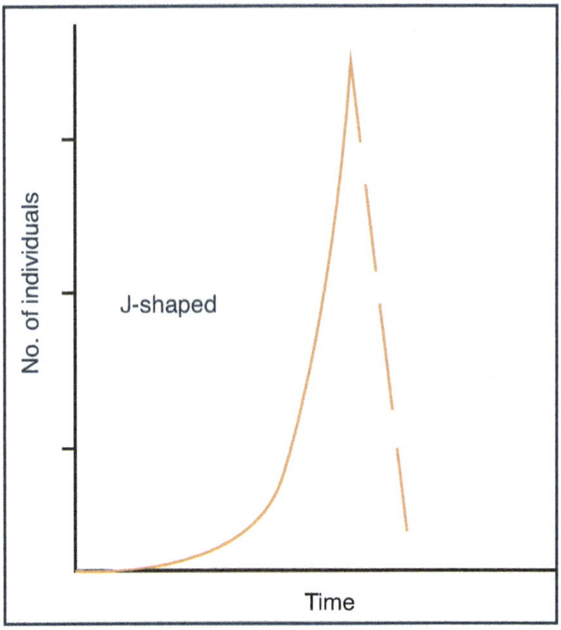

Fig. 11.2 J-Shaped Growth (Sigmoid)

This has hiked up the population density and created a great pressure and challenges on leaving space, resources, environmental condition, job opportunity, mobility and the quality of life.

The predicted scenario of world population in 2050 looks very alarming. The global population is expected to be around 9.80 billion people, which is 2 billion more than the present population as recorded on September 24, 2019, which is 7,732,640,485 at 3.09 pm IST from the site worldpopulationreview.com. It is expected that India will surpass China as the most populated country in the world with a population of 1.66 billion people compared to China who will occupy the second position with a population of 1.36 billion. Nigeria will become the world's third most populous country in 2050 with a population of 4.10 million people. The fourth position will be occupied by the United States with a population of 390 million people. Vatican City has been predicted to continue being the least populated country in the World in 2050. Presently it is having a negative growth rate in the population domain and in 2050 it will have only 800 people. The overall population will increase globally due to development in the health sector and technology. Better healthcare is the road map for reducing death rate and infant mortality. Many predictions on the world population scenario are still waiting with the advancement in computer application, artificial intelligence and environment. However, it is no exaggeration to say that if the population follows the present increasing trend then there will be an acute shortage of resources in terms of space, food and various basic needs. High-end technological development in the field of agriculture, fishery and other food items coupled with International-level policy of population transference

Table 11.1 Increase of population in 1 year (in %)

Country	2017	2018
Aruba	105,366	105,845
Afghanistan	36,296,400	37,172,386
Angola	29,816,748	30,809,762
Albania	2,873,457	2,866,376
Andorra	77,001	77,006
Arab World	411,898,965	419,790,588
United Arab Emirates	9,487,203	9,630,959
Argentina	44,044,811	44,494,502
Armenia	2,944,809	2,951,776
American Samoa	55,620	55,465
Antigua and Barbuda	95,426	96,286
Australia	24,601,860	24,992,369
Austria	8,797,566	8,847,037
Azerbaijan	9,854,033	9,942,334
Burundi	10,827,024	11,175,378
Belgium	11,375,158	11,422,068
Benin	11,175,204	11,485,048
Burkina Faso	19,193,284	19,751,535
Bangladesh	159,670,593	161,356,039
Bulgaria	7,075,947	7,024,216
Bahrain	1,494,074	1,569,439
Bahamas, The	381,761	385,640
Bosnia and Herzegovina	3,351,527	3,323,929
Belarus	9,498,264	9,485,386
Belize	375,769	383,071
Bermuda	63,874	63,968
Bolivia	11,192,854	11,353,142
Brazil	207,833,831	209,469,333
Barbados	286,233	286,641
Brunei Darussalam	424,473	428,962
Bhutan	745,568	754,394
Botswana	2,205,128	2,254,126
Central African Republic	4,596,028	4,666,377
Canada	36,540,268	37,058,856
Central Europe and the Baltics	102,738,854	102,511,922
Switzerland	8,451,840	8,516,543
Channel Islands	168,665	170,499
Chile	18,470,439	18,729,160
China	1,386,395,000	1,392,730,000
Cote d'Ivoire	24,437,469	25,069,229
Cameroon	24,566,045	25,216,237
Congo, Dem. Rep.	81,398,764	84,068,091

(continued)

Table 11.1 (continued)

Country	2017	2018
Congo, Rep.	5,110,702	5,244,363
Colombia	48,901,066	49,648,685
Comoros	813,892	832,322
Cabo Verde	537,497	543,767
Costa Rica	4,949,954	4,999,441
Caribbean small states	7,314,990	7,358,965
Cuba	11,339,259	11,338,138
Curacao	160,175	159,849
Cayman Islands	63,382	64,174
Cyprus	1,179,680	1,189,265
Czech Republic	10,594,438	10,625,695
Germany	82,657,002	82,927,922
Djibouti	944,097	958,920
Dominica	71,458	71,625
Denmark	5,764,980	5,797,446
Dominican Republic	10,513,131	10,627,165
Algeria	41,389,198	42,228,429
East Asia and Pacific (excluding high income)	2,068,155,660	2,081,651,801
Early-demographic dividend	3,207,188,541	3,249,140,605
East Asia and Pacific	2,314,202,003	2,328,220,870
Europe and Central Asia (excluding high income)	415,710,935	417,797,257
Europe and Central Asia	915,420,161	918,793,590
Ecuador	16,785,361	17,084,357
Egypt, Arab Rep.	96,442,593	98,423,595
Euro area	341,164,362	341,783,171
Eritrea		
Spain	46,593,236	46,723,749
Estonia	1,317,384	1,320,884
Ethiopia	106,400,024	109,224,559
European Union	512,191,098	513,213,363
Fragile and conflict-affected situations	504,119,229	515,215,936
Finland	5,508,214	5,518,050
Fiji	877,459	883,483
France	66,865,144	66,987,244
Faroe Islands	48,331	48,497
Micronesia, Fed. Sts.	111,459	112,640
Gabon	2,064,823	2,119,275
United Kingdom	66,058,859	66,488,991
Georgia	3,728,004	3,731,000
Ghana	29,121,471	29,767,108
Gibraltar	33,728	33,718
Guinea	12,067,539	12,414,318

(continued)

Table 11.1 (continued)

Country	2017	2018
Gambia, The	2,213,894	2,280,102
Guinea-Bissau	1,828,146	1,874,309
Equatorial Guinea	1,262,001	1,308,974
Greece	10,754,679	10,727,668
Grenada	110,874	111,454
Greenland	56,171	56,025
Guatemala	16,914,936	17,247,807
Guam	164,281	165,768
Guyana	775,221	779,004
High income	1,204,429,565	1,210,312,147
Hong Kong SAR, China	7,391,700	7,451,000
Honduras	9,429,013	9,587,522
Heavily indebted poor countries (HIPC)	759,106,221	780,234,406
Croatia	4,124,531	4,089,400
Haiti	10,982,366	11,123,176
Hungary	9,787,966	9,768,785
IBRD only	4,731,120,193	4,772,284,113
IDA and IBRD total	6,335,039,629	6,412,522,234
IDA total	1,603,919,436	1,640,238,121
IDA blend	543,525,897	555,830,605
Indonesia	264,645,886	267,663,435
IDA only	1,060,393,539	1,084,407,516
Isle of Man	83,598	84,077
India	1,338,658,835	1,352,617,328
Not classified		
Ireland	4,807,388	4,853,506
Iran, Islamic Rep.	80,673,951	81,800,269
Iraq	37,552,781	38,433,600
Iceland	343,400	353,574
Israel	8,713,300	8,883,800
Italy	60,536,709	60,431,283
Jamaica	2,920,853	2,934,855
Jordan	9,779,173	9,956,011
Japan	126,785,797	126,529,100
Kazakhstan	18,037,776	18,276,499
Kenya	50,221,473	51,393,010
Kyrgyz Republic	6,198,200	6,315,800
Cambodia	16,009,414	16,249,798
Kiribati	114,158	115,847
St. Kitts and Nevis	52,045	52,441
Korea, Rep.	51,466,201	51,635,256
Kuwait	4,056,097	4,137,309

(continued)

Table 11.1 (continued)

Country	2017	2018
Latin America and Caribbean (excluding high income)	603,254,104	609,013,934
Lao PDR	6,953,035	7,061,507
Lebanon	6,811,873	6,848,925
Liberia	4,702,228	4,818,977
Libya	6,580,724	6,678,567
St. Lucia	180,955	181,889
Latin America and Caribbean	635,372,515	641,357,515
Least developed countries: UN classification	986,365,080	1,009,662,578
Low income	687,449,530	705,417,321
Liechtenstein	37,800	37,910
Sri Lanka	21,444,000	21,670,000
Lower middle income	2,981,420,591	3,022,905,169
Low and middle income	6,306,560,891	6,383,958,209
Lesotho	2,091,412	2,108,132
Late-demographic dividend	2,276,319,334	2,288,665,963
Lithuania	2,828,403	2,789,533
Luxembourg	596,336	607,728
Latvia	1,942,248	1,926,542
Macao SAR, China	622,585	631,636
St. Martin (French part)	36,560	37,264
Morocco	35,581,294	36,029,138
Monaco	38,392	38,682
Moldova	3,549,196	3,545,883
Madagascar	25,570,540	26,262,368
Maldives	496,402	515,696
Middle East and North Africa	441,255,234	448,912,859
Mexico	124,777,324	126,190,788
Marshall Islands	58,058	58,413
Middle income	5,619,111,361	5,678,540,888
North Macedonia	2,081,996	2,082,958
Mali	18,512,394	19,077,690
Malta	467,999	483,530
Myanmar	53,382,581	53,708,395
Middle East and North Africa (excluding high income)	376,546,755	382,896,715
Montenegro	622,373	622,345
Mongolia	3,113,779	3,170,208
Northern Mariana Islands	56,562	56,882
Mozambique	28,649,007	29,495,962
Mauritania	4,282,574	4,403,319
Mauritius	1,264,613	1,265,303
Malawi	17,670,260	18,143,315
Malaysia	31,105,028	31,528,585

(continued)

Table 11.1 (continued)

Country	2017	2018
North America	361,751,263	364,290,258
Namibia	2,402,603	2,448,255
New Caledonia	280,350	284,060
Niger	21,602,472	22,442,948
Nigeria	190,873,311	195,874,740
Nicaragua	6,384,855	6,465,513
Netherlands	17,131,296	17,231,017
Norway	5,276,968	5,314,336
Nepal	27,627,124	28,087,871
Nauru	12,876	12,704
New Zealand	4,793,900	4,885,500
OECD members	1,296,225,760	1,303,529,456
Oman	4,665,935	4,829,483
Other small states	30,148,800	30,758,989
Pakistan	207,896,686	212,215,030
Panama	4,106,771	4,176,873
Peru	31,444,297	31,989,256
Philippines	105,173,264	106,651,922
Palau	17,808	17,907
Papua New Guinea	8,438,029	8,606,316
Poland	37,974,826	37,978,548
Pre-demographic dividend	894,512,725	919,485,393
Puerto Rico	3,325,001	3,195,153
Korea, Dem. People's Rep.	25,429,985	25,549,819
Portugal	10,300,300	10,281,762
Paraguay	6,867,062	6,956,071
West Bank and Gaza	4,454,805	4,569,087
Pacific island small states	2,422,086	2,457,367
Post-demographic dividend	1,106,035,186	1,109,997,273
French Polynesia	276,103	277,679
Qatar	2,724,724	2,781,677
Romania	19,587,491	19,473,936
Russian Federation	144,496,740	144,478,050
Rwanda	11,980,937	12,301,939
South Asia	1,792,835,608	1,814,388,744
Saudi Arabia	33,099,147	33,699,947
Sudan	40,813,396	41,801,533
Senegal	15,419,381	15,854,360
Singapore	5,612,253	5,638,676
Solomon Islands	636,038	652,858
Sierra Leone	7,488,431	7,650,154
El Salvador	6,388,122	6,420,744

(continued)

Table 11.1 (continued)

Country	2017	2018
San Marino	33,671	33,785
Somalia	14,589,119	15,008,154
Serbia	7,020,858	6,982,084
Sub-Saharan Africa (excluding high income)	1,050,057,829	1,078,209,758
South Sudan	10,910,759	10,975,920
Sub-Saharan Africa	1,050,153,672	1,078,306,520
Small states	39,885,876	40,575,321
Sao Tome and Principe	207,089	211,028
Suriname	570,496	575,991
Slovak Republic	5,439,232	5,447,011
Slovenia	2,066,388	2,067,372
Sweden	10,057,698	10,183,175
Eswatini	1,124,753	1,136,191
Sint Maarten (Dutch part)	40,574	40,654
Seychelles	95,843	96,762
Syrian Arab Republic	17,068,002	16,906,283
Turks and Caicos Islands	37,115	37,665
Chad	15,016,773	15,477,751
East Asia and Pacific (IDA and IBRD countries)	2,042,687,863	2,056,064,424
Europe and Central Asia (IDA and IBRD countries)	457,810,292	459,865,205
Togo	7,698,475	7,889,094
Thailand	69,209,858	69,428,524
Tajikistan	8,880,268	9,100,837
Turkmenistan	5,757,669	5,850,908
Latin America and the Caribbean (IDA and IBRD countries)	619,460,244	625,569,713
Timor-Leste	1,243,261	1,267,972
Middle East and North Africa (IDA and IBRD countries)	372,091,950	378,327,628
Tonga	101,998	103,197
South Asia (IDA and IBRD)	1,792,835,608	1,814,388,744
Sub-Saharan Africa (IDA and IBRD countries)	1,050,153,672	1,078,306,520
Trinidad and Tobago	1,384,072	1,389,858
Tunisia	11,433,443	11,565,204
Turkey	81,101,892	82,319,724
Tuvalu	11,370	11,508
Tanzania	54,663,906	56,318,348
Uganda	41,162,465	42,723,139
Ukraine	44,831,135	44,622,516
Upper middle income	2,637,690,770	2,655,635,719
Uruguay	3,436,646	3,449,299
United States	325,147,121	327,167,434
Uzbekistan	32,388,600	32,955,400
St. Vincent and the Grenadines	109,827	110,210

(continued)

Table 11.1 (continued)

Country	2017	2018
Venezuela, RB	29,390,409	28,870,195
British Virgin Islands	29,577	29,802
Virgin Islands (USA)	107,268	106,977
Vietnam	94,596,642	95,540,395
Vanuatu	285,510	292,680
World	7,510,990,456	7,594,270,356
Samoa	195,352	196,130
Kosovo	1,830,700	1,845,300
Yemen, Rep.	27,834,821	28,498,687
South Africa	57,000,451	57,779,622
Zambia	16,853,688	17,351,822
Zimbabwe	14,236,745	14,439,018

(from densely populated ecosystem to sparsely populated ecosystem) may provide a pathway to combat the threats of population explosion in 2050.

11.3 Concept of Population Control

The explosion of population in majority of the regions in the world is a matter of great concern as it is directly linked with dark chapters like poverty, job insecurity, pollution, scarcity of resources (food, space, water, etc.) and health hazards. We therefore initiated an in-depth study to scan the growth of population in different regions of the world. This data bank is sourced from CIA World Fact book and unless otherwise noted, information in this table is accurate as of January 1, 2019. Our first order analysis reflects an alarming growth of population (1 or more than 1%) in 106 countries including India (Table 11.3). This will definitely create a burden on the resource reservoir of the planet if not controlled through proper policy. Rather than simply equating population policy with family planning, the new thinking is to address some of the roots of the problem like improving women's access to education, healthcare and economic and political decisions.

Family planning is a good initiative to check the accelerating phase of population growth. The History of family planning begins from the middle of nineteenth century. The Malthusian League was founded in England in 1860 to spread the message of birth control. According to Malthus theory 'Population tends to increase geometrically while food supply increases arithmetically'. Similar efforts were also made in Sweden and France. The objective was to check/control rapid population growth and care for the health of pregnant mothers.

After World War II (1939–1945), family planning gained momentum in many countries of Asia, Africa and elsewhere. India was the first country to adopt family planning in 1951 as a National Policy. China, though initially opposed to family

Table 11.2 Country-wise area per sq. mi with population density (PD)

Country	Area (sq. mi)	PD (per sq. mi)
Afghanistan	647,500	48.0
Albania	28,748	124.6
Algeria	2,381,740	13.8
American Samoa	199	290.4
Andorra	468	152.1
Angola	1,246,700	9.7
Anguilla	102	132.1
Antigua and Barbuda	443	156.0
Argentina	2,766,890	14.4
Armenia	29,800	99.9
Aruba	193	372.5
Australia	7,686,850	2.6
Austria	83,870	97.7
Azerbaijan	86,600	91.9
Bahamas, The	13,940	21.8
Bahrain	665	1050.5
Bangladesh	144,000	1023.4
Barbados	431	649.5
Belarus	207,600	49.6
Belgium	30,528	340.0
Belize	22,966	12.5
Benin	112,620	69.8
Bermuda	53	1241.0
Bhutan	47,000	48.5
Bolivia	1,098,580	8.2
Bosnia and Herzegovina	51,129	88.0
Botswana	600,370	2.7
Brazil	8,511,965	22.1
British Virgin Is.	153	151.0
Brunei	5770	65.8
Bulgaria	110,910	66.6
Burkina Faso	274,200	50.7
Burma	678,500	69.8
Burundi	27,830	290.7
Cambodia	181,040	76.7
Cameroon	475,440	36.5
Canada	9,984,670	3.3
Cape Verde	4033	104.4
Cayman Islands	262	173.4
Central African Rep.	622,984	6.9
Chad	1,284,000	7.7

(continued)

Table 11.2 (continued)

Country	Area (sq. mi)	PD (per sq. mi)
Chile	756,950	21.3
China	9,596,960	136.9
Colombia	1,138,910	38.3
Comoros	2170	318.4
Congo, Dem. Rep.	2,345,410	26.7
Congo, Repub. of the	342,000	10.8
Cook Islands	240	89.1
Costa Rica	51,100	79.8
Cote d'Ivoire	322,460	54.8
Croatia	56,542	79.5
Cuba	110,860	102.7
Cyprus	9250	84.8
Czech Republic	78,866	129.8
Denmark	43,094	126.5
Djibouti	23,000	21.2
Dominica	754	91.4
Dominican Republic	48,730	188.5
East Timor	15,007	70.8
Ecuador	283,560	47.8
Egypt	1,001,450	78.8
El Salvador	21,040	324.3
Equatorial Guinea	28,051	19.3
Eritrea	121,320	39.5
Estonia	45,226	29.3
Ethiopia	1,127,127	66.3
Faroe Islands	1399	33.8
Fiji	18,270	49.6
Finland	338,145	15.5
France	547,030	111.3
French Guiana	91,000	2.2
French Polynesia	4167	65.9
Gabon	267,667	5.3
Gambia, The	11,300	145.3
Gaza Strip	360	3968.8
Georgia	69,700	66.9
Germany	357,021	230.9
Ghana	239,460	93.6
Gibraltar	7	3989.7
Greece	131,940	81.0
Greenland	2,166,086	0.0
Grenada	344	260.8

(continued)

Table 11.2 (continued)

Country	Area (sq. mi)	PD (per sq. mi)
Guadeloupe	1780	254.4
Guam	541	316.1
Guatemala	108,890	112.9
Guernsey	78	838.6
Guinea	245,857	39.4
Guinea-Bissau	36,120	39.9
Guyana	214,970	3.6
Haiti	27,750	299.4
Honduras	112,090	65.4
Hong Kong	1092	6355.7
Hungary	93,030	107.3
Iceland	103,000	2.9
India	3,287,590	333.2
Indonesia	1,919,440	127.9
Iran	1,648,000	41.7
Iraq	437,072	61.3
Ireland	70,280	57.8
Isle of Man	572	131.9
Israel	20,770	305.8
Italy	301,230	193.0
Jamaica	10,991	250.9
Japan	377,835	337.4
Jersey	116	785.2
Jordan	92,300	64.0
Kazakhstan	2,717,300	5.6
Kenya	582,650	59.6
Kiribati	811	130.0
Korea, North	120,540	191.8
Korea, South	98,480	496.0
Kuwait	17,820	135.7
Kyrgyzstan	198,500	26.3
Laos	236,800	26.9
Latvia	64,589	35.2
Lebanon	10,400	372.5
Lesotho	30,355	66.6
Liberia	111,370	27.3
Libya	1,759,540	3.4
Liechtenstein	160	212.4
Lithuania	65,200	55.0
Luxembourg	2586	183.5
Macau	28	16183.0

(continued)

Table 11.2 (continued)

Country	Area (sq. mi)	PD (per sq. mi)
Macedonia	25,333	80.9
Madagascar	587,040	31.7
Malawi	118,480	109.8
Malaysia	329,750	74.0
Maldives	300	1196.7
Mali	1,240,000	9.5
Malta	316	1266.5
Marshall Islands	11,854	5.1
Martinique	1100	396.5
Mauritania	1,030,700	3.1
Mauritius	2040	608.3
Mayotte	374	538.1
Mexico	1,972,550	54.5
Micronesia, Fed. St.	702	153.9
Moldova	33,843	132.0
Monaco	2	16271.5
Mongolia	1,564,116	1.8
Montserrat	102	92.5
Morocco	446,550	74.4
Mozambique	801,590	24.6
Namibia	825,418	2.5
Nauru	21	632.7
Nepal	147,181	192.2
Netherlands	41,526	397.1
Netherlands Antilles	960	231.0
New Caledonia	19,060	11.5
New Zealand	268,680	15.2
Nicaragua	129,494	43.0
Niger	1,267,000	9.9
Nigeria	923,768	142.7
N. Mariana Islands	477	172.9
Norway	323,802	14.2
Oman	212,460	14.6
Pakistan	803,940	206.2
Palau	458	44.9
Panama	78,200	40.8
Papua New Guinea	462,840	12.3
Paraguay	406,750	16.0
Peru	1,285,220	22.0
Philippines	300,000	298.2
Poland	312,685	123.3

(continued)

Table 11.2 (continued)

Country	Area (sq. mi)	PD (per sq. mi)
Portugal	92,391	114.8
Puerto Rico	13,790	284.8
Qatar	11,437	77.4
Reunion	2517	312.9
Romania	237,500	93.9
Russia	17,075,200	8.4
Rwanda	26,338	328.4
Saint Helena	413	18.2
Saint Kitts and Nevis	261	149.9
Saint Lucia	616	273.5
St Pierre and Miquelon	242	29.0
Saint Vincent and the Grenadines	389	303.0
Samoa	2944	60.1
San Marino	61	479.5
Sao Tome and Principe	1001	193.2
Saudi Arabia	1,960,582	13.8
Senegal	196,190	61.1
Serbia	88,361	106.3
Seychelles	455	179.2
Sierra Leone	71,740	83.7
Singapore	693	6482.2
Slovakia	48,845	111.4
Slovenia	20,273	99.2
Solomon Islands	28,450	19.4
Somalia	637,657	13.9
South Africa	1,219,912	36.2
Spain	504,782	80.0
Sri Lanka	65,610	308.2
Sudan	2,505,810	16.5
Suriname	163,270	2.7
Swaziland	17,363	65.5
Sweden	449,964	20.0
Switzerland	41,290	182.2
Syria	185,180	102.0
Taiwan	35,980	640.3
Tajikistan	143,100	51.2
Tanzania	945,087	39.6
Thailand	514,000	125.7
Togo	56,785	97.7
Tonga	748	153.3

(continued)

Table 11.2 (continued)

Country	Area (sq. mi)	PD (per sq. mi)
Trinidad and Tobago	5128	207.9
Tunisia	163,610	62.2
Turkey	780,580	90.2
Turkmenistan	488,100	10.3
Turks and Caicos Is	430	49.2
Tuvalu	26	454.2
Uganda	236,040	119.5
Ukraine	603,700	77.4
United Arab Emirates	82,880	31.4
United Kingdom	244,820	247.6
United States	9,631,420	31.0
Uruguay	176,220	19.5
Uzbekistan	447,400	61.0
Vanuatu	12,200	17.1
Venezuela	912,050	28.2
Vietnam	329,560	256.1
Virgin Islands	1910	56.9
Wallis and Futuna	274	58.5
West Bank	5860	419.9
Western Sahara	266,000	1.0
Yemen	527,970	40.6
Zambia	752,614	15.3
Zimbabwe	390,580	31.3

planning in the line with its communist ideology, went for it in 1962, a more aggressive policy of 'one child for one couple'. Family planning methods were facilitated by advancements in sterilization, intrauterine device, pills and condoms.

In India, family planning was given special place in the country's first 5-year plan (1951–1956). It was placed under the control of Ministry of Health and, in 1977, its nomenclature was changed to Ministry of Health and Family Welfare. All expenditure on family planning programme is borne by the Union Government. The States and Union Territories are only implementing agencies. A vast infrastructure exists in the form of community health centre, primary health centre and sub-centres.

Family planning measures include adoption of a number of contraceptive devices for preventing unwanted births. The success of these measures depends upon various socio-economic factors like education, industrial development, employments, etc.

The United Nations has played a very crucial role in popularizing family planning. This organization not only worked for creating awareness about the population problem but also provided technical and financial support for the purpose. Some of the important family planning movements are:

Table 11.3 Country-wise population growth rate (in %)

Rank	Country	Population growth rate
1	Syria	7.37
2	Angola	3.49
3	Malawi	3.31
4	Burundi	3.23
5	Chad	3.23
6	Uganda	3.18
7	Niger	3.16
8	Mali	2.98
9	Sudan	2.93
10	Zambia	2.91
11	Ethiopia	2.83
12	Burkina Faso	2.76
13	Guinea	2.75
14	Tanzania	2.74
15	Gabon	2.73
16	Benin	2.68
17	Western Sahara	2.64
18	Togo	2.61
19	Liberia	2.59
20	Cameroon	2.54
21	Nigeria	2.54
22	Iraq	2.5
23	Guinea-Bissau	2.48
24	Mozambique	2.46
25	Madagascar	2.46
26	Equatorial Guinea	2.41
27	Sierra Leone	2.4
28	Egypt	2.38
29	Afghanistan	2.37
30	Senegal	2.36
31	Congo	2.33
32	East Timor	2.32
33	Rwanda	2.3
34	Cote d'Ivoire	2.3
35	Gaza Strip	2.25
36	British Virgin Islands	2.2
37	Bahrain	2.19
38	Congo	2.17
39	Yemen	2.17
40	Ghana	2.16
41	Mauritania	2.14

(continued)

Table 11.3 (continued)

Rank	Country	Population growth rate
42	Djibouti	2.13
43	Central African Republic	2.11
44	Turks and Caicos Islands	2.09
45	Somalia	2.08
46	Jordan	2.02
47	Oman	2
48	Gambia	1.99
49	Cayman Islands	1.96
50	Qatar	1.95
51	Anguilla	1.92
52	Namibia	1.91
53	Solomon Islands	1.9
54	Luxembourg	1.9
55	West Bank	1.81
56	Vanuatu	1.81
57	Belize	1.8
58	Singapore	1.79
59	Guatemala	1.72
60	Zimbabwe	1.68
61	Papua New Guinea	1.67
62	Sao Tome and Principe	1.66
63	Saudi Arabia	1.63
64	Algeria	1.63
65	Tajikistan	1.58
66	Comoros	1.57
67	Kenya	1.57
68	Honduras	1.56
69	Brunei	1.55
70	Philippines	1.55
71	Botswana	1.52
72	Marshall Islands	1.5
73	Israel	1.49
74	Laos	1.48
75	Bolivia	1.48
76	Cambodia	1.48
77	Libya	1.45
78	United Arab Emirates	1.44
79	Pakistan	1.41
80	Kuwait	1.38
81	Malaysia	1.34
82	Cape Verde	1.32

(continued)

Table 11.3 (continued)

Rank	Country	Population growth rate
83	Haiti	1.31
84	New Caledonia	1.3
85	Ecuador	1.25
86	Aruba	1.24
87	Panama	1.24
88	Venezuela	1.21
89	Antigua and Barbuda	1.2
90	Iran	1.19
91	Paraguay	1.17
92	India	1.14
93	Costa Rica	1.13
94	Kiribati	1.12
95	Christmas Island	1.11
96	Mongolia	1.11
97	Ireland	1.11
98	Turkmenistan	1.1
99	Nepal	1.09
100	Mexico	1.09
101	Iceland	1.08
102	Bhutan	1.05
103	Bangladesh	1.02
104	Kyrgyzstan	1.02
105	Australia	1.01
106	Suriname	1
107	Malta	0.99
108	Dominican Republic	0.99
109	Kazakhstan	0.98
110	Colombia	0.97
111	Nicaragua	0.97
112	South Africa	0.97
113	Tunisia	0.95
114	Morocco	0.95
115	Norway	0.94
116	Peru	0.94
117	Uzbekistan	0.91
118	Vietnam	0.9
119	Eritrea	0.89
120	Argentina	0.89
121	Burma	0.89
122	Tuvalu	0.86
123	French Polynesia	0.85

(continued)

Table 11.3 (continued)

Rank	Country	Population growth rate
124	Indonesia	0.83
125	Azerbaijan	0.83
126	Swaziland	0.82
127	United States	0.8
128	Sweden	0.8
129	Bahamas, The	0.79
130	Liechtenstein	0.78
131	New Zealand	0.77
132	Jersey	0.76
133	Chile	0.75
134	Seychelles	0.74
135	Spain	0.73
136	Sri Lanka	0.73
137	Canada	0.72
138	Brazil	0.71
139	Macau	0.71
140	Saint Kitts and Nevis	0.7
141	San Marino	0.7
142	Switzerland	0.68
143	Belgium	0.67
144	Isle of Man	0.65
145	Samoa	0.61
146	Denmark	0.59
147	Faroe Islands	0.58
148	Mauritius	0.57
149	Fiji	0.56
150	Korea, North	0.52
151	Nauru	0.51
152	United Kingdom	0.51
153	Turkey	0.49
154	Guyana	0.48
155	Korea, South	0.44
156	Montserrat	0.43
157	Bermuda	0.43
158	Austria	0.42
159	Grenada	0.42
160	Palau	0.4
161	Netherlands	0.38
162	France	0.37
163	China	0.37
164	Finland	0.33

(continued)

Table 11.3 (continued)

Rank	Country	Population growth rate
165	Saint Lucia	0.31
166	Wallis and Futuna	0.3
167	Monaco	0.3
168	Albania	0.3
169	Hong Kong	0.29
170	Thailand	0.29
171	Guernsey	0.28
172	Uruguay	0.27
173	Barbados	0.26
174	El Salvador	0.25
175	Lesotho	0.24
176	Guam	0.23
177	Gibraltar	0.21
178	Macedonia	0.19
179	Dominica	0.17
180	Italy	0.16
181	Taiwan	0.15
182	Saint Helena	0.14
183	Czech Republic	0.1
184	Ukraine	0.04
185	Slovenia	0.03
186	Norfolk Island	0.01
187	Falkland Islands (Islas Malvinas)	0.01
188	Georgia	0.01
189	Cocos (Keeling) Islands	0
190	Pitcairn Islands	0
191	Vatican City	0
192	Tokelau	−0.01
193	Andorra	−0.01
194	Slovakia	−0.02
195	Svalbard	−0.03
196	Niue	−0.03
197	Greenland	−0.04
198	Jamaica	−0.05
199	Maldives	−0.06
200	Greece	−0.07
201	Tonga	−0.1
202	Russia	−0.11
203	Poland	−0.16
204	Germany	−0.17
205	Bosnia and Herzegovina	−0.17

(continued)

Table 11.3 (continued)

Rank	Country	Population growth rate
206	Trinidad and Tobago	−0.23
207	Saint Vincent and the Grenadines	−0.23
208	Belarus	−0.24
209	Japan	−0.24
210	Armenia	−0.25
211	Hungary	−0.26
212	Cuba	−0.27
213	Portugal	−0.27
214	Virgin Islands	−0.3
215	Montenegro	−0.34
216	Romania	−0.35
217	Serbia	−0.47
218	Croatia	−0.51
219	Northern Mariana Islands	−0.52
220	Micronesia	−0.55
221	Estonia	−0.6
222	Bulgaria	−0.63
223	Moldova	−1.06
224	Latvia	−1.1
225	Lithuania	−1.1
226	Saint Pierre and Miquelon	−1.13
227	South Sudan	−1.16
228	American Samoa	−1.35
229	Puerto Rico	−1.7
230	Cook Islands	−2.72
231	Lebanon	−3.13

1. Oral contraceptives (polls) introduced in 1960.
2. Plastic IUDs made available in 1961.
3. United Nations Fund for Population Activities (UNFPA) was created in 1961.
4. First World conference on population was organized by UNs at Bucharest (Romania) in 1974.
5. China started ‘one child for one couple’ campaign in 1979.
6. Second World War conference on population was organized by the UNs at Mexico in 1984.
7. Third World Conference on population was organized by the UNs at Cairo (Egypt) in 1994.
8. The latest World Conference on population was organized by the UNs at Cairo (Egypt) on July 11, 2012.

The net result of this family planning concept is visualized when one critically compares the average annual exponential growth rates between 1991–2001 and

2001–2011 for India. In all the States and Union Territories, the average annual exponential growth rate has reduced in 2001–2011 compared to 1991–2001 (Table 11.4).

India has also developed a much focused population policy 2020 to curb down the growth. The objective of the National Population Policy 2000 has set three objectives. One, the immediate objective, is to provide for facilities to meet the unmet needs for contraception, healthcare, infrastructure and health personnel and an integrated service delivery for basic reproductive and child healthcare. The second medium term objective is to bring the total fertility rate (TFR), i.e. the average number of children per women, in the reproductive age group, to replacement level by 2020. The third long-term objective is to achieve a stable population by 2045, a level consistent with the requirements of sustainable economic growth, social development and environmental protection.

Pointing out population growth as a major concern in India, Prime Minister Narendra Modi in his Independence Day speech on August 15, 2019, called for a deeper thought towards the issue. Apparently, the mention was an indication that the government is devising a policy or a law for curbing the population growth in country, which may also make social responsibility and stringent rules, an integral part of government family planning plans. Modi ji said that population explosion can create new problems especially for the future generations. But there is also an enlightened section of society which is aware of this challenge. We have to ponder on this issue taking along all the sections of the society.

Estimates and statistics of population in India have been showing a slightly positive picture though the country's population and remain a concern for social and economic reasons. While India's population is projected to overtake China's in less than a decade as per the United Nations 'World Population Prospects 2019' report released in June this year, the new projections for India are the lowest since the United Nations began these forecasts. The reason is the sharp decline in India's population growth rates over 10 years from 2001 to 2011. According to Census 2011, the growth rate of population has declined from 21.5% during 1991–2001 to 17.7% during 2001–2011, across all religious groups.

The population explosion has major impacts on the country ranging from health, social, environmental and economic.

In July, the government had introduced the 'Population Regulation Bill, 2019', in Rajya Sabha that calls for punitive action against people with more than two living children and making them devoid of all government services. The proposed legislation aims at disqualification from being an elected representative, denial of financial benefits and reduction in benefits under the public distribution system (PDS) for people having more than two children. The bill also suggests that government employees should give an undertaking that they will not procreate more than two children.

Public health experts are divided over the proposed 'Population Regulation Bill, 2019'. However, as a part of the disaster management arising out of population explosion, it is essential to provide food and economic security to this rising population through the lanes of alternative livelihoods as discussed in Annexure 11A.

Table 11.4 Population, percentage decadal growth and average annual exponential growth rates 1991–2001 and 2001–2011

State/UT Code	India/State/Union Territory#	Total population		Percentage decadal growth		Change in percentage decadal growth	Average annual exponential growth rate	
		2001	**2011**	**1991–2001**	**2001–2011**		**1991–2001**	**2001–2011**
1	**2**	**3**	**4**	**5**	**6**	**7**	**8**	**9**
	INDIA	**1,02,87,37,436**	**1,21,01,93,422**	**21.54**	**17.64**	**−3.9**	**1.97**	**1.64**
1	Jammu and Kashmir	1,01,43,700	1,25,48,926	29.43	23.71	−5.72	2.61	2.15
2	Himachal Pradesh	60,77,900	68,56,509	17.54	12.81	−4.73	1.63	1.21
3	Punjab	2,43,58,999	2,77,04,236	20.10	13.73	−6.37	1.85	1.3
4	Chandigarh#	9,00,635	10,54,686	40.28	17.1	−23.18	3.44	1.59
5	Uttarakhand	84,89,349	1,01,16,752	20.41	19.17	−1.24	1.87	1.77
6	Haryana	2,11,44,564	2,53,53,081	28.43	19.9	−8.53	2.53	1.83
7	NCT of Delhi#	1,38,50,507	1,67,53,235	47.02	20.96	−26.06	3.93	1.92
8	Rajasthan	5,65,07,188	6,86,21,012	28.41	21.44	−6.97	2.53	1.96
9	Uttar Pradesh	16,61,97,921	19,95,81,477	25.85	20.09	−5.76	2.33	1.85
10	Bihar	8,29,98,509	10,38,04,637	28.62	25.07	−3.55	2.55	2.26
11	Sikkim	5,40,851	6,07,688	33.06	12.36	−20.7	2.9	1.17
12	Arunachal Pradesh	10,97,968	13,82,611	27.00	25.92	−1.08	2.42	2.33
13	Nagaland	19,90,036	19,80,602	64.53	−0.47	−65	5.11	−0.05
14	Manipur	22,93,896	27,21,756	24.86	18.65	−6.21	2.25	1.72
15	Mizoram	8,88,573	10,91,014	28.82	22.78	−6.04	2.57	2.07
16	Tripura	31,99,203	36,71,032	16.03	14.75	−1.28	1.5	1.39
17	Meghalaya	23,18,822	29,64,007	30.65	27.82	−2.83	2.71	2.49
18	Assam	2,66,55,528	3,11,69,272	18.92	16.93	−1.99	1.75	1.58
19	West Bengal	8,01,76,197	9,13,47,736	17.77	13.93	−3.84	1.65	1.31
20	Jharkhand	2,69,45,829	3,29,66,238	23.36	22.34	−1.02	2.12	2.04

(continued)

Table 11.4 (continued)

State/UT Code	India/State/Union Territory#	Total population		Percentage decadal growth		Change in percentage decadal growth	Average annual exponential growth rate	
		2001	**2011**	**1991–2001**	**2001–2011**		**1991–2001**	**2001–2011**
1	**2**	**3**	**4**	**5**	**6**	**7**	**8**	**9**
21	Orissa	3,68,04,660	4,19,47,358	16.25	13.97	−2.28	1.52	1.32
22	Chhattisgarh	2,08,33,803	2,55,40,196	18.27	22.59	4.32	1.69	2.06
23	Madhya Pradesh	6,03,48,023	7,25,97,565	24.26	20.3	−3.96	2.2	1.87
24	Gujarat	5,06,71,017	6,03,83,628	22.66	19.17	−3.49	2.06	1.77
25	Daman and Diu#	1,58,204	2,42,911	55.73	53.54	−2.19	4.53	4.38
26	Dadra and Nagar Haveli#	2,20,490	3,42,853	59.22	55.5	−3.72	4.76	4.51
27	Maharashtra	9,68,78,627	11,23,72,972	22.73	15.99	−6.74	2.07	1.49
28	Andhra Pradesh	7,62,10,007	8,46,65,533	14.59	11.1	−3.49	1.37	1.06
29	Karnataka	5,28,50,562	6,11,30,704	17.51	15.67	−1.84	1.63	1.47
30	Goa	13,47,668	14,57,723	15.21	8.17	−7.04	1.43	0.79
31	Lakshadweep#	60,650	64,429	17.30	6.23	−11.07	1.61	0.61
32	Kerala	3,18,41,374	3,33,87,677	9.43	4.86	−4.57	0.9	0.48
33	Tamil Nadu	6,24,05,679	7,21,38,958	11.72	15.6	3.88	1.11	1.46
34	Puducherry#	9,74,345	12,44,464	20.62	27.72	7.1	1.89	2.48
35	Andaman and Nicobar Islands#	3,56,152	3,79,944	26.90	6.68	−20.22	2.41	0.65

= Union Territory

11.4 Take Home Messages

(A) Population is a more or less a permanent aggregation of individuals of the same species inhabiting a specific geographic area at a given time. Population ecology is the study of individuals of a species constituting population regarding various factors affecting growth, distribution, natality, mortality, etc. Population of a species at a specific place is never static. It is extremely dynamic and depicts variation in its size and density with time due to influence of various abiotic and biotic factors.

(B) The accelerating pace of population growth in the last century was not due to any undue rise in birth rate of world population, but, because in the last century there was a sharp fall in death rate, advancements in healthcare control over fatal diseases such as small pox, plague and cholera and improved food distribution system (result of food security). The average number of children born to a mother has declined from 5 to 3.5 since 1950, though the size of world population has more than doubled during the same period.

(C) Population growth is the increase in number of individuals. The rate of growth is measured by an increase in the number of individuals in a population per unit time. For controlling the growth of population, natality and mortality are important factors. The percentage ratio of natality and mortality is termed vital index. Population growth is the result of interaction of biotic potential and environmental resistance. Two distinct patterns of population growth are identified, namely, S-shaped growth (Sigmoid) and J-shaped growth.

(D) The distribution of population in the world is highly non-uniform. Some areas are thickly populated while some are sparsely populated. There are various reasons behind uneven distribution of population that can be can be categorized in to geographical factors, social and cultural factors and economic factors. The geographical factors include favourable topography, availability of mineral and freshwater resources, favourable climate soil fertility and food security that invite population and help to flourish. The social and cultural factors encompass areas of better housing, education and health facilities and religious harmony. Places of religious and cultural significance also attract people, e.g. Varanasi, Jerusalem, etc. The economic factors invite population due to presence of more industries, transportation and communication facilities, employment opportunities, etc.

(E) Family planning policy has been adopted to give a check to population explosion. The net result of this family planning concept is visualized when one critically compares the average annual exponential growth rates between 1991–2001 and 2001–2011 for India. In all the States and Union Territories, the average annual exponential growth rate has reduced in 2001–2011 compared to 1991–2001. Still to sustain this rising population food security and economic security is very vital, which can be achieved through alternative livelihood.

11.5 Brain Churners

Population is a burning issue in underdeveloped and developing nations of the world. It is intricately related with industrialization and urbanization and has severe adverse effects on food security, health security and above all economic security. The readers can consider this chapter as a knowledge box on these important issues, which are threatening the mankind today. This section is a sort of '*self-appraisal approach*' as correct answers to all the questions are marked as black in the box at the end of this section.

1. Emigration may be defined as
   (a) Permanent exit of some individuals from local population
   (b) Temporary exit of some individuals from local population
   (c) Permanent entry of some individuals into local population
   (d) Temporary entry of some individuals into local population
2. Due to immigration the size of the local population
   (a) Decreases
   (b) Increases
   (c) Is not affected
   (d) None of the above
3. In population dynamics the, term migration involves
   (a) One-way movement
   (b) Exit from local population
   (c) Two-way movement
   (d) None of the above
4. Sex ratio in population study is the ratio of
   (a) Males: Females
   (b) Females: Males
   (c) Both a and b
   (d) None of the above
5. Pre-reproduction age comprises of
   (a) Adults
   (b) Old persons
   (c) Youths
   (d) Infants/adolescents
6. Triangular age pyramid in population study represents
   (a) Zero population growth
   (b) Negative population growth
   (c) Positive population growth
   (d) None of the above

7. Vital index in population growth is the percentage ratio of
   (a) Immigration and Emigration
   (b) Natality and Mortality
   (c) Migration and Immigration
   (d) Birth rate and Emigration

8. In the equilibrium phase of S-shaped population growth curve, the slope is
   (a) 0
   (b) 1
   (c) −1
   (d) $\sqrt{3}$

9. In 2050, the global population is expected to be around
   (a) 6 billion
   (b) 8 billion
   (c) 9.5 billion
   (d) 9.8 billion

10. In bell-shaped population age pyramid, the population growth is
    (a) 0
    (b) +ve
    (c) −ve
    (d) None of the above

11. Crash phase is a part of
    (a) S-shaped growth curve
    (b) J-shaped growth curve
    (c) Age pyramid
    (d) None of the above

12. In S-shaped growth curve the population increase during
    (a) Early or lag phase
    (b) Exponential or log phase
    (c) Diminishing phase
    (d) Stationary phase

13. Population is the permanent aggregation of individuals of
    (a) Same species
    (b) Different species
    (c) Same community
    (d) Different communities

14. Natality is defined as
    (a) Average rate of reproduction/unit time
    (b) Average rate of death/unit time
    (c) Average rate of accident/unit time
    (d) None of the above
15. The nature of population is
    (a) Static
    (b) Dynamic
    (c) Both a and b
    (d) None of the above
16. Biotic community is formed by populations of
    (a) Different species
    (b) Same species
    (c) Both a and b
    (d) None of the above
17. Population explosion is due to
    (a) High birth rate and low death rate
    (b) High death rate and low death rate
    (c) Equal death and birth rates
    (d) None of the above
18. In 2050 the least populated country in the World will be (predicted)
    (a) Nigeria
    (b) Vatican City
    (c) Indonesia
    (d) Bangladesh
19. Age ratio indicates the ratio of different age groups on the basis of
    (a) Their death
    (b) Their birth
    (c) Their ability to produce
    (d) None of the above
20. Fecundity is the
    (a) Maximum natality rate achieved under ideal condition
    (b) Minimum natality rate under ideal condition
    (c) Maximum death rate
    (d) None of the above

Answer Sheet

S. No.	a	b	c	d
1.	■			
2.		■		
3.	■			
4.		■		
5.				■
6.			■	
7.		■		
8.	■			
9.				■
10.	■			
11.		■		
12.	■			
13.	■			
14.	■			
15.		■		
16.	■			
17.	■			
18.		■		
19.			■	
20.	■			

Annexure 11A: Shrimp Feed Preparation from Seaweed

Introduction

The culture of black tiger shrimp *Penaeus monodon*, Fabricius 1798 (Crustacea: Decapoda) is practiced in the brackish water system of the Indian sub-continent with the aim to increase the fish production basically for human consumption. Feed is a significant factor in increasing the productivity and profitability in the aquaculture sector (Jamu and Ayinla 2003). From economic point of view, feed cost and feed management account for at least 60% of the production cost and appear to be one of the major constraints against the greater expansion of aquaculture (Kaushik 1990). Present knowledge and understanding of the environmental impacts of shrimp feed among aquaculturists and nutritionists are very low and needs further refinement. Feed with animal ingredients generate wastes of complex character in the culture system. In aquatic animals, seaweeds have been used as a dietary supplement for sea bass (Valente et al. 2006), snakehead (Hashim and Mat-Saat 1992) and shrimp (Moss 1994; Peñaflorida and Golez 1996; Cruz-Suarez et al. 2000). In some instances, the inclusion of algae in feed formulations has resulted in improved performance, including improved feed efficiency, pellet quality and animal product quality. Most nutritional studies with seaweed have investigated low-dietary inclusion rates (less than 80 g/kg) to establish their possible usefulness as functional (binder effect), nutritional and nutraceutical (health protective effects) supplements (Cruz-Suarez et al. 2009). The optimum inclusion level varies depending on algae or consumer species (Peñaflorida and Golez 1996; Cruz-Suarez et al. 2000; Suarez-Garcia 2006). A lot of studies demonstrated the antioxidant properties of the algal carotenoids and the role they play in preventing much pathology linked to oxidative stress (Okuzumi et al. 1993; Yan et al. 1999). The main carotenoids in red seaweed are the β-carotene and α-carotene and their dihydroxylated derivatives, viz. zeaxanthin and lutein.

The present paper is the first-hand approach to study the effect of red seaweed (*Catenella repens*)-based formulated feed on water quality, shrimp quality and production, survival rate and FCR in a shrimp culture unit of central Indian Sundarbans. This mangrove-dominated deltaic complex is located at the apex of Bay of Bengal and has been declared as the World Heritage Site by UNESCO (1987) on account of its rich taxonomic diversity.

Materials and Methods

Experimental Design and Layout

The study area (Fig. 11.A.1) for culturing shrimp (*P. monodon*) was selected in the central part of Indian Sundarbans in Canning block located in South 24 Parganas district of the state of West Bengal (22°16′40.6″ N latitude & 88°38′18.4″ E longitude) during April to July 2009. The culture site is located on the bank of Matla River, which has an average salinity of 8.5 psu. Two ponds were selected in the study site out of which one was treated as control (C) and the other was treated as experimental (E). The cultured species (*P. monodon*) in the experimental pond was provided with the seaweed-based formulated feed, and the control pond was provided with traditional feed. Good quality shrimp seeds obtained from a local shrimp farm were stocked after proper acclimatization with ambient environmental conditions. The stocking density was 5 PL_{20}/m^2 in both the control and experimental ponds. The shrimps were fed initially at 15% of the biomass in each pond and the ration was then adjusted to actual consumption every day, thus reducing uneaten feed to a minimum.

Fig. 11.A.1 Map showing the location of culture pond at Canning, Indian Sundarbans

Water quality parameters were analysed fortnightly for a 90 days culture period (from April 15 to July 15, 2009). The parameters remained well within the optimum throughout the trial. Dissolved oxygen (D.O.), pH, transparency and nutrients were analysed following the standard spectrophotometric method (Strickland and Parsons 1968, 1972). Phytopigment concentration (Chl *a*) was analysed as per the method (Jeffrey and Humphrey 1975). Organic carbon content of pond bottom soil was estimated by the standard titration method (Walkey and Black 1934).

Formulation of Shrimp Feed and Analysis of Biochemical Constituents

Several studies have been carried out in the development of formulated feed for shrimp under controlled culture system (Mohanty et al. 1995; Mukhopadhyay and Ray 1999, 2001; Khan et al. 2004; Biswas et al. 2006). The red seaweed *C. repens* was selected as the candidate flora (source of astaxanthin) for preparation of shrimp feed. A comparative account of traditional feed (commonly used in the study area) and seaweed-based formulated feed is given (Table 11.A.1). The proximate composition of the seaweed meal was determined using the methods of Lowry for protein (Lowry et al. 1951), Soxhlet for lipid (Tecator 1983) and Anthrone method for carbohydrate (Trevelyan and Harrison 1952). Astaxanthin was estimated as per the standard spectrophotometric method (Schuep and Schierle 1995).

Zootechnical Parameters and Statistical Analysis

Individual weights and lengths of shrimps were taken at fortnightly interval for 90 days culture period and the relevant response variables were determined for each control and experimental ponds. Condition Index (C.I.) was analysed at fortnightly interval during the culture period as per the expression; **C**.I. = $W/L^3 \times 100$, where W = weight of the cultured species (in gm) and L = length of the cultured species (in cm). Percentage weight gain was calculated as the difference in weight from the average final weight with respect to the initial weight; weight gain = [(average individual final weight – average individual initial weight)/average individual initial weight] × 100. Feed consumption reported was the total of the consumption estimated for 90 days period. The survival rate was measured as percentage of the difference of stocking number and production volume (No.) at the end of the culture period. Feed Conversion Ratio (FCR) was analysed after the harvesting of shrimps as per the expression: FCR = $\Delta f/\Delta b$, where, Δf = Change in feed biomass and Δb = Change in body biomass of the cultured species.

Table 11.A.1 Comparative analysis of shrimp feed components

Traditional shrimp feed	Percentage (%)	Specially formulated shrimp feed	Percentage (%)
Trash shrimp dust	20	Soybean oil cake	40
Dry fish dust	30		
Soybean oil cake	10		
Rice bran	12.5	Rice bran	15
Flour	7	Wheat bran	15
Groundnut dust	5	Mustard oil cake	12.5
Maize dust	4.5		
Coconut Khail	10	Coconut oil cake	12.5
Vitamins and minerals mixture	1.0	*Catenella repens* (as a source of carotenoid)	5

Body pigmentation was assessed for each treatment on shrimp cooked for 5 min in boiling water and comparing the orange-red colouration with Roche SalmoFan™ colour score. Analysis of variance (ANOVA) was computed between all the selected parameters (indicators of our experiment) considering both control and experimental ponds to evaluate the differences caused by inclusion of seaweed in the feed.

Results

Biochemical Composition of Seaweeds and Specially Formulated Feed

The biochemical composition of *C. repens* showed protein ranges between 4.01 ± 1.28% and 15.97 ± 1.17%; lipid ranges between 0.17 ± 0.02% and 0.24 ± 0.01%; carbohydrate values between 21.52 ± 1.87% and 35.74 ± 1.55% and astaxanthin values between 87.91 ± 2.67 ppm and 188.34 ± 2.89 ppm (Banejee et al. 2009). The seaweed was used as an ingredient (5%) of specially formulated feed. The proximate composition of traditional and seaweed-based formulated feed is highlighted (Table 11.A.2). Traditional feed showed higher protein, lipid and carbohydrate values than seaweed-based feed. However, astaxanthin content was almost nil in traditional feed.

Shrimp Growth and Pigmentation

Shrimps fed with *Catenella* diet exhibited higher final weights and better weight gain (Table 11.A.3) at the end of the experiment (25.5 gm final weight and 2450% weight gain) in comparison to control pond (20 gm final weight and 1900% weight gain). C.I. values of shrimp were also higher in experimental ponds (3.41 ± 4.68) than control pond (3.08 ± 4.52) (Table 11.A.3 and Fig. 11.A.2). The FCR value for control pond was 1.45 and for experimental pond was 1.39. The survival rate was found to be 55% in the control pond and 70% in experimental pond. The production volume was also maximum in case of experimental pond (69.97 kg) compared to control pond (20.10 kg) as given (Table 11.A.4). The present pilot scale study speaks in favour of healthy pond environment, better growth, higher survival rate and low FCR values through use of seaweed-based feed.

Table 11.A.2 Proximate composition of the traditional and formulated feed

Parameters	Traditional feed	Formulated feed
Protein (%)	32 ± 2.65	28.29 ± 0.58
Lipid (%)	4.7 ± 2.05	1.62 ± 0.03
Carbohydrate (%)	22.1 ± 4.08	17.25 ± 0.50
Astaxanthin (ppm)	BDL	62.33 ± 2.78

Values are the mean ± SD

Table 11.A.3 Zootechnical parameters recorded in the culture ponds

Parameters	Control pond	Experimental pond
Condition index of shrimp	3.08 ± 4.52*	3.41 ± 4.68*
% weight gain	1900	2450
Astaxanthin in shrimp (ppm)	9.52 ± 5.45*	11.32 ± 6.37*
Roche SalmoFan™ colour score	24	29
Survival rate (%)	55	70
Δ F	27.67	72.65
Δ B	19.09	52.14
FCR	1.45	1.39

*Values are the mean ± SD

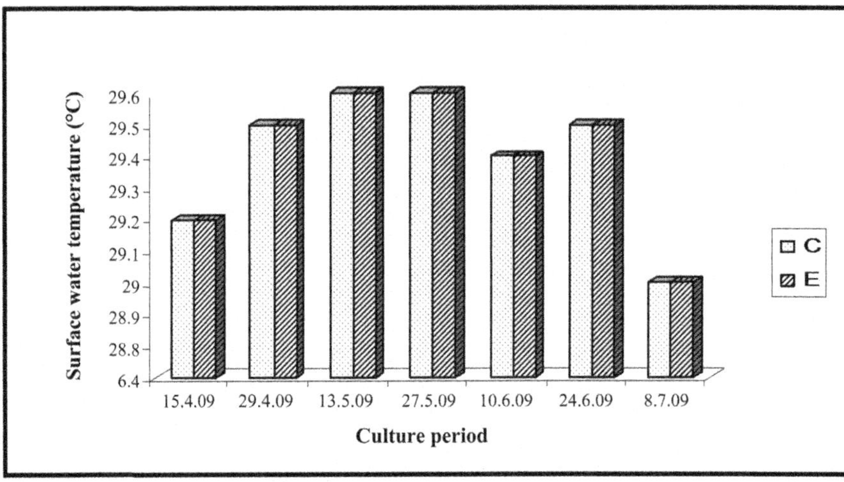

Fig. 11.A.2 Fortnightly variations of condition index of cultured shrimp from control (C) and experimental (E) ponds

An important factor governing the consumer acceptance and market value of many cultivated fish and shrimp species is the pink or red colouration of their flesh or boiled exoskeleton (Brun and Vidal 2006). In the wild, this colouration is achieved through the ingestion of carotenoid pigments particularly astaxanthin contained within invertebrate food organisms (Johnson et al. 1977; Ibrahim et al. 1984). The *Catenella*-based feed in the present study resulted in higher astaxanthin values in shrimps of experimental pond (11.32 ± 6.37 ppm) as reflected through darker orange-red colouration of shrimp exoskeleton in comparison to control pond (9.52 ± 5.45 ppm). Roche SalmoFan™ colour score showed the value of 24 in control pond, much lesser than experimental pond with a colour score of 29 (Table 11.A.3 and Fig. 11.A.3).

Table 11.A.4 Cost-benefit analysis of the culture ponds

	Items	Control pond	Experimental pond
Cost	Area (m^2)	365.40	784.08
	Stocking density @ 5 PL/ m^2	1827	3920
	Seed cost (in Rs.)	913.50	1960.00
	Feed given (in kg)	29.93	78.82
	Feed cost (in Rs.)	@ Rs. 35/– per kg = 1047.55	@ Rs. 19 per kg (excluding the cost of *C. repens*) = 1497.58
	Labour cost (supported by beneficiary himself)	Nil	Nil
	Experimental cost (in Rs.)	5000	5000
	Total cost (in Rs.)	6961.05	8457.58
	Total unit cost (in Rs./ m^2)	**19.05**	**10.79**
Benefit	Production (in kg)	@ 20 gm/ kg = 20.10	@ 25.5 gm/kg =69.97
	Return @ Rs. 350/kg	7035.00	24,489.50
	Expenditure (in Rs.)	6961.05	8457.58
	Total profit/pond (in Rs.)	73.95	16,031.92
	Profit/unit area (in Rs./ m^2)	**0.20**	**20.45**

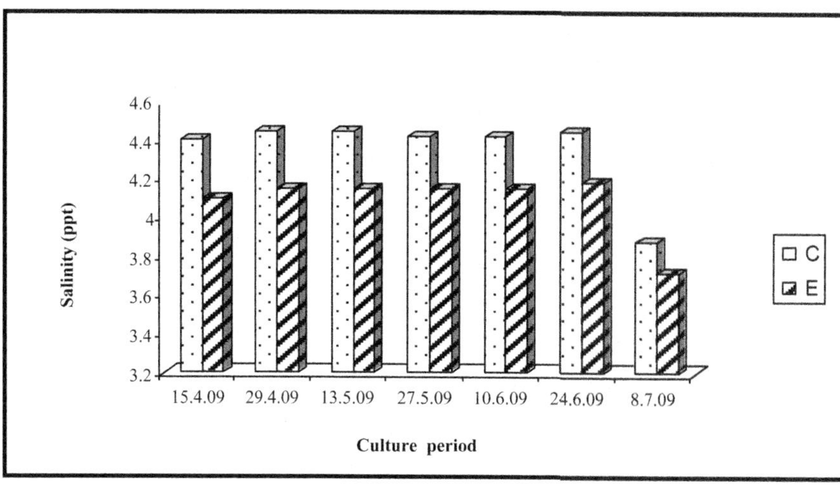

Fig. 11.A.3 Fortnightly variations of astaxanthin content of cultured shrimp from control (C) and experimental (E) ponds

Variation in Environmental Parameters

The surface water temperature during the study period ranged from 29.0 °C to 29.6 °C with a mean value of 29.4 ± 0.22 °C in both the culture ponds (Table 11.A.5 and Fig. 11.A.4). The surface water salinity ranged from 3.72 psu to 4.45 psu with

Table 11.A.5 Variation in the physico-chemical parameters of the culture ponds

Parameters	Control pond	Experimental pond
Surface water temperature (°C)	29.4 ± 0.22	29.4 ± 0.22
Surface water salinity (psu)	4.35 ± 0.2	4.08 ± 0.16
pH	7.88 ± 0.35	8.08 ± 0.03
Transparency (cm)	17.02 ± 2.02	24.3 ± 2.58
Dissolved oxygen (mgl^{-1})	4.66 ± 0.77	5.47 ± 0.12
Nitrate ($\mu gatl^{-1}$)	19.5 ± 1.21	16.6 ± 1.01
Phosphate ($\mu gatl^{-1}$)	2.22 ± 0.05	2.0 ± 0.29
Silicate ($\mu gatl^{-1}$)	64.01 ± 2.23	63.32 ± 2.67
Chlorophyll *a* (mgm^{-3})	1.74 ± 0.51	1.92 ± 0.16
Soil organic carbon (%)	1.18 ± 0.16	1.07 ± 0.1

Values are the mean ± SD

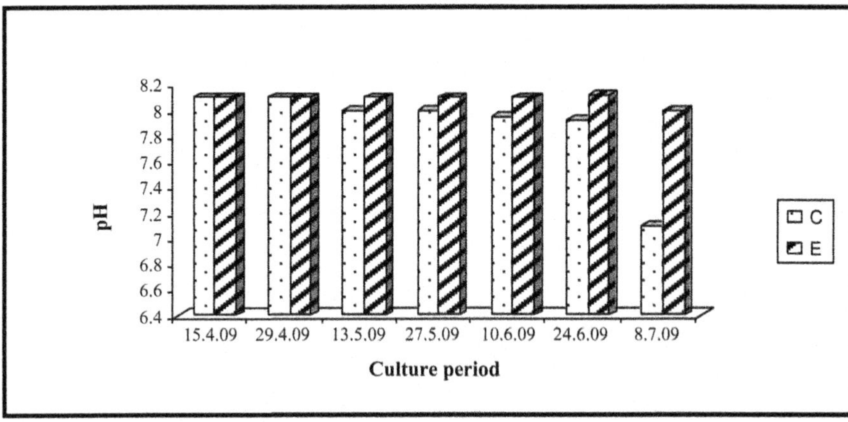

Fig. 11.A.4 Fortnightly variations of surface water temperature in control (C) and experimental (E) ponds

a mean value of 4.35 ± 0.2 psu in the control pond and 4.08 ± 0.16 psu in the experimental pond (Table 11.A.5 and Fig. 11.A.5). The surface water pH ranged from 7.10 to 8.12 with a mean value of 7.88 ± 0.35 in the control pond and 8.08 ± 0.03 in the experimental pond (Table 11.A.5 and Fig. 11.A.6). The D.O. values ranged from 3.08 mg l^{-1} to 5.58 mg l^{-1} with a mean value of 4.66 ± 0.77 mg l^{-1} in the control pond and 5.47 ± 0.12 mg l^{-1} in the experimental pond (Table 11.A.5 and Fig. 11.A.7). The water transparency during the study period ranged from 14.35 cm to 27.5 cm with a mean value of 17.02 ± 2.02 cm in the control pond and 24.3 ± 2.58 cm in the experimental pond (Table 11.A.5 and Fig. 11.A.8). The nutrient values (nitrate, phosphate and silicate) ranged from 15.12 μgat l^{-1} to 21.33 μgat l^{-1} with a mean

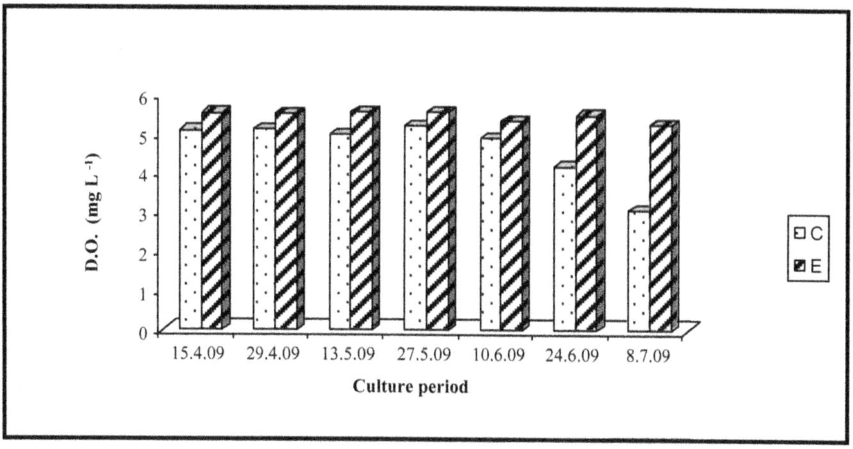

Fig. 11.A.5 Fortnightly variations of surface water salinity in control (C) and experimental (E) ponds

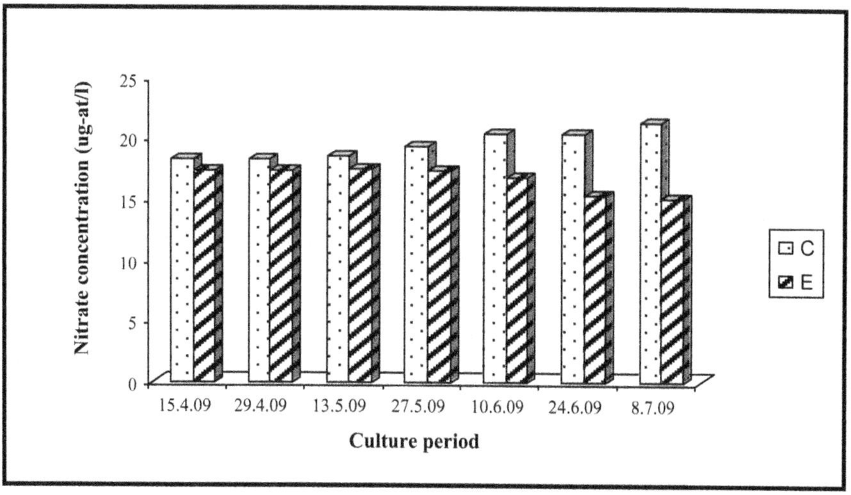

Fig. 11.A.6 Fortnightly variations of surface water pH in control (C) and experimental (E) ponds

value of 19.5 ± 1.21 μgat l^{-1} in the control pond and 16.6 ± 1.01 μgat l^{-1} in the experimental pond (Table 11.A.5 and Fig. 11.A.9) for nitrate, from 1.53 μgatl^{-1} to 2.28 μgatl^{-1} with a mean value of 2.22 ± 0.05 μgatl^{-1} in the control pond and 2.0 ± 0.29 μgatl^{-1} in the experimental pond (Table 11.A.5 and Fig. 11.A.10) for phosphate and from 59.63 μgatl^{-1} to 66.2 μgatl^{-1} with a mean value of 64.01 ± 2.23 μgatl^{-1} in the control pond and 63.32 ± 2.67 μgatl^{-1} in the experimental pond

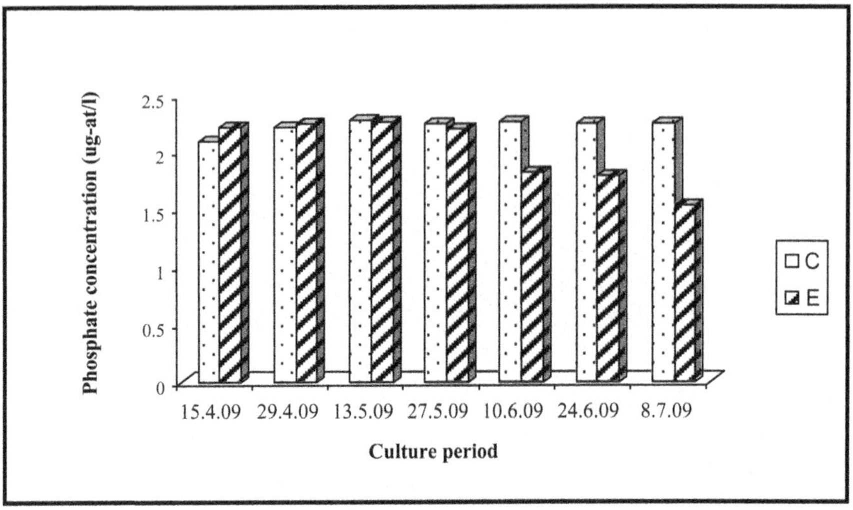

Fig. 11.A.7 Fortnightly variations of surface water dissolved oxygen in control (C) and experimental (E) ponds

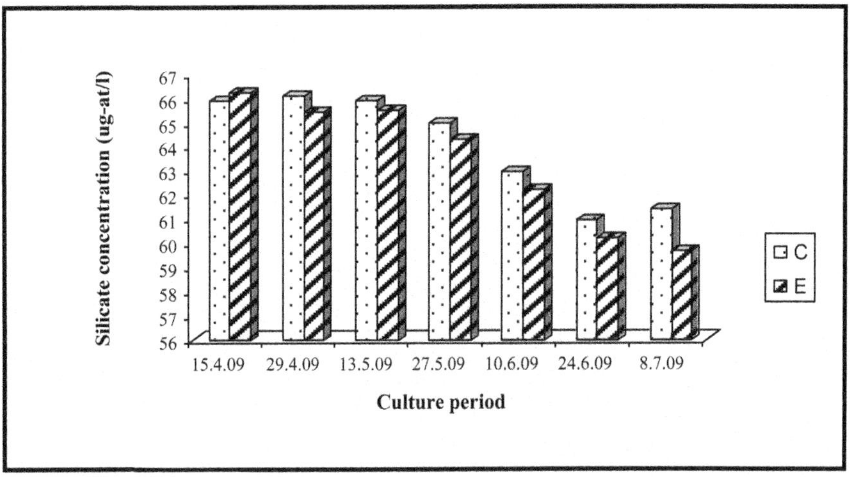

Fig. 11.A.8 Fortnightly variations of surface water transparency in control (C) and experimental (E) ponds

(Table 11.A.5 and Fig. 11.A.11) for silicate. The Chl *a* concentration during the study period ranged from 0.97 mg m^{-3} to 2.66 mg m^{-3} with a mean value of 1.74 ± 0.51 mg m^{-3} in the control pond and 1.92 ± 0.16 mg m^{-3} in the experimental pond (Table 11.A.5 and Fig. 11.A.12). The organic carbon of the pond bottom soil

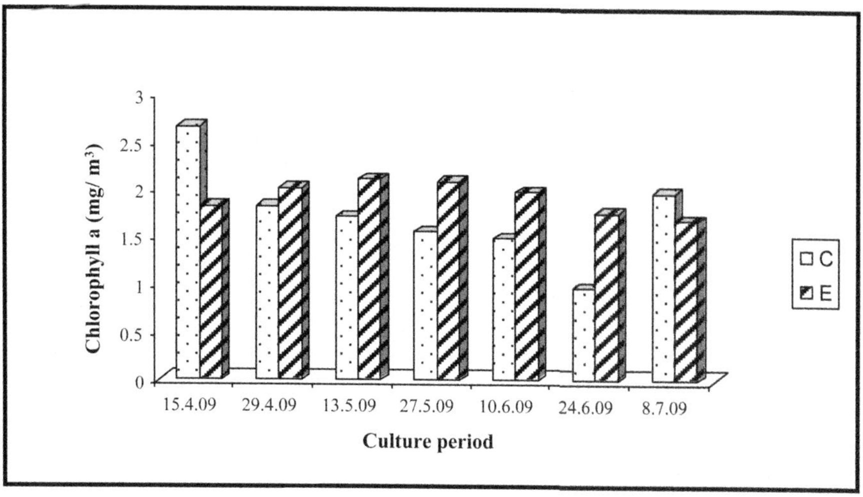

Fig. 11.A.9 Fortnightly variations of nitrate concentration in control (C) and experimental (E) ponds

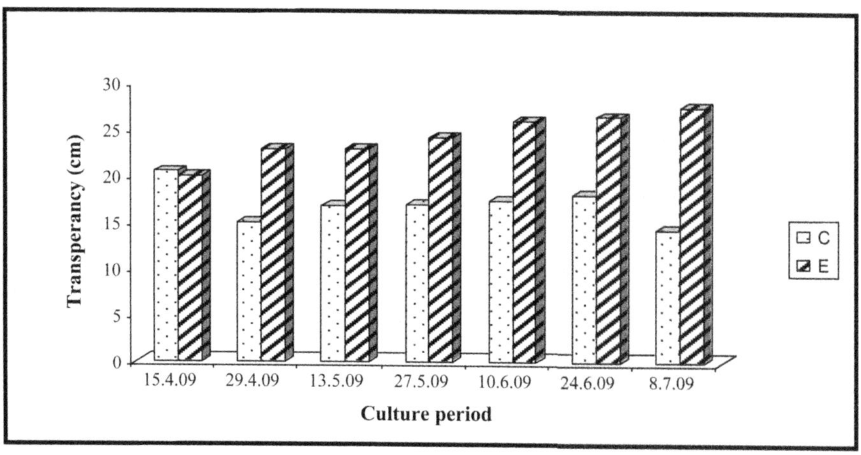

Fig. 11.A.10 Fortnightly variations of phosphate concentration in control (C) and experimental (E) ponds

during the study period ranged from 0.97% to 1.35% with a mean value of 1.18 ± 0.16% in the control pond and 1.07 ± 0.10% in the experimental pond (Table 11.A.5 and Fig. 11.A.13).

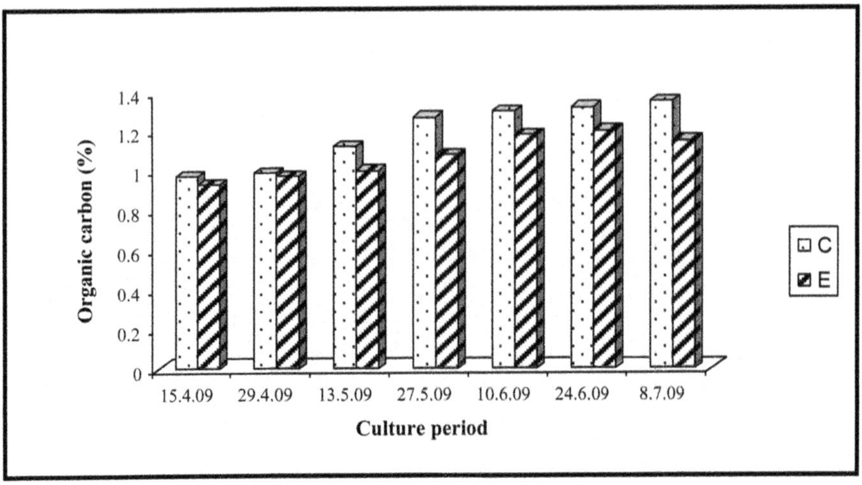

Fig. 11.A.11 Fortnightly variations of silicate concentration in control (C) and experimental (E) ponds

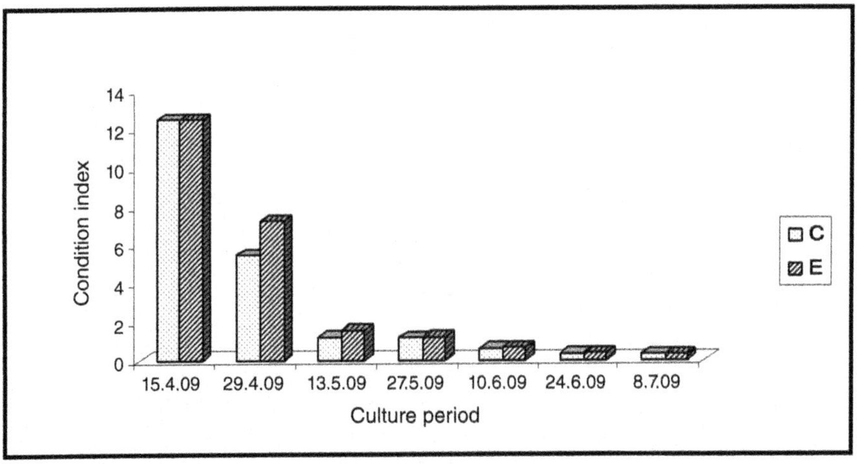

Fig. 11.A.12 Fortnightly variations of chlorophyll *a* in control (C) and experimental (E) ponds

Discussion

Feed Quality

Growth, health and reproduction of fish and other aquatic animals are primarily dependent upon an adequate supply of nutrient through feed both in terms of quality and quantity, irrespective of the culture system in which they are grown (Kader et al.

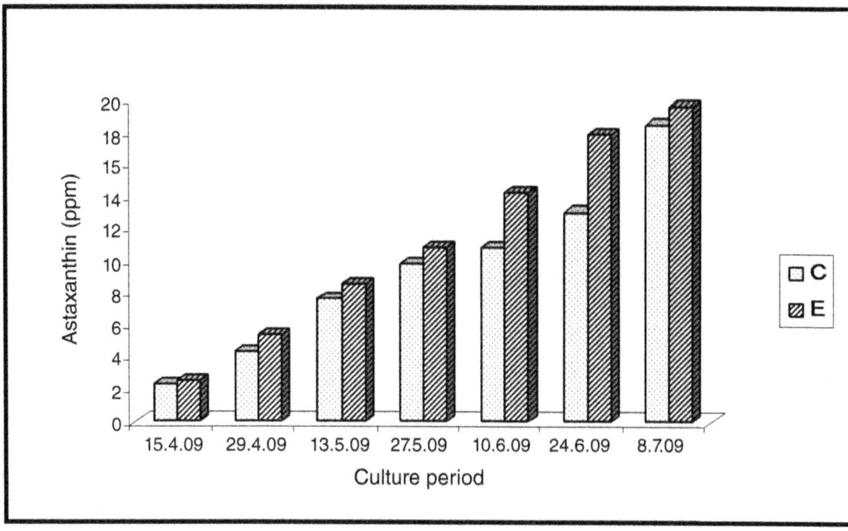

Fig. 11.A.13 Fortnightly variations of organic carbon in control (C) and experimental (E) ponds

2005). Therefore, supply of inputs (feeds, fertilizers etc.) has to be ensured so that the nutrients and energy requirements of the species under cultivation are met and the production goals of the system are achieved (Hasan 2001). The nutrient balance of feed influences feed utilization and growth of fish. It is very essential to know the nutritional requirements particularly for protein, lipid and energy for optimum growth of fish species as well as in formulating a balanced diet. Dietary protein and energy levels are known to influence the growth and body composition of fish (Lovell 1989). Improper protein and energy levels in feed increase fish production cost and deteriorates water quality. Insufficient energy in diets causes protein waste due to the increase proportion of dietary protein used for energy and the produced ammonia can reduce the water quality (Phillips 1972; Prather and Lovell 1973; Shyong et al. 1998). On the contrary, excessive energy in diets can lead to increased body lipid deposition and growth reduction because of lack of necessary nutrient for growth (Daniels and Robinson 1986; Van der Meer et al. 1997). This is often the case seen in case of traditional meal which has resulted in deterioration of water quality parameters in the control pond (Table 11.A.5). In the present study, the specially formulated feed prepared from *C. repens* had optimum protein, lipid, carbohydrate and astaxanthin content and the values were also within the recommended nutrient levels for fish (National Research Council 1983).

Protein is the major growth promoting factor in feed. The protein requirement of fish are influenced by various factors such as fish size, water temperature, feeding rate, availability and quality of natural foods and overall digestible energy content of diet (Satoh 2000; Wilson 2000). In the present study, protein level in formulated feed was found to be 28.29 ± 0.58% in comparison to 32 ± 2.65% in traditional feed. Lipids are primarily included in formulated diet to maximize their protein sparing effect (Hasan

2001) by being a source of energy. Low lipid content of the specially formulated feed results in greater binding efficiency and more pellet water stability. Lipid values were found to be lower in specially formulated feed indicating the efficiency of the present feed in maintaining water stability. Carbohydrate levels in feed are the most suitable sources of energy for shrimps. The best carbohydrate sources of shrimp feed are those derived from low-cost practical ingredients (e.g. wheat flour, rice bran, etc.). The formulated feed in the present study also supports the above view (Tables 11.A.1 and 11.A.3). Use of *C. repens* as astaxanthin supplement (source of carotenoid) has increased the antioxidative property of the feed and resulted in higher pigment concentration in shrimp species of experimental pond (Tables 11.A.3 and 11.A.2).

Growth Performance of Shrimp

The present case study exhibited significant weight gain in experimental pond compared to control pond ($p < 0.05$). The significant variation of C.I. with time is due to the growth in biomass with the passage of time. Feeding the shrimps (*P. monodon*) supplemented with *Catenella* meal resulted in better growth and survival rate in the present programme. The percentage increase in biomass was also higher in experimental pond than control pond (Table 11.A.3). This was also revealed by the low FCR values for experimental pond than control pond, which proves the acceptance of the specially formulated feed by the shrimp species. Cost-benefit analysis (CBA) has also confirmed higher profit percentage (Table 11.A.4).

Pigmentation

Colour development depends on the carotenoid content of the feed (Moretti et al. 2006) although it has been reported that dietary carotenoids are responsible for less than 20% of the flesh pigmentation in aquatic organisms (Torrisen et al. 1989; Storebakken and No 1992). Carotenoid pigmentation is affected by dietary pigment source, dosage level, duration of feeding, dietary composition, degree of carotenoid esterification, etc. (Meyers and Latscha 1997; Bjerkeng 2000; Buttle et al. 2001; Gomes et al. 2002; White et al. 2002). Animals including fish and shrimp are unable to synthesize carotenoids, but certain aquaculture species (i.e. crustaceans, omnivorous/herbivorous fish) are capable of transforming ingested carotenoids such as $\propto$ – carotene and depositing the resulting end products, usually astaxanthin in their tissues (Simpson and Chichester 1981; NRC 1983). ANOVA results showed a significant variation of astaxanthin between the control and experimental ponds ($p < 0.05$) which may be attributed to astaxanthin level in feed types. The seaweed-based feed had an astaxanthin of 62.33 ± 2.78 ppm, whereas the astaxanthin level in traditional feed was BDL. Red seaweed *C. repens* is a rich source of carotenoid (Chakraborty and Santra 2008; Banejee et al. 2009) and hence may be the cause for high astaxanthin level in the species of experimental pond. Pigmentation of muscle is a major quality attributed for shrimps (Brun and Vidal 2006; Darachai et al. 1998).

Coloration of muscles in shrimps using seaweed such as *C. repens* as a natural pigment source may enhance the potential of seaweed inclusion in shrimp feed and may perhaps replace or reduce artificial colourants currently used by the industry (Banerjee 2009; Nickell and Bromage 1998).

Water Quality of the Cultured Ponds

Aquatic parameters of the shrimp culture ponds is a reflection of the quality of feed provided to the cultured species, and the condition index values symbolize the suitability of the environment for the species (Maciena and Murphey 1988). Surface water temperature in both the culture ponds showed more or less parallel trend of variation throughout the study period. This is reflected in the ANOVA results which showed no significant variation of the parameter between the control and experimental ponds as well as between the culture periods. The uniformity in temperature profile is due to the location of both the ponds in the same site that experience similar weather and climate. Water temperature plays a major role in shrimp enzyme kinetics which may have a regulatory influence on their growth (Mitra et al. 2006). It also affects the process of moulting during the post larval stage of shrimps (WWF-India 2006). In the present study, no significant relationship was observed between condition index of shrimp and surface water temperature.

The salinity of the Hugli-Matla estuarine complex is known to exhibit intensive variations (Saha et al. Saha et al. 1995). The selected station, Canning is located in the Matla estuarine stretch, which is known for its dynamics in tidal condition. ANOVA results showed significant spatial (between ponds) and temporal (between months) variation of salinity ($p < 0.05$). The relatively higher C.I. values in the experimental pond with less salinity (4.08 ± 0.16 psu) prove the efficiency of formulated feed in combating the stress posed by salinity which in turns increases the astaxanthin level in the cultured species (Kobayashi et al. 1992; Tjahjono et al. 1994; Harker et al. 1996; Boussiba 2000; Sarada et al. 2002). Shrimp culture directly affects the pH of the pond bottom through deposition of excess feed, shrimp excreta, dead shrimps, etc. These shift the soil and overlying aquatic pH towards acidic condition. In the present study, such condition was not observed at culture site owing to the traditional practice of liming at a regular interval of time in the lower Gangetic region. ANOVA result also confirmed the above view as no significant variations of pH between the ponds was observed. High organic carbon load was observed in the control pond due to use of traditional feed resulting in low pH. ANOVA result shows significant variation of organic carbon content between the ponds ($p < 0.05$). D.O. is a vital parameter regulating the aquatic life. The shrimp health is a direct function of dissolved oxygen and its diurnal variation. Excessive organic load in the system results in lowering the D.O. value posing threat to the survival of aquatic life. In the present study, the D.O. level in the control pond showed lower value owing to deposition of organic carbon at the bottom of the pond. ANOVA result confirmed the significant variation of D.O. between the ponds. Transparency controls the phytoplankton standing stock in shrimp culture ponds due to their dependency on the solar radiation for photosynthesis. This parameter has thus a major role in regulating the phytopigment level and coloration of shrimp pond water. The formulated feed provided in the experimental pond showed increased transparency of the water due to its unique binding property. This upgraded the water quality as reflected by the high-condition index values of the shrimps in the experimental pond. The above statement was confirmed by ANOVA results which showed significant variation of transparency between the ponds ($p < 0.05$). The ready acceptance of the seaweed-based feed by the cultured species in the experimental pond may be the basis of reduced suspended particulate matter in aquatic phase that caused variation in water transparency.

Nutrients comprising of nitrate, phosphate and silicate in the aquatic phase of the culture ponds are generated through the excretory products of the cultured species, left over feed and also by the churning of the pond bed (due to runoff from the adjacent land masses). ANOVA results showed significant variation of nitrate between the ponds ($p < 0.05$) which may be due to the leaching of the feed ingredients (particularly from animal component in traditional feed) in pond water and also the faecal matter that generates ammonia (Mitra and Choudhury 1995). ANOVA results for the phosphate concentration during the study period showed no significant variation between the ponds. For silicate the variation between the ponds was significant ($p < 0.05$) which may be attributed to difference in bed materials of the ponds. Phytopigment (Chl *a*) are indicators of aquatic productivity and standing stock of phytoplankton. Although higher concentration of phytopigment signify eutrophic condition of water, but their presence in optimum level is healthy for shrimp growth as the phytoplankton constitute the natural diet of shrimp. In the present study, an optimum Chl *a* concentration and lower nutrient values in the experimental pond proves the effective utilization of nutrients by aquatic phytoplankton. ANOVA results showed no significant variation of Chl *a* between the ponds as well as between the culture periods, implying healthy growth of the cultured species.

Aquaculture has become a peak industry in the present millennium, which involves seafood farming with shrimp, cuttlefish, squid, lobster and other such culinary delights actually 'cultivated' in aquatic enclosures under scientifically controlled conditions (Rajkhowa 2005). The use of nutrient-rich feed continues to gain wide acceptance in the aquaculture industry in order to boost up the quality of the aquacultural products. The use of quality feed results in substantial reduction in the overall variable cost of an operation through improved animal performance, better FCR and improved water quality due to a reduction in the amount of nutrients and solids (i.e. faeces and uneaten food) in the waste water effluent. *C. repens*-based formulated feed showed better growth performance of the cultured species with respect to condition index values and survival rate. Body pigmentation improved in the cultured species of experimental pond and showed significantly higher astaxanthin level than the controlled pond. cost-benefit analysis (CBA) reflects high profitability of using floral-based feed instead of the traditional feed. A series of experiments are still needed for time testing the results and make the programme sustainable for the poor island dwellers of lower Gangetic delta.

Annexure 11B: Countries, Territories or Areas with Reported Laboratory-Confirmed COVID-19 Cases and Deaths; Data as of 19 March 2020

Reporting country/ territory/area	Total confirmed cases	Total confirmed new cases	Total deaths	Total new deaths	Transmission classification
Western Pacific region					
China	81,174	58	3242	11	Local transmission
Republic of Korea	8413	93	84	3	Local transmission
Japan	873	44	29	1	Local transmission
Malaysia	673	120	2	2	Local transmission
Australia	510	96	6	1	Local transmission
Singapore	313	47	0	0	Local transmission
Philippines	187	0	14	2	Local transmission
Viet Nam	66	5	0	0	Local transmission
Brunei Darussalam	56	2	0	0	Local transmission
Cambodia	35	11	0	0	Local transmission
New Zealand	20	9	0	0	Local transmission
Mongolia	5	1	0	0	Imported cases only
Territories					
Guam	5	2	0	0	Local transmission
French Polynesia	3	0	0	0	Imported cases only
European region					
Italy	35,713	4207	2978	473	Local transmission
Spain	13,716	2538	598	107	Local transmission
France	9043	0	244	0	Local transmission
Germany	8198	1042	13	0	Local transmission

(continued)

Reporting country/ territory/area	Total confirmed cases	Total confirmed new cases	Total deaths	Total new deaths	Transmission classification
Switzerland	3010	353	21	2	Local transmission
The United Kingdom	2630	672	103	0	Local transmission
Netherlands	2051	0	58	0	Local transmission
Austria	1646	314	4	1	Local transmission
Belgium	1486	0	14	0	Local transmission
Norway	1423	115	3	0	Local transmission
Sweden	1279	112	3	0	Local transmission
Denmark	1044	67	4	0	Local transmission
Portugal	642	194	2	1	Local transmission
Czechia	522	30	0	0	Local transmission
Israel	427	0	0	0	Local transmission
Greece	418	0	5	0	Local transmission
Finland	359	40	0	0	Local transmission
Ireland	292	0	2	0	Local transmission
Poland	287	0	5	0	Local transmission
Slovenia	286	0	1	0	Local transmission
Estonia	258	33	0	0	Local transmission
Iceland	250	25	0	0	Local transmission
Romania	246	62	0	0	Local transmission
Luxembourg	210	63	2	1	Local transmission
Turkey	191	51	2	1	Local transmission
Russian Federation	147	54	0	0	Imported cases only
San Marino	109	5	14	3	Local transmission

(continued)

Reporting country/ territory/area	Total confirmed cases	Total confirmed new cases	Total deaths	Total new deaths	Transmission classification
Slovakia	105	8	0	0	Local transmission
Serbia	96	11	0	0	Local transmission
Bulgaria	92	11	2	0	Local transmission
Armenia	84	32	0	0	Local transmission
Croatia	81	16	0	0	Local transmission
Latvia	71	11	0	0	Imported cases only
Albania	59	2	2	1	Local transmission
Cyprus	58	25	0	0	Local transmission
Hungary	58	8	1	0	Local transmission
Malta	48	10	0	0	Imported cases only
Belarus	46	10	0	0	Local transmission
Georgia	38	4	0	0	Imported cases only
Bosnia and Herzegovina	36	7	0	0	Local transmission
Kazakhstan	36	3	0	0	Imported cases only
North Macedonia	36	5	0	0	Local transmission
Republic of Moldova	36	0	0	0	Local transmission
Azerbaijan	34	13	1	1	Imported cases only
Lithuania	26	1	0	0	Imported cases only
Liechtenstein	25	18	0	0	Imported cases only
Ukraine	16	7	2	0	Local transmission
Uzbekistan	16	0	0	0	Imported cases only
Monaco	9	0	0	0	Under investigation

(continued)

Reporting country/ territory/area	Total confirmed cases	Total confirmed new cases	Total deaths	Total new deaths	Transmission classification
Kyrgyzstan	3	3	0	0	Under investigation
Montenegro	2	0	0	0	Imported cases only
Holy See	1	0	0	0	Under investigation
Territories					
Faroe Islands	58	11	0	0	Imported cases only
Andorra	39	23	0	0	Imported cases only
Gibraltar	8	5	0	0	Under investigation
Jersey	5	0	0	0	Imported cases only
Greenland	2	0	0	0	Under investigation
Guernsey	1	0	0	0	Imported cases only
Southeast Asia region					
Indonesia	227	55	19	14	Local transmission
Thailand	212	35	1	0	Local transmission
India	151	14	3	0	Local transmission
Sri Lanka	42	13	0	0	Local transmission
Maldives	13	0	0	0	Local transmission
Bangladesh	10	2	0	0	Local transmission
Bhutan	1	0	0	0	Imported cases only
Nepal	1	0	0	0	Imported cases only
Eastern Mediterranean region					
Iran	17,361	1192	1135	147	Local transmission
Qatar	442	0	0	0	Local transmission
Bahrain	256	5	1	0	Local transmission
Pakistan	241	54	0	0	Imported cases only

(continued)

Reporting country/ territory/area	Total confirmed cases	Total confirmed new cases	Total deaths	Total new deaths	Transmission classification
Saudi Arabia	238	67	0	0	Local transmission
Egypt	196	30	6	2	Local transmission
Iraq	164	0	12	0	Local transmission
Kuwait	142	12	0	0	Local transmission
Lebanon	133	13	4	1	Local transmission
United Arab Emirates	113	15	0	0	Local transmission
Jordan	52	13	0	0	Imported cases only
Morocco	49	11	2	0	Local transmission
Oman	33	9	0	0	Imported cases only
Tunisia	29	5	0	0	Local transmission
Afghanistan	22	0	0	0	Imported cases only
Djibouti	1	1	0	0	Under investigation
Somalia	1	0	0	0	Imported cases only
Sudan	1	0	1	0	Imported cases only
Territories					
occupied Palestinian territory	44	3	0	0	Local transmission
Region of the Americas					
United States of	7087	3551	100	42	Local transmission
America					
Canada	569	145	8	7	Local transmission
Brazil	291	57	1	1	Local transmission
Chile	238	82	0	0	Local transmission
Ecuador	155	97	2	0	Local transmission
Peru	145	59	0	0	Local transmission

(continued)

Reporting country/ territory/area	Total confirmed cases	Total confirmed new cases	Total deaths	Total new deaths	Transmission classification
Colombia	93	48	0	0	Local transmission
Mexico	93	11	0	0	Imported cases only
Panama	86	17	1	0	Local transmission
Argentina	79	14	2	0	Local transmission
Costa Rica	50	9	0	0	Local transmission
Venezuela	36	3	0	0	Imported cases only
Uruguay	29	23	0	0	Imported cases only
Dominican Republic	21	0	1	0	Local transmission
Jamaica	13	1	0	0	Local transmission
Bolivia	12	1	0	0	Imported cases only
Paraguay	11	2	0	0	Local transmission
Cuba	10	5	1	0	Local transmission
Honduras	9	1	0	0	Imported cases only
Trinidad and Tobago	7	2	0	0	Imported cases only
Guatemala	6	0	1	0	Imported cases only
Guyana	4	1	1	0	Local transmission
Bahamas	3	2	0	0	Local transmission
Barbados	2	2	0	0	Imported cases only
Saint Lucia	2	0	0	0	Imported cases only
Antigua and Barbuda	1	0	0	0	Imported cases only
Montserrat	1	1	0	0	Imported cases only
Saint Vincent and the Grenadines	1	0	0	0	Imported cases only
Suriname	1	0	0	0	Imported cases only

(continued)

Reporting country/ territory/area	Total confirmed cases	Total confirmed new cases	Total deaths	Total new deaths	Transmission classification
Territories					
Guadeloupe	33	15	0	0	Imported cases only
Martinique	23	7	0	0	Imported cases only
French Guiana	11	4	0	0	Imported cases only
Puerto Rico	5	2	0	0	Imported cases only
Aruba	4	2	0	0	Imported cases only
Saint Martin	4	2	0	0	Under investigation
Curaçao	3	0	0	0	Imported cases only
Saint Barthélemy	3	0	0	0	Under investigation
United States Virgin Islands	2	0	0	0	Imported cases only
Cayman Islands	1	0	1	0	Imported cases only
African region					
South Africa	116	54	0	0	Local transmission
Algeria	72	12	6	2	Local transmission
Senegal	36	9	0	0	Local transmission
Burkina Faso	26	6	1	1	Imported cases only
Rwanda	11	4	0	0	Local transmission
Cameroon	10	5	0	0	Local transmission
Cote d'Ivoire	9	3	0	0	Imported cases only
Ghana	9	1	0	0	Imported cases only
Nigeria	8	6	0	0	Imported cases only
Democratic Republic of the Congo	7	4	0	0	Local transmission
Kenya	7	4	0	0	Local transmission

(continued)

Reporting country/ territory/area	Total confirmed cases	Total confirmed new cases	Total deaths	Total new deaths	Transmission classification
Ethiopia	6	1	0	0	Imported cases only
Seychelles	6	2	0	0	Imported cases only
Congo	3	2	0	0	Imported cases only
Equatorial Guinea	3	2	0	0	Imported cases only
Gabon	3	2	0	0	Imported cases only
Mauritius	3	3	0	0	Under investigation
United Republic of Tanzania	3	2	0	0	Imported cases only
Liberia	2	1	0	0	Local transmission
Mauritania	2	1	0	0	Imported cases only
Namibia	2	0	0	0	Imported cases only
Zambia	2	2	0	0	Imported cases only
Benin	1	0	0	0	Imported cases only
Central African Republic	1	0	0	0	Imported cases only
Eswatini	1	0	0	0	Imported cases only
Gambia	1	1	0	0	Imported cases only
Guinea	1	0	0	0	Imported cases only
Togo	1	0	0	0	Imported cases only
Territories					
Réunion	12	3	0	0	Imported cases only
Mayotte	3	2	0	0	Imported cases only
Subtotal for all regions	**209,127**	**16,556**	**8771**	**828**	
International conveyance (Diamond Princess)	712	0	7	0	Local transmission
Grand total	**209,839**	**16,556**	**8778**	**828**	

Case classifications are based on WHO case definitions for COVID-19
Source: https://www.worldometers.info/coronavirus/

References

Banerjee K (2009) Income generation through eco-friendly pisciculture in Indian Sundarbans: an innovative approach towards sustainable green technology. Report of WOS-B Scheme, Department of Science and Technology, Government of India, pp 1–20

Banejee K, Ghosh R, Homechaudhuri S, Mitra A (2009) Seasonal variation in the biochemical composition of red seaweed (*Catenella repens*) from Gangetic delta, northeast coast of India. J Earth Syst Sci 118(5):1–10. Springer-Verlag

Biswas G, Jena JK, Singh SK, Patmajhi P, Muduli HK (2006) Effect of feeding frequency on growth, survival and feed utilization in mrigal, *Cirrhinus mrigala* and rohu, *Labeo rohita* during nursery rearing. Aquaculture 254:211–218

Bjerkeng B (2000) Carotenoid pigmentation in salmonid fishes – recent progress. In: Cruz-Suarez LE, Ricque-Marie D, Tapia-Salazar M, Olvera-Novoa MA, Cerecedo-Olvera R (eds) Avances en Nutricion Acuicola V – Memorias del Quinto Simposium Internacional de Nutricion Acuícola, Merida, Mexico, 19–22 Noviembre 2000, vol 5. Universidad Autonoma de Nuevo Leon, Monterrey, pp 71–89. ISBN 970-694-52-9

Boussiba S (2000) Carotenogenesis in the green algae *Haematococcus pluvialis*: Cellular physiology and stress response. Plant Physiol 108:111–117

Brun HL, Vidal F (2006) Shrimp pigmentation with natural carotenoids. Feed Technology, Aquaculture Asia-Pacific Magazine, pp 34–35

Buttle LG, Crampton VO, Williams PD (2001) The effect of feed pigment type on flesh pigment deposition and colour in farmed Atlantic salmon, *Salmo salar* L. Aquac Res 32:103–111

Chakraborty S, Santra SC (2008) Biochemical composition of eight benthic algae collected from Sunderban. Indian J Mar Sci 37(3):329–332

Cruz-Suarez LE, Antimo-Perez JS, Luna-Mendoza N, Tapia-Salazar M, Ricque-Marie D, (2000) Relaciones proteına renergıa y proteına vegetalranimal optimas en alimentos de engorda para Litopenaeus Õannamei y L. stylirostris. In: Cruz-Suarez, L.E., Ricque-Marie, D., Tapia-Salazar, M., Olvera-Novoa, M.A., Civera-Cerecedo, R. Eds., Advances en Nutricion Acuıcola V—Memorias del Quinto Simposio Ž. Internacional de Nutricion Acuıcola, Merida, Yucatan, 19–20 Noviembre de 2000. Universidad Autonoma de Nuevo Leon, San Nicolas de los Garza, Mexico, compact disk in press

Cruz-Suárez LE, Tapia-Salazar M, Nieto-López MG, Guajardo-Barbosa C, Ricque-Marie D (2009) Comparison of *UlvA0020clathrata* and the kelps *Macrocystis pyrifera* and *Ascophyllum nodosum* as ingredients in shrimp feeds. Aquac Nutr 15:421–430

Daniels WH, Robinson EH (1986) Protein and energy requirements of juvenile red drum (*Sciaenops ocellatus*). Aquaculture 53:243–252

Darachai J, Piyatiratitivorakul S, Kittakoop P, Nitithamyong C, Menasveta P (1998) Effect of astaxanthin on larval growth and survival of giant tiger prawn, *Penaeus monodon*. In: Flegel TW (ed) Advances in shrimp biotechnology. National Center for Genetic Engineering and Biotechnology, Bangkok, pp 117–121

Gomes E, Dias J, Silva P, Valente L, Empis J, Gouveia L, Bowen J, Young A (2002) Utilization of natural and synthetic sources of carotenoids in the skin pigmentation of gilthead seabream (*Sparus aurata*). J Eur Food Res Technol 214:287–293

Harker M, Tsavalos AJ, Young AJ (1996) Factors responsible for astaxanthin formation in the chlorophyte *Haematococcus pluvialis*. Bioresour Technol 55:207–214

Hasan MR (2001) Nutrition and feeding for sustainable aquaculture development in the third millennium. In: Aquaculture in the third millennium. Technical proceedings of the conference on aquaculture in the third millennium, Bangkok, Thailand, 20–25 February 2000. NACA, Bangkok and FAO, Rome, pp 193–219

Hashim R, Mat-Saat NA (1992) The utilization of seaweed meals as binding agents in pellet feeds for snakehead (*Channa striatus*) fry and their effects on growth. Aquaculture 108:299–308

Ibrahim A, Shimizu C, Kono M (1984) Pigmentation of culture red sea bream, *Chrysophrys major*, using astaxanthin from Antarctic krill, *Euphausia superba*, and a mysid, *Neomysis* sp. Aquaculture 38:45–57

Jamu DM, Ayinla OA (2003) Potential for the development of aquaculture in Africa. Naga 26:9–13

Jeffrey SW, Humphrey GR (1975) New spectrophotometric equations for determining chlorophyll a, b, c1 and c2 in higher plants, algae and natural phytoplankton. Biochem Physiol Pflanz Bd 167:191–194

Johnson EA, Conklin DE, Lewis MJ (1977) The yeast *Phaffia rhodoyma* as a dietary pigment source for salmonids and crustaceans. J Fish Res Board Can 34:2417–2421

Kader MA, Hossain MA, Hasan MR (2005) A survey of the nutrient composition of some commercial fish feeds available in Bangladesh. Asian Fish Sci 18:59–69

Kaushik S (1990) Use of alternative protein resources for the intensive rearing of carnivorous fish. In: Flos R, Tort L, Torres P (eds) Mediterranean aquaculture. Hellis Horwood Ltd, Chichester, pp 125–138

Khan MA, Jafri AK, Chadha NK (2004) Growth and body composition of rohu (*L. rohita*) fed compound diet: winter feeding and rearing to marketable size. J Appl Ichthyol 20:265–270

Kobayashi M, Kakizono T, Nishio N, Nagai S (1992) Effects of light intensity, light quality and illumination cycle on astaxanthin formation in a green alga *Haematococcus pluvialis*. J Ferment Bioeng 74:61–63

Lovell RT (1989) Nutrition and feeding of fish. Van Nostrand Reinhold, New York

Lowry OH, Rosebrough NJ, Farr AL, Randell RJ (1951) Protein measurements with folin-phenol reagent. J Biol Chem 193:1433–1437

Maciena MJ, Murphey BR (1988) Variation in the weight to length relationship among Florida and Northern large mough bass and their interspecific F1 hybrid. Trans Am Fish Soc 117:232–237

Meyers SP, Latscha T (1997) Carotenoids. In: DAbramo LRD, Conclin DE, Akiyama DM (eds) Crustacean nutrition, Advances in world aquaculture, vol 6. The World Aquaculture Society, Baton Rouge, pp 164–193

Mitra A, Choudhury A (1995) Causes of water pollution in prawn culture farms. Jr of Ind Ocen Stud 2(3):230–235

Mitra A, Chakraborty R, Banerjee K, Banerjee A, Mehta N, Berg H (2006) Study on the water quality of the shrimp culture ponds in Indian Sundarbans. Indian Sci Cruiser 20(1):34–43

Mohanty SN, Das KN, Sarkar S (1995) Effect of feeding varying dietary formulations on body composition of rohu fry. J Aquac 3:23–28

Moretti VM, Mentasti T, Bellagamba F, Luzzana U, Caprino F, Turchini GM, Giani I, Valfre F (2006) Determination of astaxanthin stereoisomers and colour attributes in flesh of rainbow trout (*Oncorhynchus mykiss*) as a tool to distinguish the dietary pigmentation source. J Food Addit Contam 23:1056–1063

Moss SM (1994) Growth rates, nucleic acid concentrations, and RNA/DNA ratios of juvenile white shrimp, *Penaeus vannamei* Boone, fed different algal diets. J Exp Mar Biol Ecol 182:193–204

Mukhopadhyay N, Ray AK (1999) Improvement of quality of sesame (*Seasamum indicum*) seed meal protein with supplemental amino acids in feeds for rohu, *Labeo rohita*, (Hamilton), fingerlings. Aquac Res 30:549–557

Mukhopadhyay N, Ray AK (2001) Effect of amino acids supplementation on the nutritive quality of fermented linseed meal protein in the diets for rohu, *Labeo rohita* (Hamilton), fingerlings. J Appl Ichthyol 17:220–226

National Research Council (U.S) (1983) Sub-committee on Warm water Fish Nutrition, National Research Council

Nickell DC, Bromage NR (1998) The effect of timing duration of feeding astaxanthin on the development and variation of fillet color and efficiency of pigmentation in rainbow trout *(Oncorhyncus mykiss)*. Aquaculture 169:233–246

NRC (1983) Nutrients requirements of warm water fishes and shell-fishes. National Academy Press, Washington, DC, 102 pp

Okuzumi J, Takahashi T, Yamane T, Kitao Y, Inagake M, Ohya K, Nishino H, Tanaka Y (1993) Inhibitory effects of fucoxanthin, a natural carotenoid, on N-ethyl-N'-nitro-N-nitrosoguanidine-induced mouse duodenal carcinogenesis. Cancer Lett 68:159–168

PeñafloridaV, Golez NV (1996) Use of seaweed meals from *Kappaphycus alvarezii* and *Gracilaria heteroclada* as binders in diets for juvenile shrimp *Penaeus monodon*. Aquacul 143:393–401

Phillips AM (1972) Calorie and energy requirements. In: Halver JE (ed) Fish nutrition. Academic Press, New York, pp 2–29

Prather EE, Lovell RT (1973) Response of intensively fed channel catfish to diets containing various protein energy ratio. Proc 27th South-Eastern Assoc Game Fish Comm 27:455–459

Rajkhowa I (2005) Action in aquaculture – opportunities in a quaint specialization. Business Today (May 22 Issue):131

Saha SB, Ghosh BB, Gopalakrishnan V (1995) Plankton of Hooghly estuary with special reference to salinity and temperature. J Mar Biol Assoc India 17(1):107–120

Sarada R, Tripathi U, Ravishankar GA (2002) Influence on stress on astaxanthin production in *Haematococcus pluvialis* grown under different culture conditions. Proc Biochem 37:623–627

Satoh S (2000) Common carp, *Cyprinus carpio*. In: Wilson RP (ed) Handbook of nutrient requirement of finfish. CRC Press, Boca Raton/Ann Arbor/Boston/London, pp 55–68

Schuep W, Schierle J (1995) Astaxanthin determination of stabilized, added astaxanthin in fish feeds and pre-mixes. In: Carotenoids isolation and analysis, vol 1. Birkhauser Verlag, Basel, pp 273–276

Shyong WJ, Huang CH, Chen HC (1998) Effects of dietary protein concentration on growth and muscle composition of juvenile. Aquaculture 167:35–42

Simpson KL, Chichester CO (1981) Metabolism and nutritional significance of carotenoids. Annu Rev Nutr 1:351–374

Storebakken T, No HK (1992) Pigmentation in rainbow trout. Aquaculture 100:209–229

Strickland JDH, Parsons TR (1968) A manual for seawater analysis. Bull Fish Res Board Can 167:11–131

Strickland JDH, Parsons TR (1972) A practical handbook of seawater analysis. J Fish Res Board Can Ottawa 167:311

Suarez-Garcia HA (2006) Efecto de la inclusion de alginate y harina de algas Sargassum sp y *Macrocystis pyrifera sobre la estabilidad en agua,* digestibilidad del alimento y sobre el crecimiento del camaron blanco *Litopenaeus vannamei.* Undergraduate thesis, Universidad Autonoma de Nuevo Leon, Mexico

Tecator (1983) Fat extraction on feeds with the Soxtec System HT – the influence of sample preparation and extraction media. Application note AN 67/83 (1983.06.13). Soxtec System HT Manual, Tecator AB, Sweden

Tjahjono AE, Hayama Y, Kakizono T, Terada Y, Nishio N, Nagai S (1994) Hyper-accumulation of astaxanthin in a green alga *Haematococcus pluvialis* at elevated temperatures. Biotechnol Lett 16:133–138

Torrisen OJ, Hardy RW, Shearer KD (1989) Pigmentation of salmonids – carotenoid deposition and metabolism. CRC Crit Rev Aqua Sci 1:209–225

Trevelyan WE, Harrison JS (1952) Studies on yeast metabolism. 1. Fractionation and micro-determination of cell carbohydrates. Biochem J 50:298–303

United Nations Educational (1987) Scientific and Cultural Organization Convention concerning the Protection of the World Cultural and Natural Heritage Report of the World Heritage Committee Eleventh session (UNESCO headquarters, 7–11 December 1987)

Valente LMP, Gouveia A, Rema P, Matos J, Gomes EF, Pinto IS (2006) Evaluation of three seaweeds *Gracilaria bursapatoris*, *Ulva rigida* and *Gracilaria cornea* as dietary ingredients in European seabass (*Dicentrarchus labrax*) juveniles. Aquaculture 252:85–91

Van der Meer MB, Zamora JE, Ver degem MCJ (1997) Effect of dietary lipid level on protein utilization and the size and proximate composition of body compartments of *Colossoma macropomum* (Cuvier). Aquac Res 28:405–417

Walkey A, Black IA (1934) An examination of the effect of the digestive method for determining soil organic matter and a proposed modification of the chronic and titration method. Soil Sci 37:29–38

White DA, Page GI, Swaile J, Moody AJ, Davies SJ (2002) Effect of esterification on the absorption of astaxanthin in rainbow trout, *Oncorhynchus mykiss* (Walbaum). Aquac Res 33:343–350

Wilson RP (2000) Channel catfish, *Ictalurus punctatus*. In: Wilson RP (ed) Handbook of nutrient requirement of finfish. CRC Press, Boca Raton/Ann Arbor/Boston/London, pp 35–53

WWF-India (2006) Training manual on ecofriendly and sustainable fishery, New Delhi, 130 pp

Yan X, Chuda Y, Suzuki M, Nagata T (1999) Fucoxanthin as the major antioxidant in *Hijikia fusiformis*, a common edible seaweed. Biosci Biotechnol Biochem 63:605–607

Chapter 12
Climate Change: Threat of *Era*

Abstract The present world is under the appalling shadow of climate change, which is caused both by natural and anthropogenic factors. The pulse of climate change is felt in every sector like weather, ocean characteristics, biodiversity, human health, economic profile and livelihoods. The chapter finds a mitigation mechanism to climate change through replacement of CFC and fossil fuels with appropriate technology and at the same time promoting the green cover on the Earth.

Keywords Natural factors · Anthropogenic factors · Global warming · Sea level rise (SLR) · Conference of the Parties (COP)

12.1 Causes of Climate Change

The United Nations Framework Convention on Climate Change (UNFCCC) coined the term 'climate change' for human-induced changes and 'climate variability' for other changes. Climate change may therefore be considered as any long-term significant change in the 'average weather' that a given region experiences and involves changes in the variability or average state of the atmosphere over durations ranging from decades to millions of years. The roots of these changes can be related to several dynamic processes on Earth, external forces including variations in sunlight intensity, and more recently by human activities.

For convenience of the readers, the causes of climate change have been divided into two broad domains: natural and man-made.

12.1.1 Natural Factors

The natural factors that regulate the process of climate change include:

- Variations in the Earth's orbital characteristics
- Atmospheric carbon dioxide variations
- Volcanic eruptions

A. Mitra, S. Zaman, *Environmental Science - A Ground Zero Observation on the Indian Subcontinent*, https://doi.org/10.1007/978-3-030-49131-4_12

- Variations in solar output
- Plate tectonics

(A) **Variations in the Earth's Orbital Characteristics**

According to Milankovitch theory, the normal cyclical variations in three of the Earth's orbital characteristics are probably responsible for some past climatic change. The basic idea behind this theory assumes that over time these three cyclic events result in the variation of the amount of solar radiation that is received on the surface of the planet Earth.

The first cyclical variation, known as eccentricity, controls the shape of the Earth's orbit around the Sun. The orbit gradually changes from being elliptical to being nearly circular and then back to elliptical in a period of about 100,000 years. The greater the eccentricity of the orbit (i.e., the more elliptical it is), the greater is the variation in solar energy received at the top of the atmosphere between the Earth's closest (perihelion) and farthest (aphelion) approach to the Sun. Currently, the Earth is experiencing a period of low eccentricity. The difference in the Earth's distance from the Sun between perihelion and aphelion (which is only about 3%) is responsible for approximately a 7% variation in the amount of solar energy received at the top of the atmosphere. When the difference in this distance is at its maximum (9%), the difference in solar energy received is about 20%.

The second cyclical variation results from the fact that, as the Earth rotates on its polar axis, it wobbles like a spinning top changing the orbital timing of the equinoxes and solstices (Fig. 12.1). This effect is known as the precession of the equinox. The precession of the equinox has a cycle of approximately 26,000 years. According to illustration (A), the Earth is closer to the Sun in January (perihelion) and farther away in July (aphelion) at the present time. Because of precession, the reverse will be true in 13,000 years, and the Earth will then be closer to the Sun in July (illustration B). This means, of course, that if everything else remains constant, 13,000 years from now seasonal variations in the Northern Hemisphere should be greater than at present (colder winters and warmer summers) because of the closer proximity of the Earth to the Sun.

The third cyclical variation is related to the changes in the tilt (obliquity) of the Earth's axis of rotation over a 41,000 years period. During the 41,000 year cycle, the tilt can deviate from approximately 22.5° to 24.5°. At the present time, the tilt of the Earth's axis is 23.5°. When the tilt is small, there is less climatic variation between the summer and winter seasons in the middle and high latitudes. Winters tend to be milder and summers cooler. Warmer winters allow for more snow to fall in the high-latitude regions. When the atmosphere is warmer, it has a greater ability to hold water vapour, and therefore more snow is produced at areas of frontal or orographic uplift. Cooler summers cause snow and ice to accumulate on the Earth's surface because less of this frozen water is melted. Thus, the net effect of a smaller tilt would be more extensive formation of glaciers in the polar latitudes.

Periods of a larger tilt result in greater seasonal climatic variation in the middle and high latitudes. At these times, winters tend to be colder and summers warmer. Colder winters produce more snow because of lower atmospheric temperatures. As

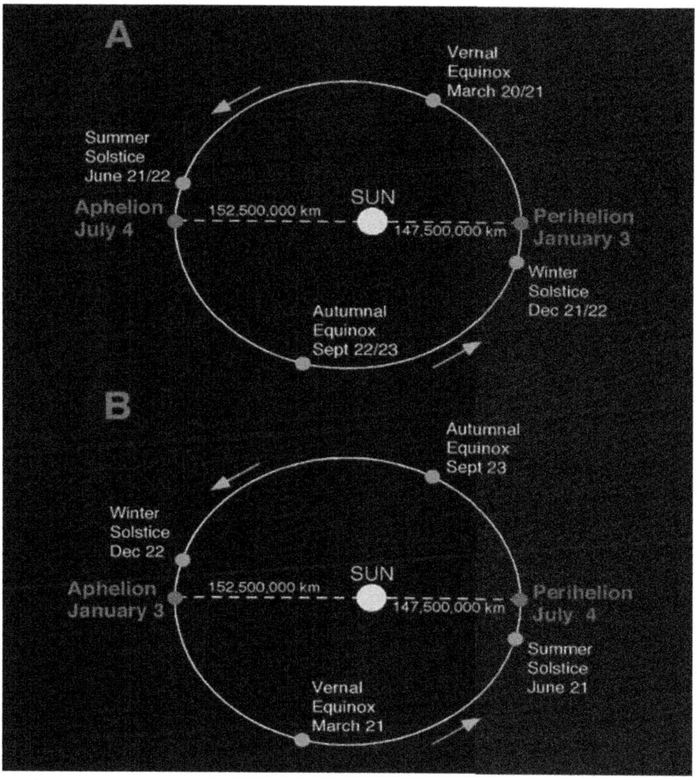

Fig. 12.1 Modification of the timing of aphelion and perihelion over time (A = today; B = 13,000 years into the future)

a result, more snow and ice accumulate on the ground surface. Moreover, the warmer summers produced by the larger tilt provide additional energy to melt and evaporate the snow that fall and accumulate during the winter months. In conclusion, glaciers in the polar regions should be generally receding, with other contributing factors constant, during this part of the obliquity cycle.

Computer models and historical evidences suggest that the Milankovitch cycles exert their greatest cooling and warming influence when the troughs and crests of all three cycles coincide with each other.

(B) **Atmospheric Carbon Dioxide Variations**

Studies on long-term climate change have revealed an inter-relationship between the concentrations of carbon dioxide in the atmosphere and mean global temperature. Carbon dioxide is one of the most important gases responsible for the greenhouse effect. Certain atmospheric gases, like carbon dioxide, water vapour and methane, are able to alter the energy balance of the Earth by their property to absorb long-wave radiation emitted from the Earth's surface. The net result of this process is the rise of Earth's temperature. Without the greenhouse effect, the average global temperature of the Earth would be around −18 °C rather than the present 15 °C.

Researchers of the 1970s CLIMAP project documented strong evidence in deep-ocean sediments regarding the variations in the Earth's global temperature during the past several hundred thousand years of the Earth's history. Other subsequent studies have confirmed these findings and have discovered that these temperature variations were closely correlated to the concentration of carbon dioxide in the atmosphere and variations in solar radiation received by the planet as controlled by the Milankovitch cycles. Measurements indicated that atmospheric carbon dioxide levels were about 30% lower during colder glacial periods. It was also theorized that the oceans were a major store of carbon dioxide and that they controlled the movement of this gas to and from the atmosphere. The amount of carbon dioxide that can be held in oceans is a function of temperature. Carbon dioxide is released from the oceans when global temperatures become warmer and diffuses into the ocean when temperatures are cooler. Initial alterations in global temperature were triggered by changes in received solar radiation by the Earth through the Milankovitch cycles. The increase in carbon dioxide then amplified the global warming by enhancing the greenhouse effect.

Over the past three centuries, the concentration of carbon dioxide has been increasing in the Earth's atmosphere because of human influences (Fig. 12.2).

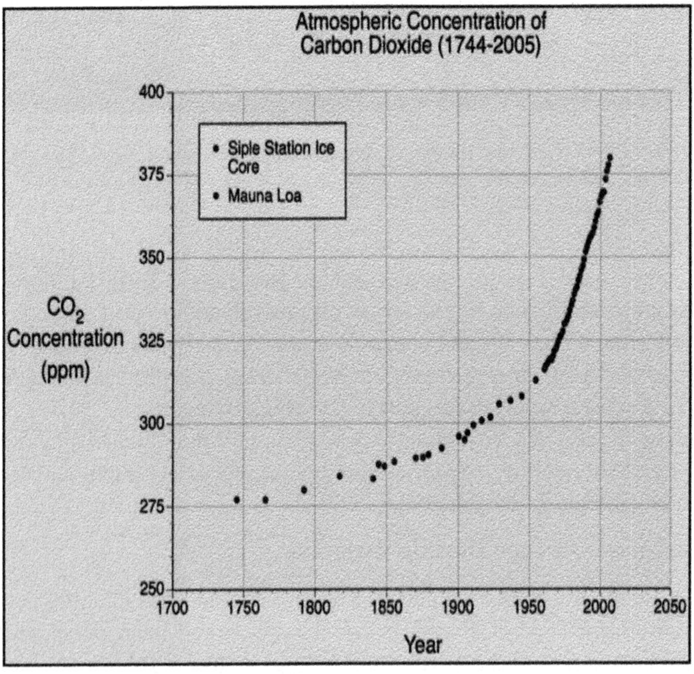

Fig. 12.2 The following graph illustrates the rise in atmospheric carbon dioxide from 1744 to 2005. Noteworthy is the exponential increase in carbon dioxide's concentration in the atmosphere. (Source: Pidwirny M. (2006), http://www.physicalgeography.net/fundamentals/7y.html)

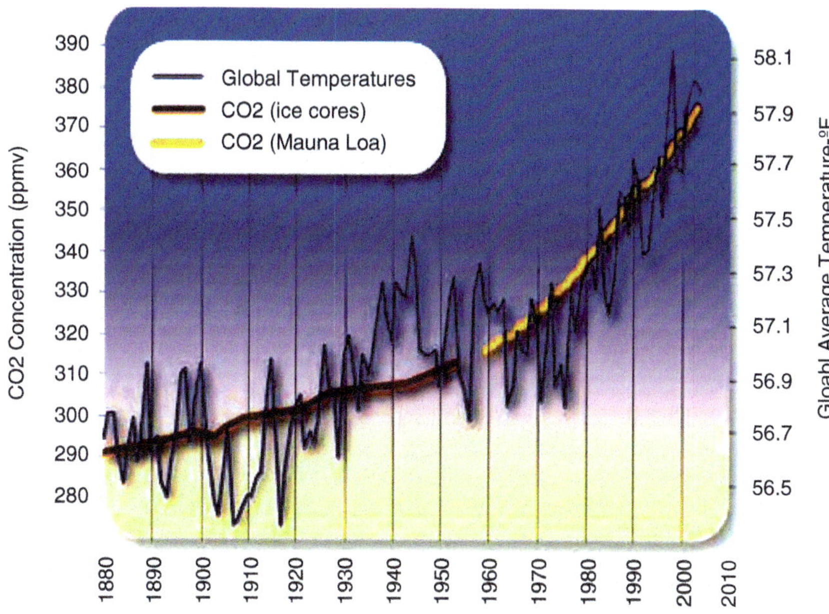

Fig. 12.3 Global average temperature and carbon dioxide concentrations during 1880 to 2004

Human activities like the burning of fossil fuels, conversion of natural prairie to farmland, intense industrialization, urbanization and deforestation have caused the release of carbon dioxide into the atmosphere. From the early 1700s, carbon dioxide has increased from 280 parts per million to 380 parts per million in 2005. Many scientists believe that higher concentrations of carbon dioxide in the atmosphere will accelerate the greenhouse effect making the planet warmer. Scientists also believe that the present *era* is already experiencing global warming due to an enhancement of the greenhouse gases. Most computer climate models suggest that the globe will warm up by 1.5–4.5 °C if carbon dioxide reaches the predicted level of 600 parts per million by the year 2050. The direct correlation between the concentrations of carbon dioxide and temperature of the Earth's atmosphere is shown in Fig. 12.3.

(C) **Volcanic Eruptions**

A connection has been established between large explosive volcanic eruptions and short-term climatic changes (Fig. 12.4) after considering long-term data of these two variables. For example, one of the coldest years in the last two centuries occurred the year following the Tambora volcanic eruption in 1815. A number of regions across the planet Earth witnessed such lowering of atmospheric temperature. Several other major volcanic events also showed a pattern of cooler global temperatures lasting 1–3 years after their eruption.

At first, scientists thought that the dust emitted into the atmosphere from large volcanic eruptions were responsible for the cooling by partially blocking the

Fig. 12.4 Explosive volcanic eruptions have been shown to have a short-term cooling effect on the atmosphere if they eject large quantities of sulphur dioxide into the stratosphere. The present figure shows the eruption of Mount St. Helens on May 18, 1980 which had a local effect on climate because of ash reducing the reception of solar radiation on the Earth's surface. Mount St. Helens had very minimal global effect on the climate because the eruption occurred at an oblique angle putting little sulphur dioxide into the stratosphere. (Source: U.S. Geological Survey; photograph by Austin Post)

transmission of solar radiation to the Earth's surface. However, measurements indicate that most of the dust thrown in the atmosphere returned to the Earth's surface within 6 months. Recent stratospheric data suggests that large explosive volcanic eruptions also eject huge amounts of sulphur dioxide gas which remains in the atmosphere for as long as 3 years. Atmospheric chemists have determined that the ejected sulphur dioxide gas reacts with water vapour commonly found in the stratosphere to form a dense optically bright haze layer that reduces the atmospheric transmission of some of the Sun's incoming radiation.

In the last century, two significant climate modifying eruptions have occurred. El Chichon in Mexico erupted in April, 1982, and Mount Pinatubo went off in the Philippines during June, 1991 (Fig. 12.5). Of these two volcanic events, Mount Pinatubo had a greater effect on the Earth's climate and ejected about 20 million tonnes of sulphur dioxide into the stratosphere (Fig. 12.6). Researchers believe that the Pinatubo eruption was primarily responsible for the 0.8 °C drop in global average air temperature in 1992. The global climatic effects of the eruption of Mount Pinatubo are believed to have peaked in late 1993. Satellite data confirmed the inter-relationship

Fig. 12.5 Ash column generated by the eruption of Mount Pinatubo on June 12, 1991. The strongest eruption of Mount Pinatubo occurred 3 days later on June 15, 1991. (Source: US Geological Survey)

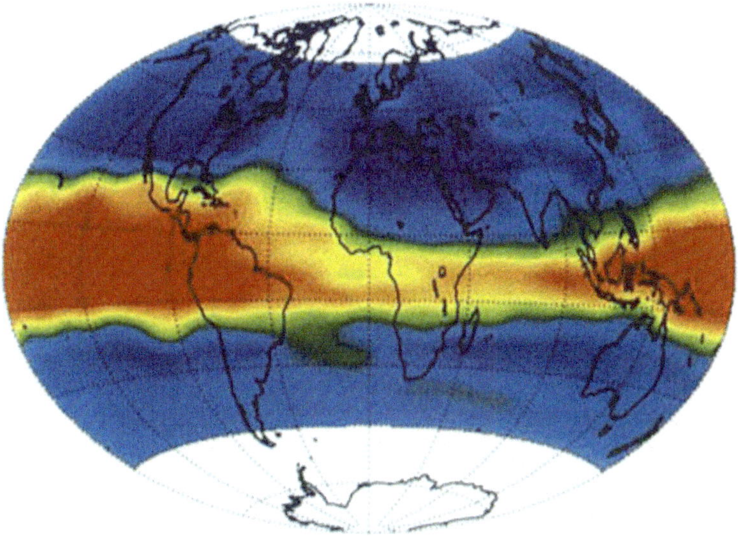

Fig. 12.6 Satellite image showing the distribution of Mount Pinatubo's sulphur dioxide and dust aerosol plume (red and yellow areas) between June 14 and July 26, 1991. (Source: SAGE II Satellite Project – NASA)

between the Mount Pinatubo eruption and reduction in global temperature in 1992 and 1993. The satellite data further indicated that the sulphur dioxide plume from the eruption caused a several percent increase in the amount of sunlight reflected by the Earth's atmosphere back to space causing the surface of the planet to cool.

(D) **Variations in Solar Output**

Until recently, many scientists thought that the Sun's output of radiation only varied by a fraction of a percent over many years. However, measurements made by satellites equipped with radiometers in the 1980s and 1990s suggested that the Sun's energy output may be more variable than was once estimated. Measurements made during the early 1980s showed a decrease of 0.1% in the total amount of solar energy reaching the Earth over just an 18-month time period. If this trend were to extend over several decades, it could influence global climate. Numerical climatic models predict that a change in solar output of only 1% per century would alter the Earth's average temperature between 0.5 °C and 1.0 °C.

Scientists have attempted to establish a linkage between sunspots and climatic change. Sunspots are huge magnetic storms that are seen as dark (cooler) areas on the Sun's surface (Fig. 12.7).

Sunspots have magnetic fields of strength up to 3000 times as great as the average magnetic field of either the Sun or the Earth. Astronomers believe the cause of sunspots is attributed to this fact. According to a standard explanation, the strong magnetic fields of the Sun have the shape of tubes just below the solar surface at the beginning of the sunspot cycle. These tubes lie perpendicular to the Sun's equator. The Sun rotates faster at its equator than at its poles, and so the tubes are stretched out in the east–west direction. Kinks then develop in the magnetic tubes and push through the solar surface. A pair of sunspots appears wherever a kink penetrates,

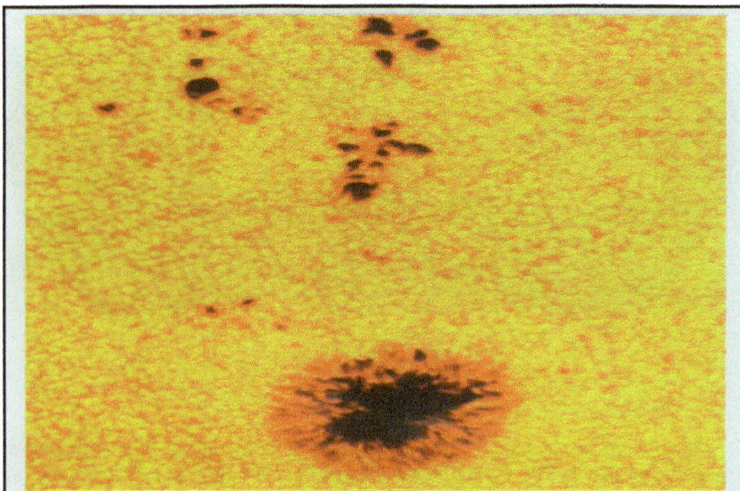

Fig. 12.7 Sunspots on the Sun's surface

because the kink both leaves and re-enters the surface. The number and size of sunspots show cyclical patterns, reaching a maximum about every 11, 90 and 180 years. The decrease in solar energy observed in the early 1980s corresponds to a period of maximum sunspot activity based on the 11-year cycle. In addition, measurements made with a solar telescope from 1976 to 1980 showed that during this period, as the number and size of sunspots increased, the Sun's surface cooled by about 6 °C. Apparently, the sunspots prevented some of the Sun's energy from leaving its surface. However, these findings tend to contradict observations made on longer time scales. Observations of the Sun during the middle of the Little Ice Age (1650–1750) indicated that very little sunspot activity was occurring on the Sun's surface. The Little Ice Age was a time of a much cooler global climate, and some scientists correlate this occurrence with a reduction in solar activity over a period of 90 or 180 years. Measurements have shown that these 90 and 180 year cycles influence the amplitude of the 11-year sunspot cycle. It is hypothesized that during times of low amplitude, like the Maunder Minimum, the Sun's output of radiation is reduced. Observations by astronomers during this period (1645–1715) revealed very little sunspot activity occurring on the Sun.

During periods of maximum sunspot activity, the Sun's magnetic field is strong. When sunspot activity is low, the Sun's magnetic field weakens. The magnetic field of the Sun also reverses every 22 years, during a sunspot minimum. Some researchers believe that the periodic droughts on the Great Plains of the United States are in some way correlated with this 22-year cycle.

(E) **Plate Tectonics**

The phenomenon of plate tectonics plays a major role in the event of climate change. On the longer time scale, plate tectonics can reorient the position of continents, shape oceans, build and tear down mountains and generally serve to provide the matrix upon which climate exists. More recently, plate motions have been implicated in the intensification of the present ice age when, approximately 3 million years ago, the North and South American plates collided to form the Isthmus of Panama and shut off direct mixing between the Atlantic and Pacific Oceans. The movement of the plate and subsequent subduction also regulate the condition of the climate. Slow subduction is associated with narrow mid-oceanic ridge and low volcanic activity. Under this condition, there is maximum exposure of marine limestone to weathering, which draws more carbon dioxide from the atmosphere to participate in the reactions resulting in the cooling effect. The opposite phenomenon of warming is linked to faster subduction process.

12.1.2 Human Influences on Climate Change

Anthropogenic factors are human activities that change the environment and influence climate. The emission of carbon dioxide due to burning of fossil fuels and the increase of greenhouse gas concentrations due to rapid urbanization, industrialization and expansion of unplanned tourism are unquestionable human influences on climate change through the event of habitat destruction or modification. However, in some cases the chain of causality is direct and unambiguous (e.g., by the effects of irrigation on temperature and humidity), while in others it is less clear. Various hypotheses for human-induced climate change have been debated for many years.

According to Church et al. (2001), the amount of anthropogenic change in land water storage systems cannot be estimated with much confidence. A number of anthropogenic factors can contribute to sea level rise. First, natural groundwater systems typically are in a condition of dynamic equilibrium where, over long time periods, recharge and discharge are in a balance. When the rate of groundwater pumping greatly exceeds the rate of recharge, as is often the case in arid and even semi-arid regions, water is removed permanently from storage. The water that is lost from groundwater storage eventually reaches the ocean through the atmosphere or surface flow, resulting in sea level rise. Second, wetland contains standing water, soil moisture and water in plants equivalent to water roughly 1 m deep. Hence, wetland destruction (for urbanization, industrialization and tourism, etc.) contributes to sea level rise. Over time scales shorter than a few years, diversion of surface waters for irrigation in the internally draining basins of arid regions results in increased evaporation. The water lost from the basin hydrologic system eventually reaches the ocean. Third, forests store water in plant tissue both above and below ground. When a forest is removed, transpiration is eliminated so that runoff is favoured in the hydrologic budget.

It has been observed by several researchers that impoundment of water behind dams removes water from the ocean and lowers sea level. Dams have led to a sea level drop over the past few decades from −0.5 to −0.7 mm $year^{-1}$ (Chao 1994; Sahagian et al. 1994). Infiltration from dams and irrigation may raise the water table by storing more water. Gornitz (2001) estimated a range of −0.33 to −0.27 mm $year^{-1}$ sea level change equivalent held by dams (not counting additional potential storage due to subsurface infiltration).

It is very difficult to provide accurate estimates of the net anthropogenic contribution to climate change due to lack of worldwide information on individual factor, although the effect caused by dams is possibly better known than other effects. According to Sahagian (2000), the sum of the anthropogenic effects could be of the order of 0.05 mm $year^{-1}$ sea level rise over the past 50 years, with an uncertainty several times as large.

In conclusion it can be advocated that the land-related contribution to sea level change has not led to a reduction in the uncertainty compared to the IPCC Technical Assessment Report (2001), which estimated the rather wide ranges of −1.1 to +0.4 mm $year^{-1}$ for 1910 to 1990 and −1.9 to +1.0 mm $year^{-1}$ for 1990. However, indirect evidence by considering other contributions to the sea level budget suggests that the land contribution either is small (<0.5 mm $year^{-1}$) or is compensated by unaccounted underestimated contributions.

Today the climate change-related researches unanimously suggest that the biggest factor of present concern is the increase in CO_2 levels due to emissions from fossil fuel combustion, followed by aerosols (particulate matter in the atmosphere), which exert a cooling effect. Cement manufacture plays a vital role in the domain of human-induced climate change. Other factors, including land use, ozone depletion, animal husbandry, agriculture and deforestation, also exert considerable influence on the climatic profile of the planet Earth. These factors are discussed here in brief.

(A) **Fossil Fuels**

Fossil fuels constitute the backbone of modern civilization. Beginning with the industrial revolution in the 1850s and accelerating ever since, the human consumption of fossil fuels has elevated CO_2 levels from a concentration of ~280 ppm to more than 380 ppm today (an increase of 35.7%). These increases are projected to reach more than 560 ppm (an increase of 100%) before the end of the twenty-first century. It is known that carbon dioxide levels are substantially higher now than at any time in the last 750,000 years. Along with rising methane levels, these changes are anticipated to cause an increase of atmospheric temperature within 1.4–5.6 °C between 1990 and 2100.

(B) **Aerosols**

Anthropogenic aerosols, particularly sulphate aerosols from fossil fuel combustion, exert a cooling influence. This, together with natural variability, is believed to account for the relative 'plateau' in the graph of twentieth-century temperatures in the middle of the century.

(C) **Cement Manufacture**

Cement manufacturing is one of the largest causes of human-induced carbon dioxide emissions. The gas is produced when calcium carbonate ($CaCO_3$) is heated to produce the calcium oxide (CaO, also called *quicklime*), which is the main ingredient for cement ($CaSiO_3$) production. The reaction steps in this manufacture are:

$$CaCO_3 \quad \rightarrow CaO + CO_2$$

$$CaO + SiO_2 \xrightarrow{\Delta} CaSiO_3$$

CO_2 produced in the first step has considerable contribution to global warming. While fossil fuel combustion and deforestation produce significantly more carbon dioxide in the Earth's atmosphere, cement production alone is responsible for approximately 2.5% of total worldwide emissions from industrial sources.

(D) **Land Use**

Prior to widespread fossil fuel use, humanity's largest effect on local climate is likely to have resulted from alteration in land-use pattern. Irrigation, deforestation and agriculture fundamentally change the environment. These activities alter the amount of water going into and out of a given location. They also may change the local albedo by influencing the ground cover and altering the amount of sunlight that is absorbed. For example, there is evidence to suggest that the climate of Greece and other Mediterranean countries was permanently changed by widespread deforestation between 700 BC and 1 AD (the wood being used for ship building, construction and fuel), with the result that the modern climate in the region is significantly hotter and drier and the species of trees that were used for ship building in the ancient world are no longer be found in the area.

A controversial hypothesis by William Ruddiman called the early anthropocene hypothesis suggests that the rise of agriculture and the accompanying deforestation led to the increases in carbon dioxide and methane during the period 5000–8000 years ago. These increases, which reversed previous declines, may have been responsible for delaying the onset of the next glacial period, according to Ruddiman's overdue-glaciation hypothesis.

In modern times, a 2007 Jet Propulsion Laboratory study found that the average temperature of California has risen about 2 degrees over the past 50 years, with a much higher increase in urban areas. The change was attributed mostly to extensive human development of the landscape.

(E) **Livestock**

According to a 2006 United Nations report, livestock is responsible for 18% of the world's greenhouse gas emissions as measured in CO_2 equivalents. This however includes land usage change, meaning deforestation in order to create grazing land. In the Amazon Rainforest, 70% of deforestation was done solely to make way for grazing land. This is the major factor in the UNFAO report (2006), which was the first agricultural report to include land usage change into the radiative forcing of livestock. In addition to CO_2 emissions, livestock produces 65% of human-induced nitrous oxide (which has 296 times the global warming potential of CO_2) and 37% of human-induced methane (which has 23 times the global warming potential of CO_2).

12.2 Effects of Climate Change

Global warming is a phase of climate change, which is associated with the increasing trend in the average temperature of the Earth's near-surface air and oceans in recent decades and its projected continuation. Global average air temperature near the Earth's surface raised 0.74 ± 0.18 °C (1.3 ± 0.32 °F) during the past century. The Intergovernmental Panel on Climate Change (IPCC) concludes, 'most of the observed increase in globally averaged temperatures since the mid-twentieth century is very likely due to the observed increase in anthropogenic greenhouse gas concentrations (http://www.ipcc.ch/SPM2feb07.pdf)', which leads to warming of the surface and lower atmosphere by accelerating the greenhouse effect. Natural phenomena such as solar variation combined with volcanoes may probably have a small warming effect from pre-industrial times to 1950 but a small cooling effect since 1950.

There is still much uncertainty over the degree by which temperatures will actually rise and what the effects of these rises will be. What is not in doubt is that the

Earth's temperature has increased by 0.5 °C over the past century and that recent years have been among the hottest on record. The likely effects of global warming have serious implications on the environment. A 20 cm rise in sea level by 2030 is expected to result from glacial melting and from the thermal expansion of the oceans as water temperatures rise. This may inundate some coastal regions and increase the risk of flooding in many areas. In the case of Indian sub-continent, population density often exceeds 1000 persons per sq. km within low elevation coastal zone (LECZ), which has placed India as one of the most vulnerable nations to climate change-induced sea level rise (Fig. 12.8).

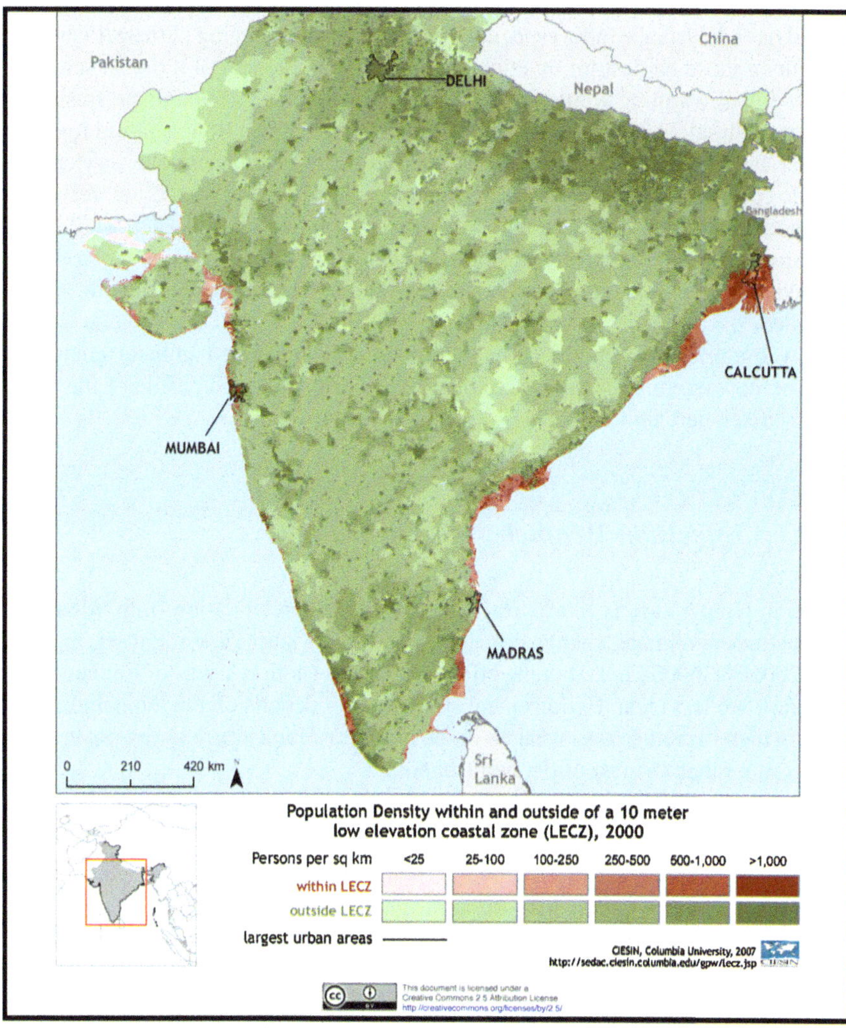

Fig. 12.8 Population density within a 10 metre low elevation coastal zone (LECZ) often exceeds 1000 persons per sq. km

Internationally, populations throughout much of China, Egypt, Denmark and Southeast Asia are most at risk – millions could lose their homes and livelihoods. In the United Kingdom, Southeast England is most vulnerable to inundation, although floods, such as that which occurred in Tywyn, mid-Wales, in 1990, are likely to increase in both frequency and severity throughout the country. Agriculture and forestry – crop failure and famine will ensue throughout marginally productive regions of the World. The effects of climate change can be felt hardest in a country like India. The fact that India has a large rural population who are dependent on the cycles of the climatic seasons for their agricultural activities, fishermen who work in the rivers and seas, farmers who need the seasonal monsoons and a large and varied indigenous population who live in harsh climatic regions of mountains, desert and river delta make India especially susceptible to a changing climate. Fishermen are among those hardest hit by climate change. The potential of a rise in sea levels due to global warming would wipe out existing shoreline, contaminate freshwater systems with salt water, diminish the minimum environment flow required for fishes in river and dry up floodplain wetlands, a rich habitat for fishes. The services that aquatic ecosystem provide to humans such as fisheries, flood protection, recreation and biodiversity conservation, which if converted in monetary terms, would be in millions of rupees. A recent global assessment of the status of aquatic ecosystems showed that their capacity to provide goods and services appears to be drastically degraded.

Few important effects of global warming and associated climate change on weather, ecosystem, biodiversity, human health and economic profile of the nation are discussed here in details.

12.2.1 Effects on Weather

A rise in temperature is likely to trigger the rate of precipitation (http://dpa.aapg.org/gac/papers/climate_change.cfm and http://www.agu.org/fora/eos/pdfs/2006EO360008.pdf), but the effects on storms (both in terms of frequency and intensity) are less clear. Extratropical storms partly depend on the temperature gradient, which is predicted to weaken in the Northern Hemisphere as the polar region warms more than the rest of the hemisphere.

Controversies also exist in the domain of correlation between sea surface temperature and Atlantic basin hurricanes.

(a) **Increased Evaporation**

Over the course of the twentieth century, evaporation rates have reduced worldwide (http://www.greenhouse.gov.au/impacts/overview/pubs/overview4.pdf); this is thought by many researchers to be explained by *global dimming*. As the climate grows warmer and the causes of global dimming are reduced, the rate of evaporation increases due to warmer oceans. Because the components of the planet Earth – its atmosphere, lithosphere and hydrosphere – comprise a closed system, this will cause heavier rainfall, more runoff and more erosion. Many scientists think that increased evaporation could result in more extreme weather as global warming progresses.

It has been predicted that each 1% increase in annual precipitation would enhance the cost of catastrophic storms by 2.8%. The Association of British Insurers (ABI) has stated that limiting carbon emissions would avoid 80% of the projected additional annual cost of tropical cyclones by the 2080s. The cost is also increasing partly because of building in exposed areas such as coasts and floodplains. The ABI claims that reduction of the vulnerability to some inevitable impacts of climate change, for example, through more resilient buildings and improved flood defences, could also result in considerable cost savings in the long term (http://www.abi.org.uk/Display/File/Child/552/Financial_Risks-of_Climate.html).

(b) **Destabilization of Local Climates**

The first recorded South Atlantic hurricane, 'Catarina/Katrina', which hit Brazil in March 2004, is an example of climatic destabilization related to climate change. In the Northern Hemisphere, the southern part of the Arctic region (home to 4,000,000 people) has experienced a temperature rise of 1 °C to 3 °C (1.8 °F to 5.4 °F) over the last 50 years. Canada, Alaska and Russia are experiencing initial melting of permafrost. This may disrupt ecosystems, and by accelerated bacterial activity in the soil, these areas can become carbon sources instead of carbon sinks (http://www.grida.no/climate/ipcc_tar/wg1/295.htm). A study conducted on changes to eastern Siberia's permafrost suggests that it is gradually disappearing in the southern regions, leading to the loss of nearly 11% of Siberia's 11,000 lakes since 1971 (http://www.gurdian.co.uk/international/story/0,1503170,00.html). At the same time, western Siberia is at the initial stage where melting permafrost is creating new lakes, which will eventually start disappearing as in the east. Furthermore, permafrost melting will also cause methane release from melting permafrost peat bogs, thus posing a positive feedback on the overall global warming phenomenon.

12.2.2 *Effects on Ocean*

(a) **Sea Level Rise**

The increase in global temperature triggers the process of expansion of ocean water volume and additional input of waters from sources like glaciers, which were locked on the land. This causes the level of seawater to rise from its normal position. The sea level has elevated more than 120 metres since the peak of the last ice age about 18,000 years ago. The bulk of that occurred before 6000 years ago. From 3000 years ago to the start of the nineteenth century, sea level was almost constant, rising at 0.1 to 0.2 mm/year. Since 1900, the level has risen at 1–2 mm/year (http://www.grida.no/climate/ipcc_tar/wg1/295.htm), and since 1992, satellite altimetry from TOPEX/

Poseidon indicates a rate of about 3 mm/year (http://www.grida.no/climate/ipcc_tar/wg1/295.htm). An increase of 1.5 to 4.5 °C temperature is estimated to lead to an increase of 15–95 cm of the sea level (IPCC 2001).

The three primary contributing factors to sea level rise often cited are (1) ocean thermal expansion; (2) glacial melt from Greenland and Antarctica (plus a smaller contribution from other ice sheets); and (3) the change in terrestrial storage. Until recently, ocean thermal expansion was expected to be the dominating factor behind the rise in sea level. However, new data sets on rates of deglaciation in Greenland and Antarctica suggest significant contribution of glacial melt, especially due to the uncertainty surrounding the dynamics of outlet glaciers (Hansen et al. 2005; Hanna et al. 2005; Howat et al. 2007; Krabill et al. 2004; Meier et al. 2007; Rignot and Kanagaratnam 2006; Stroeve et al. 2007; Velicogna and Wahr 2006). Although there is continued evidence of ice sheet growth in the eastern regions of Antarctica (Davis et al. 2005), when coupled with the measured losses in the west, and including Greenland, the evidence appears to point in the direction of increased sea level rise. The implications of these findings for sea level rise could be dramatic. The Greenland and Antarctic ice sheets contain enough water to raise the sea level by almost 70 m (Hansen et al. 2005), so even small changes in the their volume would have a significant effect.

The magnitude of contribution of freshwater to global sea level rise is highlighted in Table 12.1.

Paleoclimatic information also indicated that the warmth of the last half century is unusual in at least the past 1300 years. About 1,25,000 years ago, polar regions were relatively warmer than at present. Reductions in polar ice volume led to 4–6 m of sea level rise (IPCC 2007). As new researches on the impacts of warming are coming in the light of publications, more confidence is developing with respect to the ranges of impact, but current evidence leads to an alarming possibility that a

Table 12.1 Estimates of the various contributions to the budget of global mean sea level change for 1961 to 2003 and 1993 to 2003 compared with the observed rate of rise

Source	1961–2003	1993–2003
Thermal expansion	0.42 ± 0.12	1.6 ± 0.5
Glaciers and ice caps	0.50 ± 0.18	0.77 ± 0.22
Greenland ice sheet	0.05 ± 0.12	0.21 ± 0.07
Antarctic ice sheet	0.14 ± 0.41	0.21 ± 0.35
Sum	1.1 ± 0.5	2.8 ± 0.7
Observed	1.8 ± 0.5	3.1 ± 0.7
Difference (observed sum)	0.7 ± 0.7	0.3 ± 1.0

Sources: Willis et al. (2004), Antonov et al. (2005), Ishii et al. (2006), Lombard et al. (2006)

Remark: Ice sheet mass loss of 100 Gt/year is equivalent to 0.28 mm/year of sea level rise. A GIA correction has been applied to observations from the tide gauges and altimetry. For the sum, the error has been calculated as the square root of the sum of squared errors of the contributions. The thermostatic sea level changes are for the 0–3000 m layer of the ocean

threshold triggering many metres of sea level rise could be crossed well before the end of this century (Hansen 2006; Overpeck et al. 2006).

Several regional- and local-level case studies have been cited in favour of sea level rise due to global warming in different parts of the planet Earth. However, many case studies are not directly related to increase in temperature or expansion of ocean water mass; rather other factors like sea bed rise or tectonic reasons are also standing in queue with significant proofs behind sea level rise (Raha et al. 2012).

In context to India, *The Independent* (a news agency) reported in December 2006 that the first island claimed by rising sea levels caused by global warming in Indian Sundarbans was Lohachara Island. However, several scientists have contradicted the correlation between erosion of Lohachara Island and global warming and forwarded tectonic cause as one of the important reasons behind the loss of landmass of Lohachara Island. The Indian Sundarbans at the apex of the Bay of Bengal (between 21°13′ to 22°40′ N latitude and 88°03′ to 89°07′ E longitude) is basically a deltaic complex of approximately 426,300 ha formed by the depositional activities of the Ganges and the Brahmaputra. A group of islands and a dense network of rivers, canals and creeks comprise the area. It has been suggested that, in the past, the tidal swamps of the Sundarbans extended landwards to the base of the Rajmahal hills and filled up in the later part of the Tertiary Period (Fergusson 1863). A number of geomorphological and resultant hydrological changes have contributed to the present location and condition of the Sundarbans. Tectonic movements in the late Tertiary Period in Northwestern Punjab changed the course of the Indo-Brahma River. Some workers consider this phenomenon as the cause behind the beginning of the southeastern flow of the River Ganges (Wadia 1961). Geologists believe that there was a general southeastern slope of the Bengal Basin during the Tertiary Period. The Ganges started flowing down this slope and created a new delta through thousand years of time by silt deposition. Gradually the western part of Bengal below the Rajmahal hills rose in level and became suitable for human settlement. Later evidence suggests that the Bengal Basin tilted eastwards during the twelfth century because of neotectonic movement (Morgan and McIntire 1959). The tectonic movements of the basin in different geological periods are one of the important causes for the loss of landmass in some islands.

The frequent subsidence of some part of Sundarbans, causing the complete disappearance of forest chunk and settlement areas, is the effect of 'Swatch of no ground' phenomenon. This zone also known as Ganga trough is situated south of the Raimangal–Malancha estuary between 21°00′ and 21°22′ latitude in the Bay of Bengal region (Chaudhuri and Choudhury 1994) and has a comparatively flat floor with walls of about 12° inclination. According to Oldham (1893), 'there is a singularly deep area in the sea outside the middle of the delta, which is marked in the chart as Swatch of no ground. Here the soundings are from 5 to 10 fathoms all round that change almost abruptly to 200 and even to 300 fathoms'. All the estuaries in the Sundarbans delta (Hugli, Muriganga, Saptamukhi, Thakuran, Matla, Bidyadhari, Gosaba and Hariabhanga) are believed to be influenced by Swatch of no ground, and the unstable condition of Sundarbans is due to the existence of this submarine zone (Chaudhuri and Choudhury 1994). These discussions lead to conclude that

Table 12.2 Impacts of sea level rise across at the global level

	1 m	2 m	3 m	4 m	5 m
Land area (total = 63,332,530 sq. km)					
Impacted area	194,309	305,036	449,428	608,239	768,804
% of total area	0.31	0.48	0.71	0.96	1.21
Population (total = 4,414,030.000)					
Impacted population	56,344,110	89,640,441	133,049,836	183,467,312	245,904,401
% of total population	1.28	2.03	3.01	4.16	5.57
GDP (total = 16,890.948 million USD)					
Impacted GDP (USD)	219,181	357,401	541,744	789,569	1,022,349
% of total GDP	1.30	2.12	3.21	4.67	6.05
Urban areas (total = 1,434,712 sq. km)					
Impacted area	14,646	23,497	35,794	50,742	67,140
% of total area	1.02	1.64	2.49	3.54	4.68
Agricultural land (total = 17,975,807 sq. km)					
Impacted area	70,671	124,247	196,834	285,172	377,930
% of total area	0.39	0.69	1.09	1.59	2.10
Wetlands area (total = 4,744,149 sq. km)					
Impacted area	88,224	140,355	205,697	283,009	347,400
% of total area	1.86	2.96	4.34	5.97	7.32

Source: Dasgupta et al. (2009); The impact of sea level rise on developing countries: a comparative analysis, Climatic Change, 93:379–388

there are a lot of uncertainties to relate the erosion of islands and global warming-induced sea level rise as geological settings, geographic features and tectonic movements often generate signals of noise in such direct cause–effect relationship.

Sea level rise has considerable impact on the economic profile of the developing countries. A research analysis conducted by Dasgupta et al. (2009) for the 84 developing countries is summarized in Table 12.2. The analysis indicates that approximately 0.3% (194,000 km^2) of the territory of the 84 developing countries would be impacted by a 1 m sea level rise. This would increase to 1.2% in a 5 m sea level rise scenario. Though this remains relatively small in percentage terms, approximately 56 million people (or 1.28% of the population) in these countries would be impacted under a 1 m sea level rise scenario. This would increase to 89 million people for 2 m sea level rise (2.03%) and 245 million people (5.57%) for 5 m sea level rise. The impact of sea level rise on GDP is slightly larger than the impact on population, because GDP per capita is generally above average for coastal populations and cities. Wetland would experience significant impact even with a 1 m sea level rise. Up to 7.3% of wetlands in the 84 countries would be impacted by a 5 m sea level rise. However, these impacts are not uniformly distributed across the regions and countries of the developing world.

Impacts were also estimated for individual countries/territories. Table 12.3 summarizes the results for each indicator by presenting the top 10 impact cases (as a percentage of their national values). For this purpose, the 1 m sea level rise scenario was used. In terms of population impacted, the top 10 countries/territories worldwide

Table 12.3 Top 10 most impacted countries with a 1 m sea level rise (percentage impact in parenthesis)

Rank	Land area	Population	GDP	Urban areas	Agricultural land	Wetlands
1	The Bahamas (11.57)	Vietnam (10.79)	Vietnam (10.21)	Vietnam (10.74)	A.R. Egypt (13.09)	Vietnam (28.67)
2	Vietnam (5.17)	A.R. Egypt (9.28)	Mauritania (9.35)	Guyana (10.02)	Vietnam (7.14)	Jamaica (28.16)
3	Qatar (2.70)	Mauritania (7.95)	A.R. Egypt (6.44)	French Guiana (Fr) (7.76)	Suriname (5.60)	Belize (27.76)
4	Belize (1.90)	Suriname (7.00)	Suriname (6.35)	Mauritania (7.50)	The Bahamas (4.49)	Qatar (21.75)
5	Puerto Rico (1.64)	Guyana (6.30)	Benin (5.64)	A.R. Egypt (5.52)	Argentina (3.19)	The Bahamas (17.75)
6	Cuba (1.59)	French Guiana (Fr) (5.42)	The Bahamas (4.74)	Libya (5.39)	Jamaica (2.82)	Libya (15.83)
7	Taiwan, China (1.59)	Tunisia (4.89)	Guyana (4.64)	United Arab Emirates (4.80)	Mexico (1.60)	Uruguay (15.14)
8	The Gambia (1.33)	United Arab Emirates (4.59)	French Guiana (Fr) (3.02)	Tunisia (4.50)	Myanmar (1.48)	Mexico (14.85)
9	Jamaica (1.27)	The Bahamas (4.56)	Tunisia (2.93)	Suriname (4.20)	Guyana (1.16)	Benin (13.78)
10	Bangladesh (1.12)	Benin (3.93)	Ecuador (2.66)	The Bahamas (3.99)	Taiwan, China (1.05)	Taiwan, China (11.70)

Source: Dasgupta et al. (2009); The impact of sea level rise on developing countries: a comparative analysis, Climatic Change, 93:379–388

are Vietnam, A.R. Egypt, Mauritania, Suriname, Guyana, French Guiana (Fr), Tunisia, United Arab Emirates, the Bahamas and Benin. Around 10% of the populations of Vietnam and the A.R. of Egypt would be displaced with a 1 m sea level rise for land area. The Bahamas is by far the most impacted country, with close to 12% of its area inundated. Ten percent of the urban areas of Vietnam and Guyana would be inundated by a 1 m sea level rise, and the A.R. of Egypt's agricultural land would experience the largest percentage impact: 13% would be submerged. The research indicates that areas inundated by a 1 m sea level rise would account for 10% of GDP in Vietnam and more than 5% in Mauritania, A.R. Egypt, Suriname and Benin. Finally, nearly 28% of the wetlands in Vietnam, Jamaica and Belize would be inundated by a 1 m sea level rise. For all of the indicators used in this research, Vietnam ranks among the top 5 most impacted countries, with the A.R. of Egypt, Suriname and the Bahamas consistently ranking among the highest.

(b) **Ocean Warming**

The rise of atmospheric temperature in recent *era* has also been reflected through increase of aquatic temperature. The temperature of the Antarctic Southern Ocean rose by 0.17 °C (0.31 °F) between the 1950s and the 1980s, nearly twice the rate for the world's oceans as a whole. The temperature rise in the Gulf of Mexico is another prominent case study from the United States. As hurricanes cross the warm Loop Current coming up from South America, they can gain great strength in under a day (as did Hurricane Katrina and Hurricane Rita in 2005), with water above 85 °F seemingly promoting Category 5 storms. Hurricane season ends in November as the waters cool.

(c) **Acidification**

The world's oceans soak up much of the carbon dioxide produced by living organisms, either as dissolved gas or in the skeletons of tiny marine creatures that fall to the bottom to become chalk or limestone. Oceans currently absorb about 1 tonne of CO_2 per person per year. It is estimated that the oceans have absorbed around half of all CO_2 generated by human activities since 1800 (120,000,000,000 tonnes or 120 petagrams of carbon).

In the aquatic phase, carbon dioxide becomes a weak carbonic acid resulting in the lowering of pH value. It has been observed that the increase in the greenhouse gas since the industrial revolution has already lowered the average pH of seawater by 0.1 units. Predicted emissions could lower it further about 0.5 unit by 2100, to a level not seen for millions of years. There are concerns that lowering of aquatic pH could have a detrimental effect on corals (16% of the world's coral reefs have died from bleaching caused by warm water in 1998, which coincidentally was the warmest year ever recorded) and other marine organisms (particularly the molluscan community) due to dissolution of their calcium carbonate shells. Increased acidity may also directly affect the growth and reproduction of fish as well as the plankton on which they rely for food.

(d) **Shutdown of Thermohaline Circulation**

There is some speculation that global warming could, via a shutdown or slowdown of the thermohaline circulation, trigger localized cooling in the North Atlantic and lead to cooling, or lesser warming, in that region. This would affect in particular areas like Scandinavia and Britain that are warmed by the North Atlantic drift. More significantly, it could lead to an oceanic anoxic event.

The chances of this near-term collapse of the circulation are unclear; there is some evidence for the short-term stability of the Gulf Stream and possible weakening of the North Atlantic drift. However, whether the degree of weakening will be sufficient to shut down the circulation is under debate. There is no evidence for cooling in Northern Europe or nearby seas.

(e) **Effects on Coral Reef**

Coral reefs are colourful underwater taxonomically diverse biological units that vibrate with a wide spectrum of life forms and act as a natural protective barrier for coastal regions. Global warming affects the coral reefs in several ways. Studies show that warmer-than-normal sea temperatures are contributing to coral bleaching, which has been on the rise in recent years, weakening and killing reefs in many parts of the world. Coral bleaching occurs when the coral is stressed and expels the zooxanthellae. Without them, the coral loses its colour and becomes white. Sometimes the coral ingests new zooxanthellae, but if it fails, the coral will die. Corals thrive in water with temperature ranging between 66 and 86 °F. If the temperature changes 2 or 3 degrees outside this range, the coral faces a great stress. The trend of rising ocean water temperature in recent years due to anthropogenic-induced global warming phenomenon is thus a potential threat to coral reef community. The warmer the water, the more the ocean becomes unfavourable for the survival and growth of corals, which eventually leads to coral bleaching.

Sea level rise caused by melting sea ice and thermal expansion of the oceans could also cause problems for some reefs by making them too deep to receive adequate sunlight, which is another factor important for survival of corals.

Higher levels of carbon dioxide in the atmosphere are changing the chemistry of the oceans, making it more difficult for corals to build up calcium carbonate skeletons. The ocean absorbs almost a third of atmospheric carbon dioxide. This was, at one time, seen as positive due to the diminishing effect it has on global warming; however, high levels of carbon dioxide in the ocean lowers the pH or increase the acidity, which is harmful to the growth and survival of corals. The pH has already dropped a tenth of a unit and might drop up to a third of unit in the coming hundred years. The source of the majority of these carbon dioxide emissions is industrial exhaust, such as cars and factories. This affects corals because their skeletons are made of calcium carbonate. When the atmospheric carbon dioxide is absorbed and mixed with seawater, it forms carbonic acid. This substance complicates the building of body parts by marine animals, such as coral. Corals' growth rate is slowed, and the shells of marine creatures gradually become thin and finally dissolve. Several observations state that in 1 year alone, 16% of the world's coral reefs were wiped out. However, there are other views also. In 1997 experiment was carried out with samples of the coral *Acropora grandis* that were sampled from the hot water outlet of a nuclear power plant near Nanwan Bay, Taiwan. In 1998, the year the power plant began full operation, the coral samples were completely bleached within 2 days of exposure to a temperature of 33 °C. Two years later, however, a different picture was observed. It was found that samples taken from the same area did not even start bleaching until 6 days after exposure to 33 °C temperatures.

12.2.3 Effects on Human Health

The effects of climate change will have an enormously negative impact on human health worldwide. Increasing temperatures, drought and flooding will cause increases in starvation, disease and deaths in human populations. One of the most devastating effects of climate change on human health will be the increase in the outbreaks of diseases, such as malaria, dengue fever, encephalitis, Rift Valley fever, hantavirus pulmonary syndrome, West Nile virus, Lyme diseases, cholera and respiratory illness caused by increasing episode of fire.

The increase in outbreaks of mosquito-borne diseases in developing countries is one of the most dangerous effects associated with climate change. Malaria is one of the most prevalent and deadly of these diseases, and it is currently plaguing many developing countries, causing between one and three million deaths per year. The most dangerous form of malarial parasite is *Plasmodium falciparum* and is a widely abundant parasite in Africa. Recent evidences suggest that this form of malaria is also becoming increasingly abundant in Venezuela and Sri Lanka. Outbreaks of malaria caused by flooding and abnormally wet conditions can also be linked to drought conditions. For example, flooding in Mozambique in February and March of 2000 caused a malaria epidemic. In Brazil, abnormal droughts in the 1980s and 1990s caused crop failure and famine. Under these circumstances people were forced to migrate to other areas of the Amazon where they subsequently contracted malaria and brought it back to their home towns when the drought subsided. With increasing extreme weather events associated with climate change, it can be expected that more droughts and periods of abnormally wet conditions will result in more cases of malaria throughout the world, and, more specifically, in developing countries. This conclusion can also be drawn for other mosquito-vectored illness, such as dengue fever.

People are also experiencing increase in respiratory illnesses due to the changing climate. For instance, many people suffer from asthma, and in developing countries, asthma can be a very debilitating illness. It has been shown that increase in CO_2 causes plants to photosynthesize at a faster rate, producing more pollen. Warming temperatures have also been causing plants to release pollen earlier each year, resulting in longer growing seasons. This phenomenon has negative effects for those who suffer from asthma. Researchers also documented that droughts cause exponential increase in dust in countries like Africa and the Caribbean island, which further agitates respiratory ailments. The risks of forest fires also accelerate under drought condition as seen in Southeast Asia and the Amazon.

Heat waves are very dangerous to human health and are likely to become more frequent as climate change progresses. For instance, during the summer of 2003, Europe experienced what was considered the hottest summer since at least AD 1500. An event such as this, if it occurred in the United States, 'could cause thousands of excess deaths in the inner cities and could precipitate extensive blackouts' (Harvard Medical School Bulletin 2005; page 53). Heat waves cause death among the early, ill and very young age groups. Increased humidity associated with heat waves also has a negative effect on human health, causing discomfort, respiratory ailments and further exacerbating the problem of increasing insect populations leading to instances of insect vectored diseases.

12.2.4 Effects on Economic Profile

There exists considerable inter-relationship between climate change and economic profile of an area. The range of vulnerabilities that poor people face in different parts of the world encompasses all aspects of life that are related to climate change in some way or other. There are many ways to evaluate the relationship between climate change and vulnerability on the matrix of economic profile, but the 2001 IPCC Working Group II report on impacts, adaptation and vulnerability gives insights that may serve as a starting point of the subject. Alteration of crop production with respect to volume and pattern, salinization of coastal land, gradual vanishing of economically important fisheries and loss of mangroves are direct hit to the economic condition of the coastal people. Secondary impacts will likely include increases in food prices, frequency of fish diseases and greater problems associated with local services such as water supply and sanitation that affect the poor community.

Few case studies from the mangrove dominated deltaic complex of Indian Sundarbans revealed huge economic loss due to outbreak of viral diseases in the sector of tiger prawn culture. The frequency of the disease increases during the hot summer months (premonsoon season), when the water temperature of the culture pond exceeds ~34 °C. Warming of the Earth climate will accelerate the disease problems in the sphere of shrimp culture. This will affect a large population of coastal zone in India particularly in the maritime states of Andhra Pradesh, Orissa and West Bengal, where shrimp culture is the dominant livelihood of a considerable fraction of the local population.

The changing climate patterns, and especially the increased frequency and/or severity of extreme events, will increase vulnerability to natural disasters, both slower-onset ones such as droughts and rapid-onset disasters such as floods and cyclones. These will affect many areas, but semi-arid areas (prone to droughts) and coastal and deltaic regions (prone to floods and storms) are particularly vulnerable. The incidence of AILA (a severe tropical cyclone) on May 25, 2009 in the Gangetic delta region is a relevant example in this context. AILA was formed in the central Bay of Bengal as the net output of several factors. Around May 20, 2009, monsoon initiated at Andaman. The moisture laden south westerlies accelerated the moisture content in the winds of the Bay of Bengal. The wind speed was also variable in the northern and southern Bay of Bengal. In southern Bay of Bengal, the wind speed in the lower troposphere was around 37 km/h, whereas in northern Bay of Bengal, it was about 9 km/h. These variations led to the curling of winds, which is known as positive relative vorticity. Basically an area of depression developed in the Bay of Bengal on May 20, 2009, which transformed into a cyclone on May 23 and hit the deltaic complex of Indian Sundarbans on May 25 destroying the lives and properties of million of island dwellers.

The effect of AILA was severe resulting in the death of people, damages to properties and alterations in the physico-chemical properties of the soil and water.

Embankments were destroyed, agricultural fields and freshwater ponds lost their productivity due to intrusion of saline water and many people became homeless. 70 island dwellers died in the Indian Sundarbans region. Over 8000 people could not be traced, and about a million became homeless in India and Bangladesh. The embankments, which act as the line of defence of the villages, were completely smashed by tidal surges. Out of 3500 km long stretch of embankment, 400 km were completely broken, and another 565 became highly vulnerable. 2000 cattle died during Aila and 200 died because they drank saline water. Even honey makers' business was completely destroyed. One hundred and fifty boxes of bees were destroyed. The agricultural fields were submerged by seawater, and all standing crops were destroyed. The freshwater ponds were transformed in to brackish water bodies, and several fish species of saline water were recovered from these ponds. The entire spectrum of damages is summarized in Table 12.4.

Dangers of erosion, landslides and flash floods will also increase, particularly in many hilly and mountainous areas.

Changing climate patterns and more extreme events will also have negative impacts on several new livelihood activities (such as tourism) that will limit diversification of opportunities which, combined with damage to infrastructure and other types of physical capital, will affect the wider range of vulnerabilities (such as limited access to markets) the poor face. The weak social and political capital of many underdeveloped nations, along with extremely limited access to financial capital, means that the poor communities are least likely to be protected by investments in infrastructure or disaster mitigation and relief systems. This will ultimately disrupt the total economic spectrum of the poor class particularly of those living below the poverty line.

The predicted adverse impact on human health associated with climate change will affect the poor in particular throughout the developing countries. These risks are primarily associated with water-borne (such as dysentery or cholera) and vector-borne (such as malaria) diseases as well as heat stress morbidity and mortality. The health impacts pose a double jeopardy for poor people's livelihoods: the contribution of key productive members of the household is lost, and the cost of health care is expensive and time-consuming. Such risks will be widespread with the passage of time, but the dearth of medical care systems in many more remote, poorer areas of Africa and Asia in particular will accelerate the vulnerability and magnitude of the risks.

The deterioration of the availability or quality of water supplies in many areas (again due to wider resource stresses that climate change will exacerbate) will

Table 12.4 Damage spectrum of AILA

Affected	West Bengal	Sundarbans (Indian part)
Villages	28,359	4249
Population	670,000	254,000
Human lives lost	137	70
Livestock lost	71,196	70,811
Crop area affected (ha)	290,000	125,000

significantly increase many of these health risks, while poorer nutritional states caused by declining food security will make a large percentage of poor people more vulnerable to the effects of diseases when they do strike. Few significant changes associated with climate change are listed in Table 12.5.

Table 12.5 Twentieth century changes in the Earth's atmosphere, climate and biophysical system

Indicator characteristic	Observed change
Concentration indicator	
Atmospheric concentration of CO_2	280 ppm for the period 1000–1750 to 368 ppm in year 2000 (~31.4% increase)
Atmospheric concentration of CH_4	700 ppb for the period 1000–1750 to 1750 ppb in year 2000 (151 ± 25% increase)
Atmospheric concentration of N_2O	270 ppb for the period 1000–1750 to 316 ppb in year 2000 (17 ± 5% increase)
Tropospheric concentration of O_3	Increased by 35 ± 15% from the years 1750 to 2000; varies with region
Stratospheric concentration of O_3	Decreased over the years 1970 to 2000; varies with altitude and latitude
Atmospheric concentration of HFCs, PFCs and SF_6	Increased globally over the last 50 years
Weather indicators	
Global mean surface temperature	Increased by 0.6–0.2 °C over the twentieth century; land areas warmed more than the oceans (very likely)
Northern Hemispheric surface temperature	Increased over the twentieth century; greater than during any other century in the last 1000 years; 1990s was the warmest decade of the millennium (likely)
Diurnal surface temperature range	Decreased over the years 1950 to 2000 over land: Night time minimum temperatures increased at twice the rate of day time
Hot days/heat index	Maximum temperature (likely)
Cold/frost days	Increased (likely)
Continental precipitation	Decreased for nearly all land areas during the twentieth century (very likely). Increased by 5–10% over the twentieth century in the Northern Hemisphere (very likely) although decreased in some region (e.g., north and West Africa and parts of the Mediterranean)
Heavy precipitation	Increased at mid and high northern latitudes (likely)
Frequency and severity of droughts	Increased summer drying and associated incidence of drought in a few areas (likely)
	In some regions, such as parts of Asia and Africa, the frequency and intensity of droughts have been observed to increase in recent decades
Biological and physical indicators	
Global mean sea level	Increased at an average annual rate of 1–2 mm during the twentieth century
Duration of ice cover of rivers and lakes	Decreased by about 2 weeks over the twentieth century in mid and high latitudes of the Northern Hemisphere (very likely)
Arctic sea–ice extent and thickness	Thinned by 40% in recent decades in late summer to early autumn (likely) and decreased in extent by 10–15% since the 1950s in spring and summer

(continued)

Table 12.5 (continued)

Indicator characteristic	Observed change
Non-polar glaciers	Widespread retreat during the twentieth century
Snow cover	Decreased in area by 10% since global observation become available from satellites in the 1960s (very likely)
Permafrost	Thawed, warmed and degraded in parts of the polar, sub polar and mountainous regions
El Niño events	Became more frequent, persistent and intense during the last 20–30 years compared to the previous 100 years
Growing season	Lengthened by about 1–4 days per decade during last 40 years in the Northern Hemisphere, especially at higher latitudes
Plant and animal ranges	Shifted polewards and up in elevation for plants, insect, birds and fish
Coral reef bleaching	Increased frequency, especially during *El Niño* events
Economic indicators	
Weather-related economic losses	Global inflation; adjusted losses increased over the last 40 years. Part of the observed upward trend is linked to climatic factors

12.3 Mitigation of Climate Change

Several technological measures can be adopted to combat climate change. Some important steps are listed here:

1. Developing substitutes for chlorofluorocarbons. Chlorofluorocarbons (CFC) are the chemicals used as refrigerants. They cause ozone depletion as they release chloride radicals which on reacting with ozone cause its degradation. The chlorofluorocarbons can be substituted by hydrofluorocarbon (HFC) or hydrochlorofluorocarbon (HCFC) for using as refrigerants and in air conditioners. HFCs and HCFCs also deplete ozone but to lesser extent than CFCs.
2. Adaptation to alternative livelihoods with low carbon foot prints like farming of nutritionally rich mangrove associates species for preparing food products, seaweed culture, oyster culture, polyculture (of fishes) by promoting phytoplankton growth through iron fertilization, etc. Few alternative livelihoods are discussed in Annexure 12A.
3. Minimizing the use of nitrogen fertilizers in agriculture for reducing N_2O emissions.
4. Reducing the use of fossil fuels and promoting alternative renewable energy sources like wind energy, solar energy, tidal energy, etc.
5. Increasing the vegetation cover through plantation of selective tree species that has considerable potential to store carbon. The authors made a detailed study on this aspect in March 2020 in three cities of India, namely, Kolkata, Bhubaneswar and Jamshedpur and identified few species with high carbon sequestration potential (Table 12.6).

Table 12.6 List of dominant tree species in the selected stations with their respective above ground biomass (AGB) and above ground carbon (AGC)

Sl. No.	Species	AGB (tonnes ha^{-1})			AGC (tonnes ha^{-1})		
		A	B	C	A	B	C
1.	 *Cocos nucifera* (coconut)	1146.69	1010.38	920.75	541.24	468.82	417.10
2.	 *Azadirachta indica* (neem)	898.60	500.72	313.74	425.94	230.83	141.81
3. am	 *Mangifera indica* (mango)	1009.16	800.83	480.38	482.38	370.78	218.57
4.	 *Tamarindus indica* (tentul)	701.95	520.93	301.45	332.72	240.15	136.56
5.	 *Bombax ceiba* (shimul)	984.73	798.50	635.88	468.73	369.71	290.60
6.	 *Aegle marmelos* (bel)	379.78	246.49	107.85	179.64	113.39	48.64

(continued)

Table 12.6 (continued)

Sl. No.	Species	AGB (tonnes ha^{-1})			AGC (tonnes ha^{-1})		
		A	B	C	A	B	C
7.	 *Terminalia arjuna* (arjun)	369.40	257.49	187.09	174.36	118.45	84.38
8.	 *Delonix regia* (krishnachura)	2011.16	1261.38	860.83	955.30	586.54	392.54
9.	 *Artocarpus heterophyllus* (jackfruit)	1011.40	665.80	134.52	481.43	308.93	61.21
10.	 *Psidium guajava* (guava)	1254.28	798.64	630.77	594.53	371.37	287.00
11.	 *Acacia auriculiformis* (akashmoni)	1292.39	997.38	667.30	613.89	463.78	304.96

(continued)

Table 12.6 (continued)

Sl. No.	Species	AGB (tonnes ha^{-1})			AGC (tonnes ha^{-1})		
		A	B	C	A	B	C
12.	 *Ziziphus mauritiana* (kul)	1003.95	687.38	307.59	473.86	318.94	139.34
13.	 *Eucalyptus globulus* (eucalyptus)	1001.95	932.6	858.85	477.93	434.59	389.92
14.	 *Peltophorum pterocarpum* (radhachura)	812.9	689	609.99	385.31	319.01	276.94
15.	 *Polyalthia longifolia* (debdaru)	1010.84	796.83	347.24	482.17	368.14	157.30
16.	 *Ficus religiosa* (peepul)	2012.58	1690.50	1148.96	962.01	786.08	526.22

(continued)

Table 12.6 (continued)

Sl. No.	Species	AGB (tonnes ha^{-1})			AGC (tonnes ha^{-1})		
		A	B	C	A	B	C
17.	 *Ficus benghalensis* (banyan)	2115.45	2096.93	1101.51	1011.19	977.17	504.49

Note: A , Kolkata; B, Bhubaneswar; C, Jamshedpur

It is observed from this first-order analysis that *Ficus benghalensis* and *Ficus religiosa* are highly effective in storing carbon. The magnitude of storage is, however, site-specific showing a significant spatial trend as per the order Kolkata > Bhubaneswar > Jamshedpur. This may be attributed to edaphic factors as the trees with almost uniform age (in years) were selected for each species.

The order of AGC in Kolkata is *Ficus benghalensis* (1011.19 tha^{-1}) > *Ficus religiosa* (962.01 tha^{-1}) > *Delonix regia* (955.30 tha^{-1}) > *Acacia auriculiformis* (613.89 tha^{-1}) > *Psidium guajava* (594.53 tha^{-1}) > *Cocos nucifera* (541.24 tha^{-1}) > *Mangifera indica* (482.38 tha^{-1}) > *Polyalthia longifolia* (482.17 tha^{-1}) > *Artocarpus heterophyllus* (481.43 tha^{-1}) > *Eucalyptus globulus* (477.93 tha^{-1}) > *Ziziphus mauritiana* (473.86 tha^{-1}) > *Bombax ceiba* (468.73 tha^{-1}) > *Azadirachta indica* (425.94 tha^{-1}) > *Peltophorum pterocarpum* (385.31 tha^{-1}) > *Tamarindus indica* (332.72 tha^{-1}) > *Aegle marmelos* (179.64 tha^{-1}) > *Terminalia arjuna* (174.36 tha^{-1}).

The order of AGC in Bhubaneswar is *Ficus benghalensis* (977.17 tha^{-1}) > *Ficus religiosa* (786.08 tha^{-1}) > *Delonix regia* (586.54 tha^{-1}) > *Cocos nucifera* (468.42 tha^{-1}) > *Acacia auriculiformis* (463.78 tha^{-1}) > *Eucalyptus globulus* (434.59 tha^{-1}) > *Psidium guajava* (371.37 tha^{-1}) > *Mangifera indica* (370.78 tha^{-1}) > *Bombax ceiba* (369.71 tha^{-1}) > *Polyalthia longifolia* (368.14 tha^{-1}) > *Peltophorum pterocarpum* (319.01 tha^{-1}) > *Ziziphus mauritiana* (318.94 tha^{-1}) > *Artocarpus heterophyllus* (308.93 tha^{-1}) > *Tamarindus indica* (240.15 tha^{-1}) > *Azadirachta indica* (230.83 tha^{-1}) > *Terminalia arjuna* (118.45 tha^{-1}) > *Aegle marmelos* (113.39 tha^{-1}).

The order of AGC in Jamshedpur is *Ficus religiosa* (526.22 tha^{-1}) > *Ficus benghalensis* (504.49 tha^{-1}) > *Cocos nucifera* (417.10 tha^{-1}) > *Delonix regia* (392.54 tha^{-1}) > *Eucalyptus globulus* (389.92 tha^{-1}) > *Acacia auriculiformis* (304.96 tha^{-1}) > *Bombax ceiba* (290.60 tha^{-1}) > *Psidium guajava* (287.00 tha^{-1}) > *Peltophorum pterocarpum* (276.94 tha^{-1}) > *Mangifera indica* (218.57 tha^{-1}) > *Polyalthia longifolia* (157.30 tha^{-1}) > *Azadirachta indica* (141.81 tha^{-1}) > *Ziziphus mauritiana* (139.34 tha^{-1}) > *Tamarindus indica* (136.56 tha^{-1}) > *Terminalia arjuna* (84.38 tha^{-1}) > *Artocarpus heterophyllus* (61.21 tha^{-1}) > *Aegle marmelos* (48.64 tha^{-1}).

In addition to above technological approaches, various policy-level efforts at international levels are being carried out for the past few decades to minimize global warming by reducing the release of greenhouse gases into the environment through control of combustion of fossil fuels, deforestation and CFCs. Some of the past major efforts are as follows:

1. *Kyoto Protocol*: A group of 140 developing countries G-77 organized an International Conference during December 1–10, 1997 at Kyoto, Japan, to discuss the impact of global warming and decided to sign an agreement for controlling it. This agreement was basically a legally binding convention signed in 1992 following the Earth Summit at Rio in 1992. This protocol came into effect on February 16, 2005, and included 55 countries responsible for emitting 55% of greenhouse gases (GHGs). The United States alone accounts for 30% of world greenhouse gas emission. Following this observation, it was resolved in the meeting that greenhouse gas emissions especially carbon dioxide need to be reduced; fines would be imposed on countries flouting this protocol, and a clean development fund will need to be formed. This effort was made to curtail greenhouse gas emissions by industrialization countries.
2. *The Toronto Resolution*: A World Conference was held at Toronto, Canada, in June, 1998, where highly industrialized countries, such as the United States, the United Kingdom, Germany, France, etc., were asked for 20% voluntary reduction in the emission of carbon dioxide by the year 2005. Developed countries did not sign the Toronto Resolution. The Cancan Meet (2010), even though failed, was able to set up a green fund of $100 billion per year by 2020 for developing countries to reduce global warming. A total of 105 countries signed the treaty and agreed to stop warming the Earth and were warned that only 7 years remained to save the Earth from disastrous effects of global warming. G-8 countries meetings at Germany in July 2007 agreed to halve the production of greenhouse gases from the present level by 2050.
3. *Switching Over to Alternative Source of Energy*: Attempts were made to use alternate fuels in place of fossils. Fuels derived from fermentation of carbohydrate-rich crops for producing alcohols, such as ethanol and methanol. Solar energy should be preferred in tropical countries where sunlight is abundantly available most of the year. Biogas may be considered for fulfilment of domestic needs. In addition to this, Ocean Thermal Energy Conversion (OTEC) and tidal energies are also the roadmaps to combat the adverse impact of GHGs.

COP 21 On December 12, 2015, 196 Parties to the UN Framework Convention on Climate Change (UNFCCC) adopted the Paris Agreement, a new legally binding framework for an internationally coordinated effort to tackle climate change. The Agreement represents the culmination of 6 years of international climate change negotiations under the auspices of the UNFCCC and was reached under intense international pressure to avoid a repeat failure of the Copenhagen conference in 2009. The Agreement establishes a global warming goal of well below 2 °C on pre-industrial averages. It requires countries to formulate progressively more ambitious climate targets which are consistent with this goal. To achieve this goal, all Parties to the Paris Agreement will need to make profound changes to their economies. The Paris Agreement defines a universal, legal framework to 'strengthen the global response to the threat of climate change' (Art. 2). It establishes the obligation of all Parties to contribute to climate change mitigation and adaptation. For the first time, all countries will develop plans on how to contribute to climate change mitigation and will communicate their 'nationally determined contributions' to the Secretariat of the Convention. The Paris Agreement puts emphasis on processes rather than on defined mitigation goals. Unlike the Kyoto Protocol, the Paris Agreement does not formulate country-specific emissions targets. Instead, the Paris Agreement depends on voluntary mitigation contributions and a series of processes that seek to ensure collective and individual progress in meeting the initial and progressively more ambitious mitigation contributions.

COP 22 This made clear that the main task that lies ahead is devising rules that are inclusive and flexible in the context of environmental integrity. Ambition and transparency are essential for the successful implementation of the Paris Agreement.

- COP 22 established a timeline for negotiations on elaborating the Paris Rulebook and taking initial steps on understanding how all pieces of the Paris Agreement could fit together.
- Political momentum is building up for 2018, when it is expected that the Paris Decision's work programme will be concluded and a complete set of guidelines, modalities and procedures adopted.
- Although little progress was achieved on finance, Parties agreed that the Adaptation Fund should serve the Paris Agreement. The increased profile of the Adaptation Fund can also be partly attributed to the true African focus given by Morocco to COP 22, placing the initiative for the Adaptation of African Agriculture at the heart of the discussions.
- Workshops and events outside the negotiating rooms took centre stage, bringing together a range of stakeholders and signaling that climate change is increasingly becoming a global, multi-level and cross-sectoral regime.

COP 22 produced little news on climate finance. Parties initiated a process to identify quantitative and qualitative information on the provision of financial resources (and which must be communicated biennially by developed countries under the Paris Agreement). Parties did not make progress on defining modalities for the accounting of financial resources provided and mobilized by developed countries. Negotiations will resume in May at the 46th meeting of the Subsidiary Body for Technical Advice (SBSTA). In addition to promoting transparency, accuracy, consistency and comparability in reporting support, issues to be considered will include clarity on what counts as climate finance, the need to avoid double-counting of finance and how to integrate these modalities into the enhanced transparency framework established by Article 13 of the Paris Agreement.

COP23 COP23 was a largely procedural meeting concerned with defining the rules of implementation of the Paris Agreement.

The meeting once again highlighted the interdependence of the different agenda items under the Paris Agreement. Negotiations will have to result in a 'package deal' in 2018, requiring talks on all articles to progress in coordination and at similar speeds. Progress on the main negotiation items was captured through a series of 'informal notes'. These lay the groundwork for a draft negotiation text on the 'Paris Rulebook', which summarizes implementation guidance for various aspects under the Paris Agreement.

There is reasonable confidence that Parties are on track to conclude the deliberations on the Paris Rulebook at COP24 in Katowice this year. Parties did however recognize the need for an additional negotiation session between May and December 2018.

COP23 hosted a record number of side events, giving a voice to a wide range of actors and showcasing a large number of initiatives. Negotiations took place in the Bula zone, while the Bonn zone brought together observers that presented concrete solutions for implementing the Paris Agreement. The physical distance between the official negotiations and the side events posed challenges for negotiators, who had to invest considerable time to commute between the Bula and Bonn zones. It also limited interactions between observers and negotiators. The presentation of concrete climate actions in the Bonn Zone was among the most exciting aspects of COP23 and for many provided cause for hope even against the backdrop of warnings that existing pledges are grossly insufficient for meeting the objectives of the Paris Agreement. NGOs and forward-looking businesses presented countless initiatives and projects. Over 200 local and regional leaders adopted the Bonn–Fiji Agreement, which commits them to the inclusive implementation of the Paris Agreement at the local level and to raising climate ambition.

12.4 Take Home Messages

(A) The United Nations Framework Convention on Climate Change (UNFCCC) uses the term 'climate change' for human-induced changes and 'climate variability' for other changes. Climate change is therefore any long-term significant change in the 'average weather' that a given region experiences and involves changes in the variability or average state of the atmosphere over durations ranging from decades to millions of years.

(B) The natural factors that trigger climate change are variations in the Earth's orbital characteristics, atmospheric carbon dioxide variations, volcanic eruptions, variations in solar output and plate tectonics. The human-induced factors that cause climate change are burning of fossil fuels, intense industrialization (preferably cement manufacturing), changing land-use pattern, livestock promotion, deforestation, etc.

(C) The effects of climate change can be felt hardest in a country like India. The fact that India has a large rural population who are dependent on the cycles of the climatic seasons for their agricultural activities, fishermen who work in the rivers and seas, farmers who need the seasonal monsoons and a large and varied indigenous population who live in harsh climatic regions of mountains, desert and river delta, makes India especially susceptible to a changing climate. Fishermen are among those hardest hit by climate change. The potential of a rise in sea levels due to global warming would wipe out existing shoreline, contaminate freshwater systems with salt water, diminish the minimum environment flow required for fishes in river and dry up floodplain wetlands, a rich habitat for fishes. The services that aquatic ecosystem provide to humans such as fisheries, flood protection, recreation and biodiversity conservation, which if converted in monetary terms, would be in millions of rupees.

(D) Some technological steps to combat climate change are developing substitutes for chlorofluorocarbons; adaptation to alternative livelihoods with low carbon foot prints; minimizing the use of nitrogen fertilizers in agriculture for reducing N_2O emissions, reducing the use of fossil fuels; promoting alternative renewable energy sources like wind energy, solar energy, tidal energy, etc.; and planting selective floral species with carbon storage potential.

(E) In addition to technological approaches, various policy-level efforts at international levels are being carried out for the past few decades to minimize global warming by reducing the release of greenhouse gases into the environment through control of combustion of fossil fuels, deforestation and CFCs. Kyoto Protocol, Toronto Resolution, Conference of the Parties (COP) are some justified steps in this context.

12.5 Brain Churners

This chapter provides the readers a basic idea on environment. The chapter has immense importance on the backdrop of natural resources, which is gradually depleting with the rise of human population. It is expected that the readers will find interest in answering the questions asked on various types of natural resources and their conservation. The section will also help the readers to face various competitive examinations as environment has been given thrust in recent times. The readers can consider this section as a sort of *self-appraisal approach* as correct answers to all the questions are filled in black in the Answer Box at the end of this section.

1. At the present time, the tilt of the Earth's axis is

   (a) 23.5°
   (b) 27.5°
   (c) 29.5°
   (d) None of these

2. Without the greenhouse effect, the average global temperature of the Earth would be around

   (a) −19 °C
   (b) −18 °C
   (c) −20 °C
   (d) None of these

3. Most computer climate models suggest that the globe will warm up by 1.5–4.5 °C if carbon dioxide reaches the predicted level of

   (a) 800 parts per million by the year 2050
   (b) 900 parts per million by the year 2050
   (c) 700 parts per million by the year 2050
   (d) 600 parts per million by the year 2050

4. Researchers believe that the Pinatubo eruption was primarily responsible for the 0.8 °C drop in global average air temperature in

   (a) 1993
   (b) 1992
   (c) 1995
   (d) None of the above

5. The North and South American plates collided to form the Isthmus of Panama and shut off direct mixing between the

   (a) Indian and Arctic Oceans
   (b) Arctic and Atlantic Oceans
   (c) Atlantic and Pacific Oceans
   (d) None of the above

6. Dams have led to a sea level drop over the past few decades

   (a) From −0.5 to −0.7 mm $year^{-1}$
   (b) From −0.7 to −0.8 mm $year^{-1}$
   (c) From −0.6 to −0.7 mm $year^{-1}$
   (d) From −0.4 to −0.5 mm $year^{-1}$

7. Reductions in polar ice volume led to

   (a) 5–7 m of sea level rise
   (b) 7–8 m of sea level rise
   (c) 4–6 m of sea level rise
   (d) None of the above

8. The frequent subsidence of some part of Sundarbans, causing the complete disappearance of forest chunk and settlement areas, is the effect of
   (a) 'Dredging of no ground' phenomenon
   (b) 'Digging of no ground' phenomenon
   (c) 'Drilling of no ground' phenomenon
   (d) 'Swatch of no ground' phenomenon.
9. Ganga trough is situated south of the
   (a) Raimangal – Malancha estuary
   (b) Hooghly – Matla estuary
   (c) Saptamukhi – Thakurani estuary
   (d) None of the above
10. In the aquatic phase, carbon dioxide becomes a weak carbonic acid resulting in the
    (a) Increasing of pH value
    (b) Lowering of pH value
    (c) Neutralizing of pH value
    (d) None of the above
11. Corals thrive in water with temperature ranging between 66 and 86 °F
    (a) 77–86 °F
    (b) 96–86 °F
    (c) 66–86 °F
    (d) None of the above
12. Chlorofluorocarbons (CFC) are the chemicals used as
    (a) Preservatives
    (b) Pesticides
    (c) Antioxidants
    (d) Refrigerants
13. The Earth's orbit gradually changes from being elliptical to being nearly circular and then back to elliptical in a period of about
    (a) 100,000 years
    (b) 150,000 years
    (c) 120,000 years
    (d) None of the above
14. The precession of the equinox has a cycle of approximately
    (a) 24,000 years
    (b) 25,000 years

(c) 26,000 years
(d) None of the above

15. The third cyclical variation is related to the changes in the tilt (obliquity) of the Earth's axis of rotation over a

(a) 42,000 years period
(b) 41,000 years period
(c) 43,000 years period
(d) None of the above

16. Computer models and historical evidences suggest that the Milankovitch cycles exert their greatest cooling and warming influence when the troughs and crests of all three cycles

(a) Coincide with each other
(b) Alter with each other
(c) Separate from each other
(d) None of the above

17. Variations in solar radiation received by the planet as controlled by the

(a) Milankovitch cycles
(b) Geothermal cycles
(c) Hydrological cycles
(d) None of the above

18. Recent stratospheric data suggests that large explosive volcanic eruptions also eject huge amounts of

(a) Carbon dioxide gas
(b) Nitrogen dioxide gas
(c) Sulphur dioxide gas
(d) All of the above

19. Sunspots are huge

(a) Sand storms
(b) Magnetic storms
(c) Wind storms
(d) All of the above

20. Anthropogenic aerosols, particularly sulphate aerosols from fossil fuel combustion, exert

(a) Warming influence
(b) Both warming and cooling influence
(c) Cooling influence
(d) None of the above

Answer Sheet

S. No.	a	b	c	d
1.	■			
2.		■		
3.				■
4.		■		
5.			■	
6.	■			
7.			■	
8.				■
9.	■			
10.		■		
11.			■	
12.				■
13.	■			
14.			■	
15.		■		
16.	■			
17.	■			
18.			■	
19.		■		
20.			■	

Annexure 12A: Value Addition to Rasgulla (A Traditional Indian Sweet) Through Tulsi (*Ocimum sanctum*) Leaf Extract: An Innovative Livelihood Programme

Introduction

Ocimum sanctum, commonly known as holy basil, belongs to Lamiaceae family (Fig. 12.A.1). The plant is used in Indian sub-continent since long time back to treat various types of human diseases (Pattanayak et al. 2010; Saharkhiz et al. 2015). The leaf is taken in many forms such as in herbal tea, dried powder, fresh leaf mixed with honey, etc. to protect against cough and cold. It has also been documented that tulsi may be a COX-2 inhibitor like several model pain killers due to its significant amount of eugenol (Rao et al. 2013; Pattanayak et al. 2010; Gaber et al. 2015). The plant contains several bioactive compounds (Devendran and Balasubramanian 2011; Triveni Kumar et al. 2013). Recent studies have shown the presence of several chemicals of pharmaceutical importance in this species (Rao et al. 2013; Pattanayak et al. 2010; Gupta et al. 2002). Considering the meritorious background of the species, the present pilot-scale experiment was conducted by the researchers and sweet

Fig. 12.A.1 Preparation of rasgulla from tulsi extract at Satyanarayan Mistanna Bhandar, Singur, Hooghly district, West Bengal (India)

makers of Satyanarayan Mistanna Bhandar, Singur, Hooghly district and West Bengal (India) to prepare special type of rasgulla by mixing formulated proportion of tulsi leaf extract in place of water in sugar syrup (Fig. 12.A.1). The primary vision of this study is to upgrade the quality of the sweet in terms of nutritional value and orient the local level traditional track of sweet making through the lane of pharmaceutical science.

Material and Methods

Analysis of *Ocimum sanctum* (Tulsi) Leaf

Fresh leaves of tulsi were collected from Singur region (22°81′50″ N; 88°23′45″ E) of Hooghly district in West Bengal from different patches. The dried leaf samples were grinded by mortal pastel and protein content was measured by Lowry's method using spectrophotometer as per the standard procedure (Lowry et al. 1951). The total carbohydrate content was estimated by phenol sulphuric acid method using D-Glucose as standard (Dubois et al. 1956). The total fat in the leaf sample was estimated by Soxhlet method using petroleum ether (80 °C) as per the procedure (AOAC 2000). The ascorbic acid of leaf sample was quantified as per the standard protocol (Zvaigzne et al. 2009; Nielsen 1998) and expressed in mg/100 gm. Minerals like Ca, K and Mg were analysed as per the standard method of flame photometry.

Preparation of *Ocimum sanctum* Leaf Extract

The collected tulsi leaves were shade dried and powdered. The dried leaf dusts were soaked overnight in sterilized water and filtered through fine loin cloth. The filtrate was centrifuged at 5000 rpm for 20 min using a REMI cold centrifuge. The supernatant, thus obtained, was filtered through loin cloth and the collected filtrate was preserved at −20 °C in clean Tarson bottles.

Preparation of Rasgulla with Tulsi Extract

Rasgulla is prepared from soft and freshly made channa. Cow milk is preferred for the production of rasgulla because it gives soft, spongy and juicy product.

For rasgulla production, channa is kneaded into smooth paste, and then small balls of about 15–20 mm diameter and 10–15 grams in weigh are made. The surface of balls should be smooth and free from of any cracks. One kg of channa yields about 40–60 rasgullas (depending on the size of the rasgulla). Rasgulla balls are cooked in sugar syrup having 50–60% sugar concentration for about 15–20 min. During cooking, a small amount of water is continuously added to maintain sugar concentration. This makes up for the loss of water due to evaporation. About 10% of cooking solution is replaced by fresh one, every time it is reused to cook another batch or rasgulla. After cooking rasgulla, balls are soaked in 40–45% sugar syrup for about 1–23 h. Our specially formulated rasgulla replaced the water of sugar syrup with *O. sanctum* leaf extract and mixed with equivalent proportion of sugar as that of normal (control). The entire marketing of this formulated rasgulla was done by Satyanarayan Mistanna Bhandar, Singur, Hooghly district, West Bengal (India) (Fig. 12.A.2).

Analysis of Rasgulla (A Comparative Analysis of Control and Experimental Samples)

The success of any experimental approach is monitored on the basis of a comparative study between control and experimental samples (Fig. 12.A.3) in terms of few parameters (variables) that are relevant indicators of human health. In the present programme, we have used the values of protein, carbohydrate, fat, ascorbic acid, Ca, K and Mg as indicators of comparative study. These indicators (compounds/elements) are intricately related to human health.

Fig. 12.A.2 Sticker and signboard (both in English and local language Bengali) for marketing tulsi extract based rasgulla

Fig. 12.A.3 View of control (without tulsi extract) rasgulla and experimental (with tulsi extract)

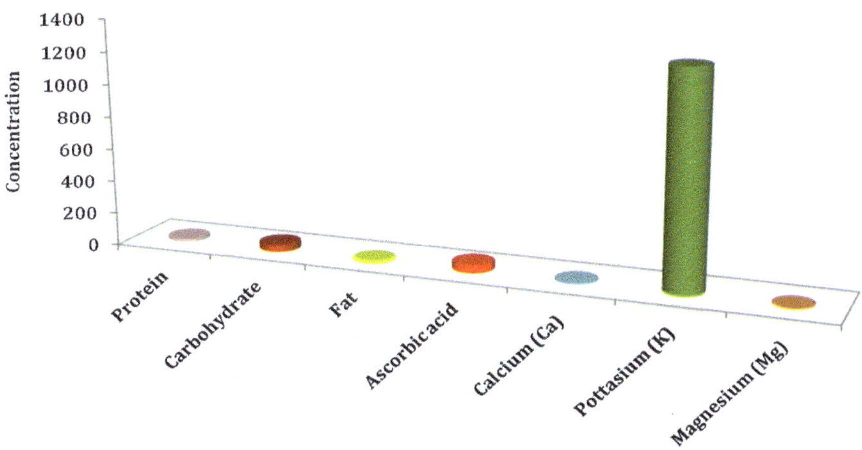

Fig. 12.A.4 Composition of *O. sanctum* leaf; units of protein, carbohydrate and fat are in percentage; for ascorbic acid the unit is mg/100 gm, and for Ca, K and Mg, the units are mg/100 g dry matter

Result

Figure 12.A.4 presents the composition of *O. sanctum* leaf.

Figure 12.A.5 represents a comparative picture of control and experimental samples, which clearly depicts the improvement of the experimental rasgulla (prepared from tulsi extract), from the nutritional point of view.

Discussion

Rasgulla is undoubtedly the most popular traditional Indian milk sweet prepared from channa. The soft ball of channa soaked in sugar syrup delights the taste buds of sweet lovers. However, to many customers, this sweet ball is a cause of concern

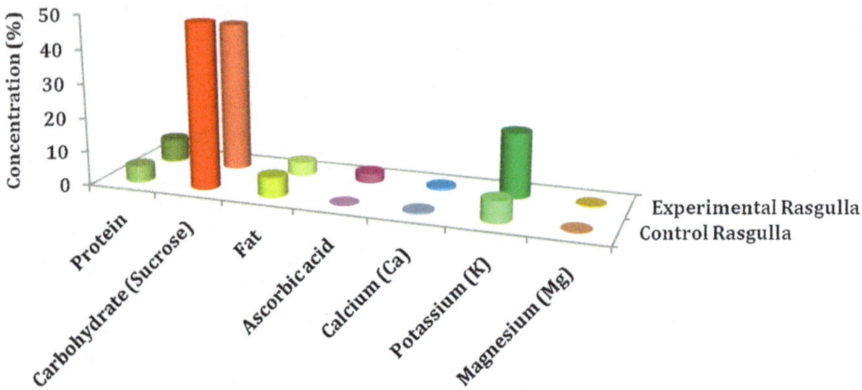

Fig. 12.A.5 Comparative study of nutritional value between control (without tulsi extract) and experimental (with tulsi extract) rasgullas

as the sweet sugar is intricately related with diabetes. Hence, the diabetic patients usually avoid the consumption of this delicious sweet. There are various types of rasgulla sold in the market. This includes ordinary sugar mixed with rasgulla, spongy rasgulla and even diabetic rasgulla. Spongy rasgulla differs in terms of taste, texture body and succulence as compared to ordinary rasgulla. The diabetic rasgulla is prepared by replacing sucrose with low-calorie sweeteners or alcoholic sugar such as sorbitol to cater the need of the people suffering from diabetes to health conscious consumers. In this pilot programme, we have introduced *O. sanctum* leaf extract with the aim to introduce the medicinal value of the species to the consumer through mouth-watering rasgulla. We observe that there is an increase of 48.68% protein in our formulated rasgulla compared to control samples. Interestingly the fat and sucrose have decreased by 32.74% and 9.43%, respectively. Also the minerals like Ca and K have increased. The ascorbic acid has also increased in the experimental rasgulla which strongly confirms the role of this formulated sweet to fight against cough and cold. It can be concluded from the current work that the consumption of this specially formulated rasgulla from *O. sanctum* leaf extract has the capability to provide the protection against several diseases preferably cough and cold, which is very common among the population of the tropical countries. We expect that the pharmaceutical benefits of tulsi (Annexure 12B) can be introduced in human system through this effort of preparing rasgulla from the leaf extract of the species, provided a backup nursery which is maintained for a regular flow of this unique medicinal raw material.

Annexure 12B

1. With high levels of beta-carotene, vitamin A, cryptoxanthin, lutein and zeaxanthin in tulsi, the plant parts act as protective scavengers against oxygen-derived free radicals and reactive oxygen species (ROS), thereby preventing premature ageing and various diseases. The high level of vitamin C also helps to fight against cough and cold.
2. Tulsi is rich in zeaxanthin. Zeaxanthin, a yellow flavonoid carotenoid compound, helps in filtering harmful UV rays by reaching the retina, thereby protecting against age-related muscular degeneration (ARMD), particularly in elderly people.
3. Tulsi contains exceptionally high amounts of vitamin A, which is essential for good vision due to its antioxidant properties. Besides, it is also necessary for maintaining a healthy mucus membrane and skin.
4. Tulsi contains significant amounts of vitamin K, which plays an important role in strengthening bones and assisting the mineralization process of the bones.
5. Potassium, manganese, copper and magnesium are vital for the proper functioning of the body. While potassium is required for controlling heart rate and blood pressure, manganese is important for the antioxidant enzyme superoxide dismutase. Tulsi has all these compounds in considerable level and, hence, should form a part of a healthy diet.
6. Tulsi is a good source of volatile oils, like eugenol, cineole, estragole, limonene, linalool, myrcene, sabinene and others, which have strong antibacterial properties. These oils combine to fight against many bacterial infections, such as enterococcus, staphylococcus, etc.
7. With anti-inflammatory effects, tulsi is recommended to people suffering from arthritis. Basil oil, in particular, contains eugenol, a substance that blocks the activity of an enzyme in the body called cyclooxygenase which causes swelling.
8. Magnesium, present in considerable quantity in tulsi, relaxes the muscles and blood vessels, thereby improving blood flow and reducing the risk of irregular heart rhythm of heart muscles and blood vessels.
9. Tulsi oil makes unique skin and hair moisturizer, as it enhances the lustre and shine of dull-looking skin and hair. Tulsi is widely used for curing skin problems, such as acne and psoriasis.
10. Constipation, stomach cramps, indigestion and flatulence can be treated by consuming tulsi tea. As such, it provides immediate relief from gas in the stomach and intestines.
11. The essential oil of tulsi acts as an anti-vomiting agent in motion sickness and various other vomiting cases. Besides, it is highly recommended in effectively curing cold, influenza, whooping cough, asthma, bronchitis and sinus infections.

(continued)

12. For treating stress-related problems like migraines and depression, tulsi oil finds a significant place in aromatherapy due to its calming effects and promotion of mental strength and clarity.
13. Tulsi, when consumed in the form of herbal tea, helps in alleviating the symptoms of urinary infections, colics, anorexia, gastric ulcers and diarrhoea.
14. Tulsi is widely used in Ayurvedic medicines for curing a variety of common ailments, such as diabetes, respiratory disorders, impotence, allergies and infertility.
15. With a powerful compound called cinnamanic acid present, tulsi has been proved to stimulate circulation, stabilize blood sugar and improve respiration.
16. Effective treatment of skin rashes, eczema, wounds and insect bites constitutes other valuable health benefits of tulsi.

References

Antonov JI, Levitus S, Boyer TP (2005) Steric variability of the world ocean, 1955–2003. Geophys Res Lett 32(12):L12602. https://doi.org/10.1029/2005GL023112

Association of Official Analytical Chemists (AOAC) (2000) Official method of analysis, 17th edn. Association of Official Analytical Chemists, Washington, DC

Church JA, et al, (2001) Changes in sea level. Climate Change 2001: The Scientific Basis. Contribution of Working Group I to the Third Assessment Report of the Intergovernmental Panel on Climate Change [J. T. Houghton, Y. Ding, D. J. Griggs, M. Noquer, P. J. van der Linden, X. Dai, K. Maskell and C. A. Johnson (eds.)]. Cambridge University Press, Cambridge, United Kingdom and New York, NY, USA, pp. 639–693

Chao B (1994) Man-made lakes and global sea level. Nature 370:258

Chaudhuri AB, Choudhury A (1994) Mangroves of the Sundarbans, vol I, India. IUCN – The World Conservation Union, Bangkok

Davis CH, Li Y, McConnell JR, Frey MM, Hannah E (2005) Snowfall-driven growth in East Antarctica ice sheet mitigates recent sea-level rise. Science 308(5730):1898–1901

Dasgupta S, Laplante B, Meisner C, Wheeler D, Yan J (2009) The impact of sea level rise on developing countries: a comparative analysis. Clim Chang 93:379–388

Devendran G, Balasubramanian U (2011) Qualitative phytochemical screening and GC-MS analysis of *Ocimum sanctum* L. leaves. Asian J Plant Sci 1:44–48

Dubois M, Gilles KA, Hamilton JK, Rebers PA, Smith F (1956) Colorimetric methods for determination of sugars and related substances. J Anal Chem 28:350–356

Fergusson J (1863) Recent changes in the delta of the Ganges. Quat J Geol Soc 19:321–354

Gaber A, Hassan MM, El-Desoky SE, Attia OA (2015) *In vitro* antimicrobial comparison of Taif and Egyptian pomegranate peels and seeds extracts. J Appl Biol Biotechnol 3(2):012–017

Gornitz V (2001) Impoundment, groundwater mining, and other hydrologic transformations: impacts on global sea level rise. In: Douglas BC, Kearney MS, Leatherman SP (eds) Sea level rise: history and consequences. *Academic*, San Diego, pp 97–119

Gupta SK, Prakash J, Srivastava S (2002) Validation of traditional claim of Tulsi, *Ocimum sanctum* Linn. as a medicinal plant. Indian J Exp Biol 40:765–773

Hanna E, Huybrechts P, Janssens I, Cappelen J, Steffen K, Stephens A (2005) Runoff and mass balance of the Greenland ice sheet: 1958–2003. J Geophys Res 110:D13108

Hansen J (2006) Can we still avoid dangerous human-made climate change?. In: Presentation on December 6, 2005 to the American Geophysical Union in San Francisco. http://www.columbia.edu/~jeh1/newschool–text–and–sides.pdf

Hansen J, Nazarenko L, Ruedy R, Sato M, Willis J, Del Genio A, Koch D, Lacis A, Lo K, Menon S, Novakov T, Perlwitz J, Russell G, Schmidt G, Tausnev N (2005) Earth's energy imbalance: confirmation and implications. Science 308:1431–1435

Harvard Medical School/Oxford University Press (2005) 287–323. Monbiot. The Guardian, 29 November

Howat I, Joughin I, Scambos T (2007) Rapid changes in ice discharge from Greenland outlet glaciers. Science 315:1559–1561

IPCC (2001) The scientific basis. Contribution of working group I to the third assessment report of the Intergovernmental Panel on Climate Change (IPCC). Cambridge University Press, Cambridge, UK

IPCC (2007) Climate change 2007: the physical science basis. Summary for policymakers. IPCC Secretariat, Geneva

Ishii M, Kimoto M, Sakamoto K, Iwasaki SI (2006) Steric sea level changes estimated from historical ocean subsurface temperature and salinity analyses. J Oceanogr 62(2):155–170

Krabill W, Hanna E, Huybrechts P, Abdalati W, Cappelen J, Csatho B, Frederick E, Manizade S, Martin C, Sonntag J, Swift R, Thomas R, Yunge J (2004) Greenland ice sheet: increased coastal thinning. Geophys Res Lett 31:L24402

Lombard A, Cazenave A, Traon PYL, Guinehut S, Cecile C (2006) Perspectives on present-day sea level change: a tribute to Christial le provost. Ocean Dyn 56(5–6):445–451

Lowry OH, Rosebrough NJ, Farr AL, Randall RJ (1951) Protein measurement with the Folin-Phenol reagents. J Biol Chem 193:265–275

Meier M, Dyurgerov M, Rick U, O'Neel S, Preffer W, Anderson R, Anderson S, Glazovsky A (2007) Glaciers dominate eustatic sea level rise in the 21st century. Science 317:1064–1067

Morgan JP, McIntire WG (1959) Quaternary geology of the Bengal Basin, East Pakistan and Burma. Bull Geol Soc Am 70:319–342

Nielsen D (1998) Food analysis. Aspen Publishers, Inc. Gaithersburg, Maryland United State Department of Agriculture Nutrient base 2004

Oldham RD (1893) A manual of geology of India. Government of India, Calcutta

Overpeck J, Otto- Bliesner B, Miller G, Muhs D, Alley R, Kichl J (2006) Paleo-climatic evidence for future ice sheet instability and rapid sea level rise. Science 311:1064–1067

Pattanayak P, Behera P, Das D, Panda SK (2010) *Ocimum sanctum* Linn. A reservoir plant for therapeutic applications: an overview. Pharmacogn Rev 4(7):95–105

Pidwirny M (2006) Causes of climate change. In: Fundamentals of physical geography, 2nd edn

Raha A, Das S, Banerjee K, Abhijit M (2012) Climate change impacts on Indian Sunderbans: a time series analysis (1924–2008). Biodivers Conserv 21(5):1289–1307. https://doi.org/10.1007/s10531-012-0260-z

Rao SA, Vijay Y, Deepthi T, Lakshmi CS, Rani V, Rani S (2013) Antidiabetic effect of ethanolic extract of leaves of *Ocimum sanctum* in alloxan induced diabetes in rats. Int J Basic Clin Pharmacol 2:613–616

Rignot E, Kanagaratnam P (2006) Changes in the velocity structure of the Greenland ice sheet. Science 311:986–990

Sahagian DL (2000) Global physical effects of anthropogenic hydrological alterations: sea level and water redistribution. Glob Planet Chang 25:39–48

Sahagian DL, Schwartz FW, Jacobs DK (1994) Direct anthropogenic contributions to sea level rise in the twentieth century. Nature 367:54–56

Saharkhiz MJ, Alam Kamyab A, Kazerani NK, Zomorodian K, Pakshir K, Rahimi MJ (2015) Chemical compositions and antimicrobial activities of *Ocimum sanctum* L. essential oils at different harvest stages. Jundishapur J Microbiol 8(1):e13720

Stroeve J, Holland M, Meier W, Scambos T, Serreze M (2007) Arctic sea ice decline: faster than forecast. Geophys Res Lett 34:L09501

Triveni Kumar K, Singh AK, Kumar R, Gupta V, Tripathi K (2013) *Ocimum sanctum* Linn: a review on phytopharmacology and therapeutic potential of Tulsi. Int J Pharm Phytopharmacol Res 3:148–151

Velicogna I, Wahr J (2006) Measurements of time variable gravity show mass loss in Antarctica. Science 311:1754–1756

Wadia DN (1961) Geology of India. M.C. Muller & Co Ltd, London

Willis JK, Roemmich D, Cornuelle B (2004) International variability in upper-ocean heat content, temperature and thermosteric expansion on global scales. J Geophys Res 109:C12036. https://doi.org/10.1029/2003JC002260

Zvaigzne G, Karklina D, Seglina D, Krasnova I (2009) Antioxidants in various citrus fruit juices. Chem Technol 3(52):56–61

The manufacturer's authorised representative in the EU is Springer Nature Customer Service Centre GmbH, Europaplatz 3, 69115 Heidelberg, Germany. If you have any concerns regarding our products, please contact ProductSafety@springernature.com

Printed and bound by CPI Group (UK) Ltd, Croydon, CR0 4YY
23/09/2026
02221536-0002